Advanced Materials for Energy Storage and Conversion

Advanced Materials for Energy Storage and Conversion

Editor

Ning Sun

Basel • Beijing • Wuhan • Barcelona • Belgrade • Novi Sad • Cluj • Manchester

Editor
Ning Sun
Beijing University of
Chemical Technology
Beijing, China

Editorial Office
MDPI
St. Alban-Anlage 66
4052 Basel, Switzerland

This is a reprint of articles from the Special Issue published online in the open access journal *Coatings* (ISSN 2079-6412) (available at: https://www.mdpi.com/journal/coatings/special_issues/material_energy_storage_conversion).

For citation purposes, cite each article independently as indicated on the article page online and as indicated below:

Lastname, A.A.; Lastname, B.B. Article Title. *Journal Name* **Year**, *Volume Number*, Page Range.

ISBN 978-3-0365-9995-3 (Hbk)
ISBN 978-3-0365-9996-0 (PDF)
doi.org/10.3390/books978-3-0365-9996-0

© 2024 by the authors. Articles in this book are Open Access and distributed under the Creative Commons Attribution (CC BY) license. The book as a whole is distributed by MDPI under the terms and conditions of the Creative Commons Attribution-NonCommercial-NoDerivs (CC BY-NC-ND) license.

Contents

About the Editor vii

M. I. Rodríguez-Tapiador, J. M. Asensi, M. Roldán, J. Merino, J. Bertomeu and S. Fernández
Copper Nitride: A Versatile Semiconductor with Great Potential for Next-Generation Photovoltaics
Reprinted from: *Coatings* 2023, 13, 1094, doi:10.3390/coatings13061094 1

Dong-Min Shin, Hyunsu Son, Ko Un Park, Junyoung Choi, Jungdon Suk, Eun Seck Kang, et al.
Al_2O_3 Ceramic/Nanocellulose-Coated Non-Woven Separator for Lithium-Metal Batteries
Reprinted from: *Coatings* 2023, 13, 916, doi:10.3390/coatings13050916 19

Ziyang Guo, Xiaodong Tian, Yan Song, Tao Yang, Zihui Ma, Xiangjie Gong, et al.
Hard Carbons Derived from Phenyl Hyper-Crosslinked Polymers for Lithium-Ion Batteries
Reprinted from: *Coatings* 2023, 13, 421, doi:10.3390/coatings13020421 35

Yue Xin, Zhaoxin Yu, Razium Ali Soomro and Ning Sun
Facile Synthesis of Polyacrylic Acid/Graphene Oxide Composite Hydrogel Electrolyte for High-Performance Flexible Supercapacitors
Reprinted from: *Coatings* 2023, 13, 382, doi:10.3390/coatings13020382 47

Refka Sai and Rasha A. Abumousa
Impact of Iron Pyrite Nanoparticles Sizes in Photovoltaic Performance
Reprinted from: *Coatings* 2023, 13, 167, doi:10.3390/coatings13010167 59

Stoyan I. Karakashev, Nikolay A. Grozev, Svetlana Hristova, Kristina Mircheva and Orhan Ozdemir
Electrostatic Forces in Control of the Foamability of Nonionic Surfactant
Reprinted from: *Coatings* 2023, 13, 37, doi:10.3390/coatings13010037 73

Hongming Zhang, Jiahe Zhuang, Xiangrui Feng and Ben Ma
$Co_{0.6}Ni_{0.4}S_2$/rGO Photocatalyst for One-Pot Synthesis of Imines from Nitroaromatics and Aromatic Alcohols by Transfer Hydrogenation
Reprinted from: *Coatings* 2022, 12, 1799, doi:10.3390/coatings12121799 83

Yuqi Zhang, Zhonghui Sun, Dongyang Qu, Dongxue Han and Li Niu
Recent Advances in $CoSe_x$ and $CoTe_x$ Anodes for Alkali-ion Batteries
Reprinted from: *Coatings* 2023, 13, 1588, doi:10.3390/coatings13091588 97

Raksan Ko, Dong Hyun Lee and Hocheon Yoo
Annealing and Doping Effects on Transition Metal Dichalcogenides—Based Devices: A Review
Reprinted from: *Coatings* 2023, 13, 1364, doi:10.3390/coatings13081364 125

Ruxin Yu, Gang Liu, Linbo Xu, Yanqiang Ma, Haobin Wang and Chen Hu
Review of Degradation Mechanism and Health Estimation Method of VRLA Battery Used for Standby Power Supply in Power System
Reprinted from: *Coatings* 2023, 13, 485, doi:10.3390/coatings13030485 145

Yanan Mei, Yuling He, Haijiang Zhu, Zeyu Ma, Yi Pu, Zhilin Chen, et al.
Recent Advances in the Structural Design of Silicon/Carbon Anodes for Lithium Ion Batteries: A Review
Reprinted from: *Coatings* 2023, 13, 436, doi:10.3390/coatings13020436 165

About the Editor

Ning Sun

Ning Sun received her Ph.D. degree in 2019 from Beijing University of Chemical Technology (BUCT). She is now working as an associate professor at BUCT. Her main research interests include the synthesis of carbon materials and 2D MXene for application in energy storage, such as in supercapacitors and secondary batteries.

Article

Copper Nitride: A Versatile Semiconductor with Great Potential for Next-Generation Photovoltaics

M. I. Rodríguez-Tapiador [1,*], J. M. Asensi [2,3], M. Roldán [4], J. Merino [5], J. Bertomeu [2,3] and S. Fernández [1,*]

1. Energy Department, Center for Energy, Environmental and Technological Research (CIEMAT), Av. Complutense 40, 28040 Madrid, Spain
2. Departament de Física Aplicada, Universitat de Barcelona, 08028 Barcelona, Spain; jmasensi@ub.edu (J.M.A.); jbertomeu@ub.edu (J.B.)
3. Institute of Nanoscience and Nanotechnology (IN2UB), Universitat de Barcelona, 08028 Barcelona, Spain
4. National Fusion Laboratory, CIEMAT, Av. Complutense 40, 28040 Madrid, Spain; marcelo.roldan@ciemat.es
5. Technology Support Center (CAT), University Rey Juan Carlos, Tulipán, s/n, Móstoles, 28039 Madrid, Spain; jesus.merino@urjc.es
* Correspondence: mariaisabel.rodriguez@ciemat.es (M.I.R.-T.); susanamaria.fernandez@ciemat.es (S.F.); Tel.: +34-913466591 (M.I.R.-T.); +34-913460923 (S.F.)

Abstract: Copper nitride (Cu_3N) has gained significant attention recently due to its potential in several scientific and technological applications. This study focuses on using Cu_3N as a solar absorber in photovoltaic technology. Cu_3N thin films were deposited on glass substrates and silicon wafers via radio-frequency magnetron sputtering at different nitrogen flow ratios with total pressures ranging from 1.0 to 5.0 Pa. The thin films' structural, morphology, and chemical properties were determined using XRD, Raman, AFM, and SEM/EDS techniques. The results revealed that the Cu_3N films exhibited a polycrystalline structure, with the preferred orientation varying from 100 to 111 depending on the working pressure employed. Raman spectroscopy confirmed the presence of Cu-N bonds in characteristic peaks observed in the 618–627 cm^{-1} range, while SEM and AFM images confirmed the presence of uniform and smooth surface morphologies. The optical properties of the films were investigated using UV-VIS-NIR spectroscopy and photothermal deflection spectroscopy (PDS). The obtained band gap, refractive index, and Urbach energy values demonstrated promising optical properties for Cu_3N films, indicating their potential as solar absorbers in photovoltaic technology. This study highlights the favourable properties of Cu_3N films deposited using the RF sputtering method, paving the way for their implementation in thin-film photovoltaic technologies. These findings contribute to the progress and optimisation of Cu_3N-based materials for efficient solar energy conversion.

Keywords: Cu_3N films; reactive magnetron sputtering; photothermal deflection spectroscopy (PDS); solar energy conversion

1. Introduction

Copper nitride (Cu_3N) is a compound that has attracted increasing attention in recent years due to its unique properties and potential applications in several fields [1], as shown in Figure 1.

Regarding its crystal structure, Cu_3N is a binary compound made up of copper (Cu) and nitrogen (N) that displays a cubic anti-ReO$_3$ crystal system (space group Pm3m). The bond angle of Cu-N-Cu reported is approximately 180 degrees, and the lattice parameter and density are 3.817 Å and 6.1 g/cm^3, respectively [2]. This material is metastable and exhibits thermal stability close to 250 °C [3]. Cu_3N displays insulator-to-conductor transition behaviour, with electrical conductivity values ranging between 10^{-3} and 10^{-2} Scm^{-1} [4]. Moreover, the electrical properties of the Cu_3N films can be modified by doping them with metal or non-metal elements such as fluorine or gold, resulting in either p-type or n-type behaviour [5,6]. To switch between these electric characters, doping concentration, synthesis conditions, surface modifications, and interface engineering can be adjusted to alter charge carriers, crystal structure, defect density, and electrical properties. By controlling these factors, copper nitride's resistivity can be fine tuned to achieve the desired conductivity [7–9]. This approach can potentially enhance and broaden the scope of applications for Cu_3N films. As a result, it is a promising candidate for future technological advancements, developments, and innovations in various fields, including electronic devices such as thin-film transistors (TFTs) and complementary metal-oxide-semiconductor (CMOS) circuits [10].

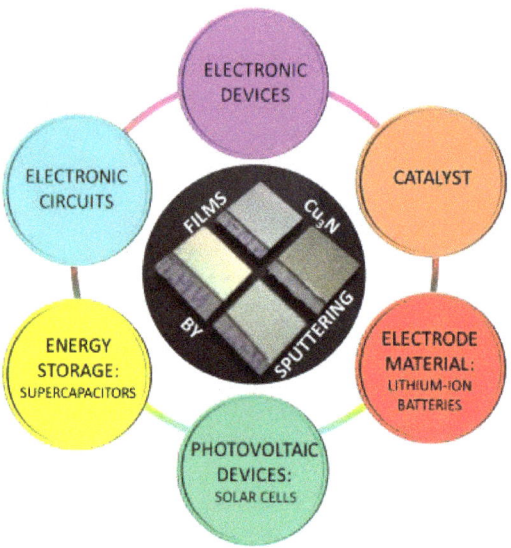

Figure 1. Some applications of Cu_3N films in different technological fields.

Additionally, Cu_3N has been investigated as a potential material for energy storage applications, such as batteries and other energy storage devices, due to its thermal stability and chemical stability [11]. Recent works have reported that Cu_3N can be used as an electrode material in lithium-ion batteries, exhibiting excellent performance in capacity and cycle stability [12]. Moreover, copper nitride has demonstrated high electrochemical and catalytic activity for various important reactions, making it a promising material for numerous applications [13–15].

From the optical point of view, experimental studies have shown that this material has indirect and direct band gap values ranging from 1.17 to 1.69 eV and 1.72 to 2.38 eV, respectively [16,17]. Considering these values, Cu_3N can be considered a promising light absorber material for solar cells. Its crystal structure and chemical composition can be optimised by controlling the technical parameters during preparation, resulting in an optimised optical bandgap and maximum photovoltaic voltage [4,18,19]. By developing p-type and n-type Cu_3N (100) thin films via different technologies and adjusting the Cu/N chemical composition via reasonable control methods, bipolar doping can be added under Cu defects, leading to materials ready for photovoltaic applications thanks to their excellent indirect bandgap values. This approach can significantly increase the conversion efficiency of solar energy [4,20].

There are several chemical and physical methods to prepare copper nitride films. Among the chemical fabrication techniques, chemical vapour deposition (CVD) and atomic layer deposition (ALD) are commonly used to fabricate it in a gas phase. These methods involve the use of precursors such as $Cu(hfac)_2$ and $[Cu(sBu-Me-amd)]_2$ [21] to determine its resulting phase composition and morphology and to establish the growth rate [22,23]. Ammonolysis reactions can also be used to prepare bulk Cu_3N powder samples [24], as well as thin films [25]. In addition, recent works have shown that Cu$_3$N nanowire arrays can be synthesised by an ammonolysis reaction from copper (II) oxide precursors grown on copper surfaces deposited by electro or PVD in an ammonia solution [26,27].

Regarding the physical ones, sputtering is one of the most popular physical vapour deposition (PVD) methods used to fabricate films of metals, alloys, oxides, and nitrides [28–31]. Since the pioneering work of Terada et al. (1989) on epitaxial growth of copper nitride [32], reactive RF magnetron sputtering has become the most widely used mode for the fabrication of binary nitride. This method involves using a vacuum chamber in which a copper target is bombarded with high-energy ions, causing copper atoms to be ejected from the target and deposited onto a substrate to form a thin film. By introducing nitrogen gas, with or without argon gas, into the chamber during this process, Cu$_3$N can be grown on a substrate. This method's ease, simplicity, low cost, reproducibility and sustainability make it a very attractive choice for the growth of Cu$_3$N thin films. Previous studies have already reported that by modifying the bias voltage [33], the type of substrate [32], the working pressure [34], and the RF power [35], the film properties can be adjusted, allowing the variation in optical, electrical, structural, and morphological features to suit them to the desired ones depending on the application field. In addition, in our previous works, we have demonstrated the strong effect of RF power on modifying the morphological, structural, and optical Cu$_3$N characteristics, intending to use them as solar absorbers for next-generation photovoltaics [19].

In this work, we investigated the impact of the process gas and its pressure on the properties of Cu$_3$N films prepared via reactive RF magnetron sputtering at room temperature (RT). Two different atmospheres were studied: an environment based on the mixture of N$_2$ and Ar gases and another one based on a pure N$_2$ gas, while the working pressures ranged from 1.0 to 5.0 Pa. The changes in crystalline nature, chemical composition, morphology, and electrical and optical properties were examined in depth. We aim to determine what material and sputtering deposition conditions lead to the most suitable properties and the highest possible absorption coefficient to be used as an absorber in a solar cell.

2. Materials and Methods

Cu$_3$N thin films were deposited on different substrates, <100> polished n-type floating zone crystalline silicon (c-Si) wafers and 1737F Corning glass (Corning Inc., New York, NY, USA) via reactive RF magnetron sputtering in a commercial MVSystem LLC (Golden, CO, USA) mono-chamber sputtering system. The 3-inch diameter Cu target, with a purity of 99.99%, was from Lesker company (St. Leonards-on-Sea, East Sussex, UK). Before the sputtering deposition, the surface of the silicon wafer was prepared by removing the native silicon dioxide layer using a solution of 1% hydrofluoric acid (HF) in a mixture of deionised water and isopropyl alcohol. The wafer was immersed in this solution for 5 min. Next, the glass substrates were subjected to ultrasonic cleaning with ethanol and deionised water for 3 min. Then, they were submerged in isopropyl alcohol. Afterwards, all substrates were dried by blowing nitrogen gas over them.

The sputtering chamber was initially pumped to a base pressure of 2.6×10^{-5} Pa, and the distance between the target and substrate was set to 10 cm. A pre-sputtering process was performed for 5 min to clean the target surface. Then, the deposition was conducted for 30 min at room temperature (RT) and 50 W of RF power. The process gases used were N$_2$ (99.9999%) and Ar (99.9999%), with flow rates of 20 sccm and 10 sccm, respectively, controlled using mass flow controllers from MKS Instruments (MKS Instruments, Andover, MA, USA). The total gas pressure was varied between 1.0 and 5.0 Pa by adjusting the position of the "butterfly" valve in the magnetron system. The thickness of the films was measured using a Dektak 8 profilometer (Bruker, San José, CA, USA). A tip force of 68.67 µN and a scan size of 2000 µm were used in all cases. To determine the crystallinity of the Cu$_3$N films, X-ray diffraction (XRD) was performed using a commercial system (model PW3040/00 X'Pert MPD/MRD) (Malvern Panalytical Ltd., Malvern, UK) with Cu-kα radiation (λ = 0.15406 nm). The 2θ range scanned was 10–60°, with a step size of 0.01° and a time of 20 s per step. The topography was analysed with a multimode nanoscope atomic force microscopy (AFM) model IIIA (SPM; Veeco Digital Instrument) in tapping mode using a silicon nitride AFM tip (OTR8, Veeco, Santa Clara, CA, USA). The surface roughness was quantified using mean root square (RMS) analysis, and the grain size from the two-dimensional (2D) AFM 1×1 µm^2 images, using the Gwyddion software (Gwyddion software, http://gwyddion.net/ accessed on 20 March 2023).

In addition, the variation of surface morphology as a function of the gaseous environment used was determined using a JEOL JSM 7600F scanning microscope, equipped with a field emission Schottky electron gun (FESEM), in-lens secondary electron detector, and elemental analysis system for chemical composition EDS (energy-dispersive X-ray spectroscopy). Several surface regions were analysed at an acceleration voltage of 15 kV to quantitatively determine the amount of Cu and N in the thin film.

The molecular structure was determined using a dispersive spectrometer Confocal Raman microscope with capabilities for obtaining XYZ 3D confocal Raman images, equipped with a 532 nm laser, two diffraction gratings (600 and 1800 g/nm), three objectives ($5\times$, $75\times$ and $100\times$), and option to obtain photocurrent mappings (Horiba LabRam soleil, Longjumeau Cedex, France). This measurement provides valuable information about the sample's vibrational modes and helps characterise its molecular structure. Finally, to determine the suitability of Cu_3N as a solar absorber, the optical transmittance spectra were measured at normal incidence using a UV/VIS/NIR Perkin Elmer Lambda 1050 spectrophotometer. The optical band gap energies (E_g) were calculated from these spectra for indirect and direct transitions.

The optical properties of Cu_3N thin films deposited on glass were analysed using UV-VIS-NIR optical spectroscopy and photothermal deflection spectroscopy (PDS). The transmittance (T_{opt}) and reflectance (R_{opt}) spectra were obtained with a PerkinElmer Lambda 950 UV-Vis-NIR spectrometer equipped with an integrating sphere. While the transverse PDS setup used to measure weaker absorption consists of a 100 W tungsten halogen lamp, PTI 01-0002 monochromator (spectral range of 400–2000 nm), and Thorlabs MC1000 optical chopper (4 Hz light modulation frequency). A Signal Recovery 7265 lock-in amplifier was connected to a Hamamatsu C10442-02 PSD position-sensitive detector to measure the deflection of an MC6320C 10 mW laser probe beam. Samples were put in a quartz cell filled with Fluorinert TM FC-40. T_{opt} and R_{opt} measurements allow for determining the optical absorbance ($A_{opt} = 1 - T_{opt} - R_{opt}$) in the strong absorption region. In addition, the interference fringes observed in both spectra can be used to estimate the film thickness and their refractive index (n_∞).

On the other hand, the PDS measurement is very effective for determining absorbance (APDS) in the weak (and very weak) absorption region. Thus, it is possible to determine the absorbance over a broad spectrum range by combining optical measurements with PDS. The absorption coefficient (α) is obtained by a fit based on the calculation of absorbance using the transfer matrix method (TMM). These measurements allow the determination of the most suitable sputtering conditions to achieve a more efficient solar absorber material.

3. Results

The Cu_3N films in this study exhibited excellent physical stability and good adhesion to the respective substrate, even after exposure to ambient air; therefore, no evidence of cracking or peeling off was observed after the deposition process. Table 1 details the sputtering deposition conditions and the corresponding measured film thickness.

As observed, the deposition rate varied depending on the deposition conditions used, obtaining similar values to those reported by other authors [32,36]. These data reveal that the working pressure is essential in thin film fabrication. As the working pressure increases, the number of impacts between the sputtered species and gas atoms increases in sputtering processes. Thus, we observed that the deposition rate decreases with higher working pressure. This effect leads to a decrease in the deposition rate due to a reduction in the mean free path of the species within the plasma [3]. This trend was observed for the samples deposited in the N_2 pure atmosphere, as shown in Figure 2. It is well known that there is an abrupt decrease in deposition rate due to the nitridation process that is more favoured at high gas pressures [37]. At the same time, the sputtering ratio Cu/N began to be higher when the N_2 gas pressure decreased. This would favour the metallic regime and, hence, obtain a gradual increase in the sputtering yield at low working pressures. In addition, this would indicate that, under such values of N_2 gas pressure, the target "poisoning" effect would not have started yet; hence, a gradual decrease in the deposition rate with the gas pressure was achieved.

Table 1. Deposition conditions of Cu_3N films, varying N_2 flow ratios (0.7 and 1.0) and different total pressures.

Total Pressure (Pa)	N_2 Flux	Ar Flux	N_2 Flow Ratio	Deposition Rate (nm/s)	Thickness (nm)
1.0	20	10	0.7	0.053	95 ± 5
2.0	20	10	0.7	0.054	96 ± 7
3.5	20	10	0.7	0.115	207 ± 18
5.0	20	10	0.7	0.055	98 ± 2
1.0	20	0	1.0	0.091	164 ± 12
2.0	20	0	1.0	0.065	118 ± 20
3.5	20	0	1.0	0.060	109 ± 16
5.0	20	0	1.0	0.058	104 ± 20

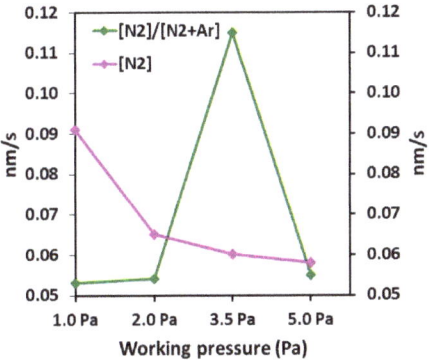

Figure 2. Deposition rate to different N_2 flow ratios of 0.7 and 1.0 and to different total working pressure (Pa). Deposition conditions of Cu_3N films: RT and P_{rf} = 50 W.

On the other hand, it can be noticed that the films prepared under the N_2/Ar gas mixture environment presented lower deposition rates than those prepared in N_2 pure atmosphere. In the case where the sputtering process was carried out in a N_2/Ar atmosphere, the decrease in the deposition rate was not so evident, remaining almost constant at 0.055 nm/s (see Figure 2). This could be attributed to a poorer nitridation occurring on the surface target due to the lower presence of N_2 in the gas mixture. It should be pointed out that a significant increase was observed in the deposition rate at 3.5 Pa, reaching a value of 0.115 nm/s. This was attributed to the change in the preferred crystal structure orientation, specifically to the (111) plane, a plane of lower density that would lead to a rougher film with a lower refractive index, as will be shown later [3].

Figure 3 illustrates the X-ray diffraction (XRD) patterns of the Cu_3N deposited at the operating pressure range of 1.0–5.0 Pa in different environments: an N_2 pure atmosphere (Figure 3a) and a gas mixture of N_2/Ar (Figure 3b). All of the films exhibited a polycrystalline nature, characterised by an anti-ReO_3 crystal structure, typical of cubic Cu_3N

(card number 00-047-1088), and hence, dominated with the Cu$_3$N phase. Regardless of the working gas pressure used, the samples deposited in the N$_2$ pure atmosphere showed the (100) plane as the preferred orientation (Figure 3a).

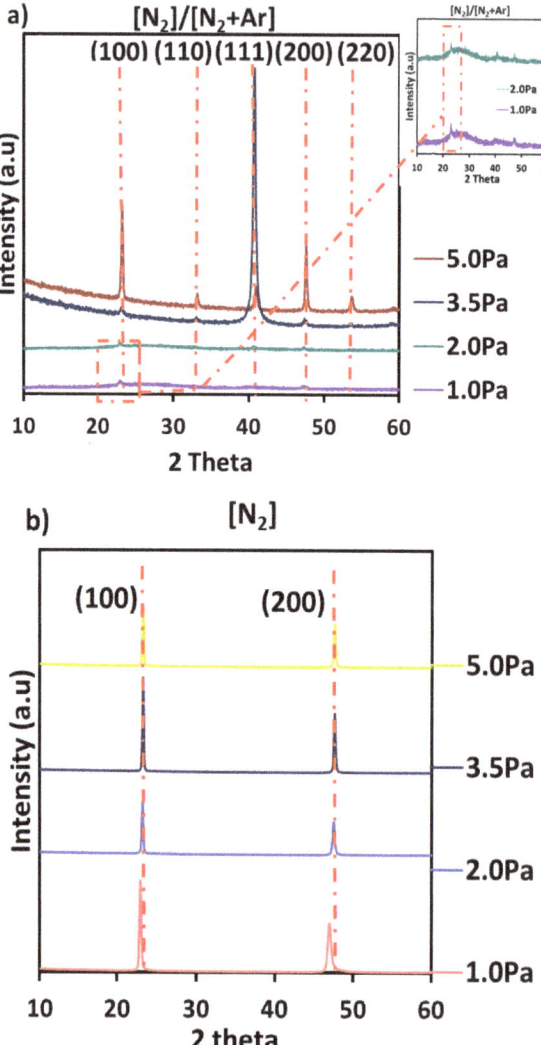

Figure 3. X-ray diffraction spectra of Cu$_3$N films at different gas pressures (Pa) and nitrogen flow ratios. (a) [N$_2$]/[N$_2$ + Ar] r = 0.7 and (b) [N$_2$] r = 1.0.

On the other hand, the patterns of the samples deposited in the N$_2$/Ar gas mixture (Figure 3b) showed the appearance of (100), (110), (111), (200), and (220) diffraction peaks. In this case, and at low gas pressures, a weak (100) peak emerged over an amorphous hump caused by the glass substrate. At the pressure of 3.5 Pa, the film showed a (111) preferred orientation, while in the diffraction pattern of the sample deposited at 5.0 Pa, the (100) plane appeared as the most substantial peak. Other authors previously observed this change obtained in orientation. It could be attributed to the higher density of the N atoms that reached the substrate and reacted with the Cu atoms, leading to high-density Cu-N bonds for the preferential growth along the (100) direction. This can indicate that the nitridation was more effective when the gas pressure increased and, hence, the growth was favoured along the N-rich planes of Cu$_3$N [38].

On the other hand, the constant lattice a was determined by calculating the interplanar spacing using Bragg's law [39], expressed as the following Equation (1):

$$d(hkl) = \frac{a}{\sqrt{(h^2 + k^2 + l^2)}} \quad (1)$$

where d is the interplanar spacing, and h, k, and l are the Miller indices. Furthermore, the grain size (τ) was determined using the Debye–Scherrer Equation (2) [40] as follows:

$$\tau = \frac{k\lambda}{\beta \cdot \cos\theta_B} \quad (2)$$

where k is a constant (0.9), λ is the X-ray wavelength (0.154 nm), θ is the diffraction angle, and β is the full width at half maximum (FWHM) of the predominant peak.

Tables 2 and 3 summarise the FWHM of the main diffraction peak, the lattice constant, the predominant plane, the 2θ value, and the grain sizes derived from the XRD patterns of all the samples, depending on the N$_2$ flow ratio.

Table 2. XRD data extracted from the XRD spectra of the Cu$_3$N films fabricated on glass via RF magnetron sputtering in a mixed N$_2$/Ar atmosphere.

N$_2$ Flow Ratio: 0.7				
Working pressure (Pa)	1.0 *	2.0 *	3.5	5.0
2θ (°)	23.03	22.97	40.73	23.23
Predominant direction *	-	-	(111)	(100)
Lattice parameter a (nm)	0.3858	0.3872	0.3813	0.3814
FWHM (°)	0.24	0.24	0.53	0.31
Grain size (nm)	34	34	16	27

* Poor crystalline quality. Preferential orientation is not easy to identify.

Table 3. XRD data extracted from the XRD spectra of the Cu$_3$N films fabricated on glass via RF magnetron sputtering in a pure N$_2$ atmosphere.

N$_2$ Flow Ratio: 1.0				
Working pressure (Pa)	1.0	2.0	3.5	5.0
2θ (°)	22.77	23.17	23.29	23.28
Predominant direction	(100)	(100)	(100)	(100)
Lattice parameter a (nm)	0.3903	0.3840	0.3810	0.3813
FWHM (°)	0.20	0.15	0.21	0.21
Grain size (nm)	40	55	39	40

It can be observed that the FWHM values for the samples deposited in the N$_2$/Ar gas mixed atmosphere were superior to those for the samples deposited in the pure N$_2$ atmosphere, indicating an improved quality for these last films. The same trend was obtained for the grain size, reaching values as high as 55 nm when the samples were fabricated in a pure N$_2$ atmosphere. Concerning the lattice parameter, at the low pressures of 1.0 and 2.0 Pa, greater values than the theoretical one (0.38170 nm) were achieved, regardless of the N$_2$ flow ratio used. This fact could indicate a move away from the stoichiometry condition for such samples [41].

Figure 4 pictures the plan-view FESEM and AFM 1 × 1 µm^2 2D micrographs of the Cu$_3$N thin films deposited in N$_2$/Ar gas mixed atmosphere (Figure 4a) and pure N$_2$ environment (Figure 4b). FESEM analysis revealed the presence of smooth and uniform surfaces, composed mainly of columnar grains, which are characteristic of the sputtering method [32,42,43]. These findings align with the results obtained from the AFM analysis [34,44,45], as shown in Figure 4.

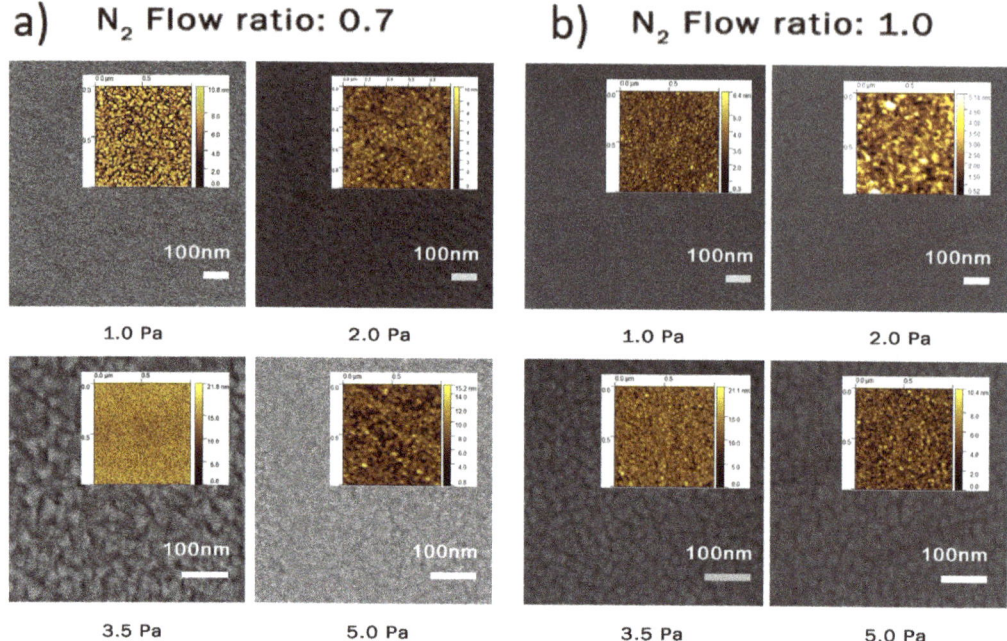

Figure 4. Top view FESEM images: Magn. 90,000×; 100 nm, and AFM 1 × 1 µm² 2D micrographs. (**a**) [N$_2$]/[N$_2$ + Ar] r = 0.7 and (**b**) [N$_2$] r = 1.0.

The size of the grains was influenced by both the environment and the total pressure applied during the deposition process. It was observed that lower working pressures resulted in larger grain sizes. This phenomenon can be attributed to the formation of Cu$_3$N crystallites and the adhesion of tiny copper crystals, possibly caused by a decrease in N$_2$ density. These observations are supported by grain size calculations performed using the commercial software Gwyddion. This result could be reinforced with the grain size values calculated with the Debye–Scherrer equation in XRD.

Tables 4 and 5 summarise the grain size and surface roughness RMS of the Cu$_3$N films, calculated from the 2D AFM micrographs using the Gwyddion software, depending on the N$_2$ flow ratio.

Table 4. Surface roughness RMS and grain size calculated from AFM 1 × 1 µm² images of the Cu$_3$N films fabricated on glass via RF magnetron sputtering in a mixed N$_2$/Ar atmosphere.

N$_2$ Flow Ratio: 0.7				
Working pressure (Pa)	1.0	2.0	3.5	5.0
RMS (nm)	1.39 ± 0.15	1.26 ± 0.14	2.77 ± 0.23	1.77 ± 0.21
Grain size (nm)	40 ± 2	35 ± 3	30 ± 3	34 ± 3

Table 5. Surface roughness RMS and grain size calculated from AFM 1 × 1 µm² images of the Cu$_3$N films fabricated on glass via RF magnetron sputtering in a pure N$_2$ atmosphere.

N$_2$ Flow Ratio: 1.0				
Working pressure (Pa)	1.0	2.0	3.5	5.0
RMS (nm)	0.90 ± 0.15	1.15 ± 0.50	2.21 ± 0.27	1.34 ± 0.12
Grain size (nm)	19 ± 2	32 ± 2	33 ± 2	35 ± 2

Based on these findings, it can be concluded that regardless of the total working pressure and N_2 flow ratio used, the surfaces were very flat, presenting RMS values that did not exceed 3 nm. In this sense, it can be noted that even though the RMS was very low, there was a difference between the films deposited using pure N_2 gas and those deposited in the N_2/Ar gas mixture, obtaining slightly smoother surfaces in the first case. On the other hand, the grain size varied depending on the process parameters. Larger sizes as working pressure increased. The larger grain sizes obtained for the films deposited at 1.0 Pa and 2.0 Pa in a mixture of N/Ar gases may be due to the formation of small agglomerates due to the amorphous character seen in XRD. It should be pointed out that the grain size values estimated from XRD patterns were slightly higher than the obtained from AFM measurements. This can be explained because the first ones were an average value of a larger area analysed, while the second ones were calculated in a small area at a specific point.

The chemical composition of the samples was determined qualitatively using EDS data (Table 6). The analysis revealed a Cu/N ratio below three, indicating the non-stoichiometry of the deposited material. Interestingly, an increased Cu/N ratio was observed at higher working pressures. This phenomenon can be attributed to the increased energy of the nitrogen atoms at higher working pressures, enhancing the formation of bonds with the copper atoms. Moreover, all Cu_3N films exhibited the presence of trace amounts of oxygen. This observation may be associated with exposure to ambient oxygen, as observed in the Raman analysis but not detected in XRD patterns.

Table 6. The EDS analysis provided insights into the relative surface composition of the examined films.

Working Pressures	1.0 Pa	2.0 Pa	3.5 Pa	5.0 Pa
Cu/N ratio $[N_2]/[N_2 + Ar]$ $r = 0.7$	1.79	2.08	2.10	2.17
Cu/N ratio $[N_2]$ $r = 1.0$	1.87	1.88	2.07	2.13

Figure 5 displays the Raman spectra of the Cu_3N films deposited at different working pressures and N_2 flow ratios of 0.7 (Figure 5a) and 1.0 (Figure 5b). Cu_3N has a crystal structure belonging to the Pm-3m space group where the unit cell contains one formula unit. As a result, no first-order Raman signal is expected for a perfect cubic Cu_3N. Although theoretical calculations suggest the absence of active Raman modes, the possibility of modes arising due to the breakdown of the selection rule cannot be ruled out. This is due to $Cu_xN_{(1-x)}$'s highly non-stoichiometric nature and the breakdown in crystal symmetry caused by defects in the structure. The prominent Raman peak around 619–627 cm^{-1} in Figure 5 corresponds to the stretching of the Cu-N bond, characteristic of Cu_3N [41,46,47]. A slight shift of that Raman peak was observed as the working pressure varied, while the Raman shift value tends to move to lower ones when using the N_2/Ar gas mixture, compared to using the pure N_2 gas in the deposition process.

Furthermore, Raman shifts of CuO_2 and CuO also appeared in the spectra at 94 cm^{-1}, 150 cm^{-1}, and 295 cm^{-1}, respectively. The samples prepared in the N_2/Ar gas mixture showed a higher presence of different copper oxides that may have formed on the film surface upon contact with atmospheric air and/or within the crystal structure. In order to analyse the Raman signal derived from the presence of these types of oxides, these measurements were complemented with the XRD data. As a result, it was confirmed that the characteristic 2θ peaks of CuO_2 and CuO at 36.5° and 35.5°, respectively [48], did not appear in any diffraction patterns, reinforcing our prediction that the oxidation process would be happening due to environmental causes. However, the role of oxygen impurities cannot be ignored, as oxygen always remains an unintentional impurity in nitride-based materials [49]. Therefore, a more detailed theoretical analysis is required to interpret and assign the active Raman peak appropriately. Tables 7 and 8 show the prepared samples' Raman shift values and FWHM.

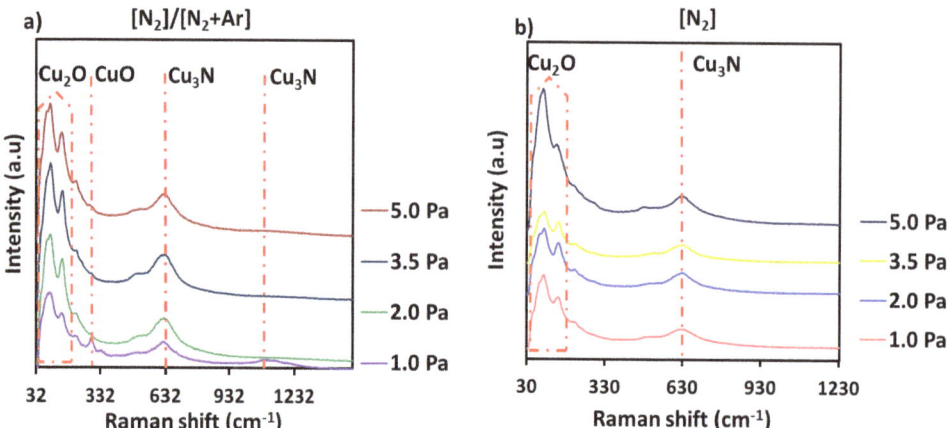

Figure 5. Raman spectra for Cu$_3$N films deposited at different pressures on glass. (a) [N$_2$]/[N$_2$ + Ar] r = 0.7 and (b) [N$_2$] r = 1.0.

Table 7. Raman peak position and full width at half-maximum (FWHM) of the Cu$_3$N films at different pressures on glass in a mixed N$_2$/Ar atmosphere.

N$_2$ Flow Ratio: 0.7				
Working pressure (Pa)	1.0	2.0	3.5	5.0
FWHM (cm^{-1})	78.9 ± 2.8	76.8 ± 1.8	82.9 ± 3.4	68.1 ± 2.7
Peak position (cm^{-1})	618.0 ± 1.2	621.0 ± 0.4	619.0 ± 0.6	621.0 ± 0.6

Table 8. Raman peak position and full width at half-maximum (FWHM) of the Cu$_3$N films at different pressures on glass in a pure N$_2$ atmosphere.

N$_2$ Flow Ratio: 1.0				
Working pressure (Pa)	1.0	2.0	3.5	5.0
FWHM (cm^{-1})	82.9 ± 1.7	74.9 ± 2.0	68.0 ± 3.6	65.2 ± 1.3
Peak position (cm^{-1})	623.0 ± 0.3	625.0 ± 0.4	626.0 ± 0.5	627.0 ± 0.4

The position of the main peak and the FWHM were calculated by simulating via the OriginLab program (OriginPro 8, OriginLab Corporation, Northampton, MA, USA). As observed, the FWHM value decreased as the working pressure increased, suggesting that the lower the FWHM value, the lower the nitrogen concentration in the sample. There is an exception for the sample prepared in the gas mixture at 3.5 Pa, where the FWHM did not exhibit that tendency, attributed to its structural change to the (111) preferential plane revealed by its XRD pattern. However, this sample does not follow that trend. In summary, these results obtained from XRD and Raman show that the films with superior quality that could serve as solar absorbers are those prepared in the pure N$_2$ atmosphere and at pressures of 3.5 Pa and 5.0 Pa. Furthermore, these films exhibited a less variety of grain orientations, with a predominant (100) plane and the Raman shift values were closer to the formation of the theoretical bonding structure.

Optical properties were determined from transmittance and reflectance spectra obtained using UV-Vis-NIR spectroscopy and shown in Figure 6. The transmittance spectra showed high transmittance in the NIR region (>700 nm), which gradually decreased in the VIS range (450–700 nm), reaching very low values in the UV range (300–400 nm).

For most samples, a minimum transmittance was observed in the transparent region (>700 nm), corresponding to a maximum reflectance. This is an apparent effect due to interference from multiple reflections inside the films because of the samples' homogeneity and flatness, and, thanks to that. The light maintains its coherence in the internal reflections. It should be noted that the thin thickness of the films (<200 nm) limits the number of interferences observed in the spectra (in fact, the maximum transmittance beyond 2500 nm, which should be about 92% according to the refractive index of the glass, is not observed in any case). However, the detection of the minimum of T_{opt} (or a maximum of R_{opt}) is sufficient to determine, by the fit of T_{opt} (or R_{opt}) in the transparent region, the thickness and refractive index (n_∞) of the films.

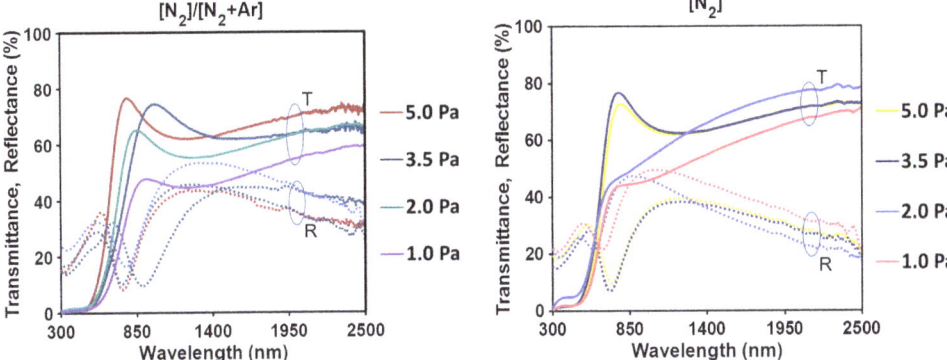

Figure 6. Transmittance and reflectance spectra of Cu_3N films deposited at different pressures and atmospheres.

Table 9 summarises the optical parameters obtained. It can be observed that the thickness values obtained from the adjustments of the optical spectra were slightly superior to the measured with profilometry (data in Table 1). This disparity in thickness could be attributed to how this value is determined because profilometry provides direct thickness measurements, whereas PDS relies on indirect one.

Table 9. Obtained values of the optical properties of Cu_3N films deposited on glass analysed via UV-VIS-NIR optical spectroscopy and PDS at different N_2 flow ratios.

	N_2 Flow Ratio: 0.7			
Working Pressure (Pa)	1.0	2.0	3.5	5.0
Optical fit parameters				
Film thickness (nm)	110	123	173	125
Refractive index n_∞	3.05	2.65	2.44	2.45
Band gap fit parameters				
Transition energy E_2 (eV)	2.18	2.27	2.38	2.46
Direct band gap E_g^d (eV)	1.91	2.02	2.09	2.22
Indirect band gap E_g^i (eV)	1.10	1.28	1.22	1.53
Urbach fit parameters				
Transition energy E_1 (eV)	1.55	1.69	1.64	1.90
Urbach energy E_U (meV)	261	233	243	205
Absorption coefficient at E_1 (cm^{-1})	6.3×10^4	5.1×10^4	4.0×10^4	5.2×10^4
	N_2 flow ratio: 1.0			
Working Pressure (Pa)	1.0	2.0	3.5	5.0
Optical fit parameters				
Film thickness (nm)	109	65	126	123
Refractive index n_∞	2.85	2.79	2.43	2.43
Band gap fit parameters				
Transition energy E_2 (eV)	2.30	2.25	2.42	2.37
Direct band gap E_g^d (eV)	2.05	2.06	2.21	2.15
Indirect band gap E_g^i (eV)	1.30	1.48	1.55	1.49
Urbach fit parameters				
Transition energy E_1 (eV)	1.74	1.82	1.96	1.88
Urbach energy E_U (meV)	253	186	229	217
Absorption coefficient at E_1 (cm^{-1})	8.0×10^4	6.5×10^4	6.0×10^4	5.3×10^4

The refractive index values were in the range of 2.4–3.0, which is consistent with the values typically reported for Cu_3N [50]. It can be noticed that higher refractive index values were obtained for the samples deposited at lower total pressures, regardless of the atmosphere used in the film deposition. This phenomenon can be attributed to the higher Cu content in the samples deposited at lower working pressures. This assumption can also be supported by examining the XRD diffractograms and EDS analysis.

The absorption coefficient (α) in the strong absorption region can be calculated with reasonable accuracy from the absorbance A_{opt}, reflectance R_{opt} and thickness d of the film as follows:

$$\alpha \approx -\frac{1}{d}\ln\left(1 - \frac{A_{opt}}{1 - R_{opt}}\right) \tag{3}$$

In the weak absorption region, Equation (3) is unsuitable for estimating α because of multiple reflections. However, in this case, assuming that the refractive index is practically constant ($n \approx n_\infty$), it is possible to derive α a wide range of the spectrum (from about 3.5 eV to 0.5 eV).

Figure 7a shows the A_{opt} and A_{pds} spectra for the Cu_3N samples deposited at 5.0 Pa and N_2/Ar gas mixture. As can be seen, the determination of A_{opt} at wavelengths longer than 800 nm is unreliable (the error in the measurement of T_{opt} and R_{opt} is of the order of A_{opt}). However, the PDS measurement in this region allows us to obtain the absorbance, A_{pds}. Figure 7b shows the absorption coefficient spectrum obtained by combining the calculation according to Equation (3) for the intense absorption region ($\alpha > 1/d$) and the TMM model fit for the weak absorption region.

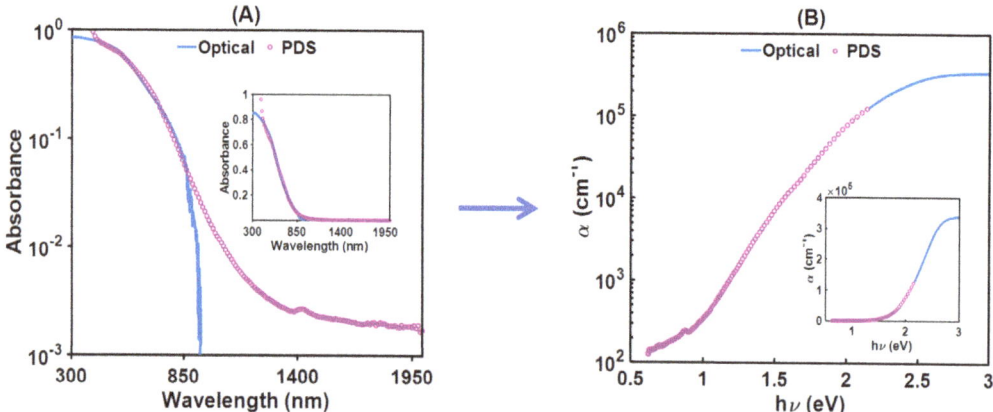

Figure 7. Absorbance obtained from optical measurements and PDS for the Cu_3N film deposited at 5.0 Pa pressure in pure N_2 atmosphere. In (A), the absorbance spectra are shown, and in (B), the absorption coefficient is calculated from the fit based on the transfer matrix method. The insets show the same graphs on a linear scale.

Once α was obtained, the indirect and direct optical band gaps of Cu_3N were obtained using the Tauc plot:

$$(\alpha h\nu)^{1/m} = B(h\nu - E_g) \tag{4}$$

where $h\nu$ is the photon energy, E_g is the band gap energy, B is a constant, and m is a factor, which depends on the nature of the electron transition (2 for indirectly allowed transitions and 1/2 for direct allowed transitions). Figure 8 shows the Tauc plots with $m = 2$ and $m = 1/2$ of the different Cu_3N samples deposited at different pressures in the two environments. The decrease in total pressure implies a decrease in both band gaps. This effect was more significant in the samples deposited on the N_2/Ar mixture.

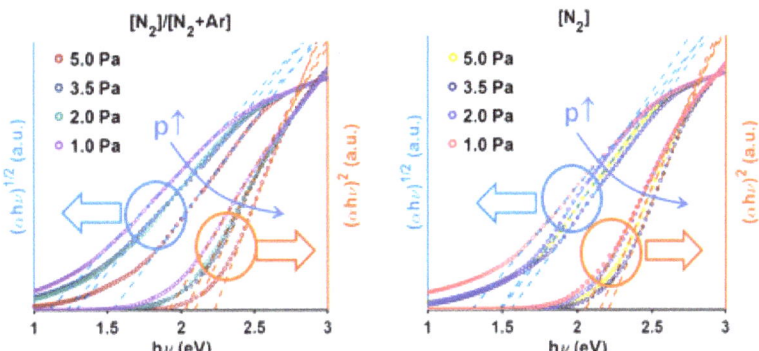

Figure 8. A plot of $(\alpha h\nu)^{1/2}$ and $(\alpha h\nu)^2$ vs. photon energy (eV) for the Cu$_3$N films at different gas pressures and atmospheres. For ease of comparison, normalised values of the absorption coefficient are considered.

Trying to further analysis, different energy ranges were distinguished according to the behaviour of α. Thus, the electronic transitions associated with the direct band gap took place for photon energies in a narrow range from 2.1 to 2.4 eV, as a lower limit, to approximately 2.5 eV, as an upper limit. On the other hand, the electronic transitions associated with the indirect gap cover a larger range extending about 0.5 eV toward lower photon energies. In addition, at lower energies, exponential Urbach absorption (α_U) can be observed, which is related to electronic transitions involving the band tails.

To accurately determine the direct and indirect band gaps, we performed a fine least-squares fit of the spectral dependence α according to the following model:

$$\text{At } h < E_1: \quad \alpha_U(h\nu) = \alpha_0 \exp(h\nu/E_U) \tag{5}$$

where E_1 is the transition energy, E_U is the Urbach energy (the slope of the exponential tail), and α_0 is the absorption prefactor.

$$\text{At } h > E_2: \quad \alpha_d(h\nu) = B_d\sqrt{h\nu - E_g^d} \quad \text{(direct Tauc model)} \tag{6}$$

E_2 is the transition energy, B_d is a constant, and E_g^d is the direct band gap energy.

$$\text{At } E_1 < h < E_2 \quad \alpha_i(h\nu) = B_i\left(h\nu - E_g^i\right)^2 \quad \text{(indirect Tauc model)} \tag{7}$$

B_i is a constant, and E_g^i is the indirect band gap energy. On the other hand, it can be observed that α must vary continuously and smoothly through the different regions. This implies imposing continuity conditions for α and its derivative at the two transition energies:

$$\begin{array}{ll} \alpha_i(E_1) = \alpha_U(E_1) & \frac{d\alpha_i}{dE}(E_1) = \frac{d\alpha_U}{dE}\alpha_U(E_1) \\ \alpha_i(E_2) = \alpha_d(E_2) & \frac{d\alpha_i}{dE}(E_2) = \frac{d\alpha_d}{dE}\alpha_U(E_2) \end{array} \tag{8}$$

which reduces the eight parameters of the model described by Equations (5)–(7) to only four independent ones. The graphs in Figure 9 depict the fit of the experimental data with the model described by Equations (5)–(8). A correlation between these energies and total pressure was observed, the effect more measurable for the samples deposited in the mixed N$_2$/Ar atmosphere. The inset in Figure 8 pictures the fit in the sub-gap region at the edge of the indirect band gap at E_1, plotted in a logarithmic scale. This is the region where the absorption coefficient is described by the Urbach exponential, related to the states in the band tails. The same as was deduced from the Tauc plots in Figure 7, the samples with the narrowest band gap (direct and indirect) were obtained at the lowest total pressures.

Therefore, by modifying the deposition parameters, such as total working pressure and N_2 flow ratio, the optical properties of Cu_3N thin films can be tailored to achieve desirable properties, making them promising for applications in optoelectronics and photonics.

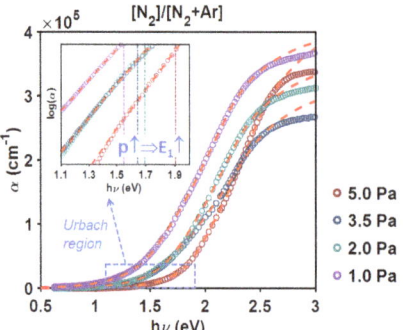
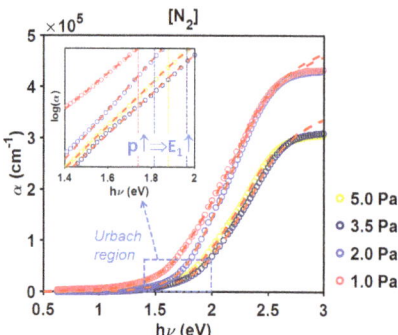

Figure 9. Absorption coefficient spectra of Cu_3N layers at different pressures and atmospheres. The red dashed lines show the fit according to the model equations described (5)–(8). In the logarithmic scale, the insets show the exponential tail of the Urbach region; note the effect of pressure on the transition energy E_1.

Finally, the shaded regions in Figure 10 represent the range of photon energies ($E_1 < h\nu < E_2$) associated with the electronic transition by the indirect gap. In general, there is an increasing tendency for the band gap with the total pressure.

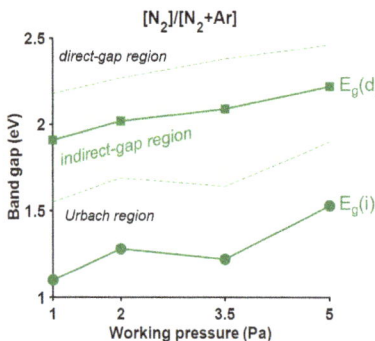
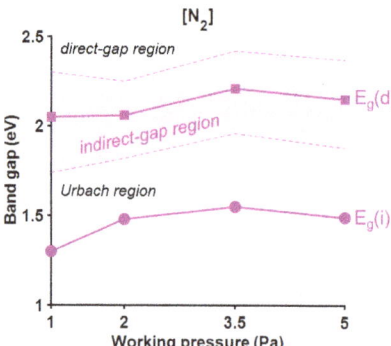

Figure 10. Evolution with a working pressure of the energies of the direct gap Eg(d) and the indirect gap Eg(i) for the two deposition atmospheres. The shaded region indicates the range of photon energies ($E_1 < h\nu < E_2$) associated with electronic transitions according to the indirect gap: for higher energies ($h\nu > E_2$), the transitions are associated with the direct gap and for lower energies ($h\nu < E_1$) with the Urbach tail.

Regarding the Urbach energy (E_U), increasing values were obtained as the total pressure decreased. These values were superior to those of other semiconductor materials [20]. It is known that high E_U values indicate a higher internal defect density, while lower Eu ones suggest a lower internal defect density. Hence, the Urbach band tail is related to impurity adsorption [51,52]. According to the literature, Cu_3N shows E_U values ranging from 105 to 238 meV [20,51] depending on the substrate temperature. Therefore, it can be concluded that the samples deposited at higher total pressures feature reasonable band gap energies and E_U values to be used in photovoltaic applications.

4. Conclusions

In this study, the Cu_3N films were successfully deposited on glass substrates via RF magnetron sputtering using a power of 50 W and different N_2 flow ratios to determine their effect on the film properties. The results provide valuable insights into these films' optical and structural properties. The analysis of the film structure revealed polycrystallinity, with a preferred growth orientation along the (100) plane when the N_2 flow ratio was $r = 1.0$. However, an amorphous matrix was observed for an N_2 flow ratio of 0.7 and low total working pressures up to 2.0 Pa; at working pressures above 2.0 Pa, a tendency for growth along the (111) plane was obtained. These findings and the lattice parameter values extracted from the XRD patterns suggest that the films deposited in the N_2/Ar gas mixture tend to have a higher concentration of copper. Raman characterisation confirmed the formation of Cu-N bonds, as evidenced by the characteristic peak observed in these spectra. The presence of oxygen in the Raman spectra and in the EDS analysis was attributed to environmental factors due to no Cu-O bond-related structures detected in the XRD patterns. The Cu/N ratio demonstrated an increase in the total working pressure. FESEM and AFM analysis showed a film morphology consisting of columnar grains with a very smooth and homogeneous surface. Through a combined analysis of the optical properties using conventional UV-VIS-NIR and PDS spectroscopies the absorption coefficient over a wide range of photon energies (from 0.5 eV to 3.5 eV) was determined. A model with two band gaps, indirect and direct, and the Urbach exponential tail in the sub-gap region described the complete absorption coefficient spectrum. Depending on the deposition conditions, the energy of the direct gap varied in the range of 1.1 to 1.5 eV and the direct gap in the range of 1.9–2.2 eV. Generally, as the working pressure decreased, the energies of the two gaps tended to decrease, with the effect being most evident in the layers deposited in the N_2/Ar gas mixture environment. The samples deposited with the lowest working pressure (1 Pa) presented the highest value of the Urbach energy (>250 meV). The minimum value of the Urbach energy of 183 meV was found for the sample deposited at 2 Pa in a pure N_2 atmosphere. Finally, these films exhibited desirable structural, morphological, and optical properties, making them promising candidates as solar absorbers. The findings could contribute to developing and optimising Cu_3N-based materials for efficient solar energy conversion.

Author Contributions: Conceptualization, M.I.R.-T. and S.F.; methodology, M.I.R.-T. and S.F.; software, J.M.A.; validation, M.I.R.-T., M.R and J.M.A.; formal analysis, M.R., J.M.A., J.M. and M.I.R.-T.; investigation, M.I.R.-T. and S.F.; resources, J.B. and S.F.; data curation, M.I.R.-T., M.R. and J.M.A.; writing—original draft preparation, M.I.R.-T., M.R. and J.M.A.; writing—review and editing, M.I.R.-T., M.R., J.M.A., J.B. and S.F.; visualization, M.I.R.-T., J.M.A. and M.R.; supervision, S.F.; project administration, J.B. and S.F.; funding acquisition, J.B. and S.F. All authors have read and agreed to the published version of the manuscript.

Funding: This research was funded by MCIN/AEI/10.13039/501100011033, grant number PID2019-109215RB-C42 and PID2019-109215RB-C43. M.I.R.-T. also acknowledges partial funding from MEDIDA C17.I2G: CIEMAT. Nuevas tecnologías renovables híbridas, Ministerio de Ciencia e Innovación, Componente 17 "Reforma Institucional y Fortalecimiento de las Capacidades del Sistema Nacional de Ciencia e Innovación". Medidas del plan de inversiones y reformas para la recuperación económica funded by the European Union—NextGenerationEU.

Institutional Review Board Statement: Not applicable.

Informed Consent Statement: Not applicable.

Data Availability Statement: Not applicable.

Acknowledgments: The authors would like to thank A. Soubrié from Centro de Microscopía Electrónica "Luis Bru" for her help and advice in AFM measurements.

Conflicts of Interest: The authors declare no conflict of interest.

References

1. Evans, H.A.; Wu, Y.; Seshadri, R.; Cheetham, A.K. Perovskite-related ReO$_3$-type structures. *Nat. Rev. Mater.* **2020**, *5*, 196–213. [CrossRef]
2. Zachwieja, U.; Jacobs, H. Ammonothermalsynthese von kupfernitrid, Cu$_3$N. *J. Less Common Met.* **1990**, *161*, 175–184. [CrossRef]
3. Yue, G.H.; Yan, P.X.; Liu, J.Z.; Wang, M.X.; Li, M.; Yuan, X.M. Copper nitride thin film prepared by reactive radio-frequency magnetron sputtering. *J. Appl. Phys.* **2005**, *98*, 103506. [CrossRef]
4. Matsuzaki, K.; Okazaki, T.; Lee, Y.-S.; Hosono, H.; Susaki, T. Controlled bipolar doping in Cu$_3$N (100) thin films. *Appl. Phys. Lett.* **2014**, *105*, 222102. [CrossRef]
5. Matsuzaki, K.; Harada, K.; Kumagai, Y.; Koshiya, S.; Kimoto, K.; Ueda, S.; Sasase, M.; Maeda, A.; Susaki, T.; Kitano, M.; et al. High-Mobility p-Type and n-Type Copper Nitride Semiconductors by Direct Nitriding Synthesis and In Silico Doping Design. *Adv. Mater.* **2018**, *30*, e1801968. [CrossRef]
6. Lu, N.; Ji, A.; Cao, Z. Nearly constant electrical resistance over large temperature range in Cu$_3$NM$_x$ (M = Cu, Ag, Au) compounds. *Sci. Rep.* **2013**, *3*, 3090. [CrossRef]
7. Chen, S.-C.; Huang, S.-Y.; Sakalley, S.; Paliwal, A.; Chen, Y.-H.; Liao, M.-H.; Sun, H.; Biring, S. Optoelectronic properties of Cu$_3$N thin films deposited by reactive magnetron sputtering and its diode rectification characteristics. *J. Alloys Compd.* **2019**, *789*, 428–434. [CrossRef]
8. Mukhopadhyay, A.K.; Momin, M.A.; Roy, A.; Das, S.C.; Majumdar, A. Optical and Electronic Structural Properties of Cu$_3$N Thin Films: A First-Principles Study (LDA + U). *ACS Omega* **2020**, *5*, 31918–31924. [CrossRef]
9. Tanveer, Z.; Mahmood, K.; Ikram, S.; Ali, A.; Amin, N. Modulation of thermoelectric properties of thermally evaporated copper nitride thin films by optimizing the growth parameters. *Phys. B Condens. Matter* **2021**, *605*, 412712. [CrossRef]
10. Matsuzaki, K.; Katase, T.; Kamiya, T.; Hosono, H. Symmetric ambipolar thin-film transistors and high-gain CMOS-like inverters using environmentally friendly copper nitride. *ACS Appl. Mater. Interfaces* **2019**, *11*, 35132–35137. [CrossRef]
11. Ścigała, A.; Szłyk, E.; Dobrzańska, L.; Gregory, D.H.; Szczęsny, R. From binary to multinary copper based nitrides–Unlocking the potential of new applications. *Coord. Chem. Rev.* **2021**, *436*, 213791. [CrossRef]
12. Chen, W.; Zhang, H.; Yang, B.; Li, B.; Li, Z. Characterization of Cu$_3$N/CuO thin films derived from annealed Cu$_3$N for electrode application in Li-ion batteries. *Thin Solid Film.* **2019**, *672*, 157–164. [CrossRef]
13. Wu, H.; Chen, W. Copper nitride nanocubes: Size-controlled synthesis and application as cathode catalyst in alkaline fuel cells. *J. Am. Chem. Soc.* **2011**, *133*, 15236–15239. [CrossRef] [PubMed]
14. Lee, B.S.; Yi, M.; Chu, S.Y.; Lee, J.Y.; Kwon, H.R.; Lee, K.R.; Kang, D.; Kim, W.S.; Lim, H.B.; Lee, J. Copper nitride nanoparticles supported on a superparamagnetic mesoporous microsphere for toxic-free click chemistry. *Chem. Commun.* **2010**, *46*, 3935–3937. [CrossRef] [PubMed]
15. Yin, Z.; Yu, C.; Zhao, Z.; Guo, X.; Shen, M.; Li, N.; Muzzio, M.; Li, J.; Liu, H.; Lin, H.; et al. Cu$_3$N Nanocubes for Selective Electrochemical Reduction of CO(2) to Ethylene. *Nano Lett.* **2019**, *19*, 8658–8663. [CrossRef]
16. Ghoohestani, M.; Karimipour, M.; Javdani, Z. The effect of pressure on the physical properties of Cu$_3$N. *Phys. Scr.* **2014**, *89*, 35801. [CrossRef]
17. Sahoo, G.; Meher, S.; Jain, M.K. Room temperature growth of high crystalline quality Cu$_3$N thin films by modified activated reactive evaporation. *Mater. Sci. Eng. B* **2015**, *191*, 7–14. [CrossRef]
18. Noh, H.; An, H.; Lee, J.; Song, J.; Hong, H.J.; Seo, S.; Jeong, S.Y.; Kim, B.-J.; Ryu, S.; Lee, S. Large enhancement of the photocurrent density in N-doped Cu$_3$N films through bandgap reduction. *J. Korean Ceram. Soc.* **2020**, *57*, 345–351. [CrossRef]
19. Rodríguez-Tapiador, M.I.; Merino, J.; Jawhari, T.; Muñoz-Rosas, A.L.; Bertomeu, J.; Fernández, S. Impact of the RF Power on the Copper Nitride Films Deposited Nitrogen Environment for Applications as Eco-Friendly Solar Absrober. *Materials* **2023**, *16*, 1508. [CrossRef]
20. Birkett, M.; Savory, C.N.; Fioretti, A.N.; Thompson, P.; Muryn, C.A.; Weerakkody, A.D.; Mitrovic, I.Z.; Hall, S.; Treharne, R.; Dhanak, V.R.; et al. Atypically small temperature-dependence of the direct band gap in the metastable semiconductor copper nitrideCu$_3$N. *Phys. Rev. B* **2017**, *95*, 115201. [CrossRef]
21. Pinkas, J.; Huffman, J.C.; Baxter, D.V.; Chisholm, M.H.; Caulton, K.G. Mechanistic Role of H$_2$O and the Ligand in the Chemical Vapor Deposition of Cu, Cu$_2$O, CuO, and Cu$_3$N from Bis (1, 1, 1, 5, 5, 5-hexafluoropentane-2, 4-dionato) copper (II). *Chem. Mater.* **1995**, *7*, 1589–1596. [CrossRef]
22. Fallberg, A.; Ottosson, M.; Carlsson, J.-O. CVD of Copper(I) Nitride. *Chem. Vap. Depos.* **2009**, *15*, 300–305. [CrossRef]
23. Modin, A.; Kvashnina, K.O.; Butorin, S.M.; Werme, L.; Nordgren, J.; Arapan, S.; Ahuja, R.; Fallberg, A.; Ottosson, M. Electronic structure of Cu$_3$N films studied by soft x-ray spectroscopy. *J. Phys. Condens. Matter.* **2008**, *20*, 235212. [CrossRef]
24. Paniconi, G.; Stoeva, Z.; Doberstein, H.; Smith, R.I.; Gallagher, B.L.; Gregory, D.H. Structural chemistry of Cu$_3$N powders obtained by ammonolysis reactions. *Solid State Sci.* **2007**, *9*, 907–913. [CrossRef]
25. Szczesny, R.; Hoang, T.K.A.; Dobrzanska, L.; Gregory, D.H. Solution/Ammonolysis Syntheses of Unsupported and Silica-Supported Copper(I) Nitride Nanostructures from Oxidic Precursors. *Molecules* **2021**, *26*, 4926. [CrossRef]
26. Scigala, A.; Szlyk, E.; Rerek, T.; Wisniewski, M.; Skowronski, L.; Trzcinski, M.; Szczesny, R. Copper Nitride Nanowire Arrays-Comparison of Synthetic Approaches. *Materials* **2021**, *14*, 603. [CrossRef]
27. Xia, Y.; Fan, G.; Chen, K.; Chen, Y.; He, Z.; Ou, J. Preparation and anti-corrosion performances of grass-like microstructured superhydrophobic surface on copper via solution-immersion. *Mater. Lett.* **2022**, *323*, 132482. [CrossRef]

28. Kashin, A.; Ananikov, V. A SEM study of nanosized metal films and metal nanoparticles obtained by magnetron sputtering. *Russ. Chem. Bull.* **2011**, *60*, 2602–2607. [CrossRef]
29. Padamata, S.K.; Yasinskiy, A.; Yanov, V.; Saevarsdottir, G. Magnetron Sputtering High-Entropy Alloy Coatings: A Mini-Review. *Metals* **2022**, *12*, 319. [CrossRef]
30. Tadjine, R.; Houimi, A.; Alim, M.M.; Oudini, N. Oxygen flow rate effect on copper oxide thin films deposited by radio frequency magnetron sputtering. *Thin Solid Film.* **2022**, *741*, 139013. [CrossRef]
31. Hu, D.-C.; Kuo, D.-H.; Kao, J.-Y.; Chen, C.-S.; Tsao, C.-C.; Hsu, C.-Y. Fabrication of nitride films by co-sputtering of high-entropy alloys and tungsten. *J. Aust. Ceram. Soc.* **2022**, *59*, 105–115. [CrossRef]
32. Islam, M.M.; Georgiev, D.G. Stable stoichiometric copper nitride thin films via reactive sputtering. *Appl. Phys. A* **2022**, *128*, 579. [CrossRef]
33. Kong, Q.; Ji, L.; Li, H.; Liu, X.; Wang, Y.; Chen, J.; Zhou, H. Influence of substrate bias voltage on the microstructure and residual stress of CrN films deposited by medium frequency magnetron sputtering. *Mater. Sci. Eng. B* **2011**, *176*, 850–854. [CrossRef]
34. Figueira, C.; Rosario, G.D.; Pugliese, D.; Rodríguez-Tapiador, M.; Fernández, S. Effect of Argon on the Properties of Copper Nitride Fabricated by Magnetron Sputtering for the Next Generation of Solar Absorbers. *Materials* **2022**, *15*, 8973. [CrossRef]
35. Rodríguez-Tapiador, M.; Merino, J.; Jawhari, T.; Muñoz-Rosas, A.; Bertomeu, J.; Fernández, S. Power effect on the properties of copper nitride films as solar absorber deposited in pure nitrogen atmosphere. *Authorea Prepr.* 2022; preprint. [CrossRef]
36. Ghosh, S.; Singh, F.; Choudhary, D.; Avasthi, D.; Ganesan, V.; Shah, P.; Gupta, A. Effect of substrate temperature on the physical properties of copper nitride films by rf reactive sputtering. *Surf. Coat. Technol.* **2001**, *142*, 1034–1039. [CrossRef]
37. Abe, T.; Yamashina, T. The deposition rate of metallic thin films in the reactive sputtering process. *Thin Solid Film.* **1975**, *30*, 19–27. [CrossRef]
38. Dorranian, D.; Dejam, L.; Sari, A.H.; Hojabri, A. Structural and optical properties of copper nitride thin films in a reactive Ar/N$_2$ magnetron sputtering system. *Eur. Phys. J. Appl. Phys.* **2010**, *50*, 20503. [CrossRef]
39. Hadian, F.; Rahmati, A.; Movla, H.; Khaksar, M. Reactive DC magnetron sputter deposited copper nitride nano-crystalline thin films: Growth and characterization. *Vacuum* **2012**, *86*, 1067–1072. [CrossRef]
40. Xiao, J.; Li, Y.; Jiang, A. Structure, optical property and thermal stability of copper nitride films prepared by reactive radio frequency magnetron sputtering. *J. Mater. Sci. Technol.* **2011**, *27*, 403–407. [CrossRef]
41. Nowakowska-Langier, K.; Chodun, R.; Minikayev, R.; Okrasa, S.; Strzelecki, G.W.; Wicher, B.; Zdunek, K. Phase composition of copper nitride coatings examined by the use of X-ray diffraction and Raman spectroscopy. *J. Mol. Struct.* **2018**, *1165*, 79–83. [CrossRef]
42. Alyousef, H.A.; Hassan, A.; Zakaly, H.M. Exploring the Impact of Substrate Placement on Cu$_3$N Thin Films as a Solar Cell Window Layer: Structural and Optical Attributes. *Mater. Today Commun.* **2023**, *35*, 106183. [CrossRef]
43. Paredes, P.; Rauwel, E.; Rauwel, P. Surveying the Synthesis, Optical Properties and Photocatalytic Activity of Cu$_3$N Nanomaterials. *Nanomaterials* **2022**, *12*, 2218. [CrossRef]
44. Chen, Y.-H.; Lee, P.-I.; Sakalley, S.; Wen, C.-K.; Cheng, W.-C.; Sun, H.; Chen, S.-C. Enhanced Electrical Properties of Copper Nitride Films Deposited via High Power Impulse Magnetron Sputtering. *Nanomaterials* **2022**, *12*, 2814. [CrossRef]
45. Kumar, K.; Kumar, A.; Devi, S.; Tyagi, S.; Kaur, D. Relevant photovoltaic effect in N-doped CQDs/MoS$_2$ (0D/2D) quantum dimensional heterostructure. *Ceram. Int.* **2022**, *48*, 14107–14116. [CrossRef]
46. Fallberg, A.; Ottosson, M.; Carlsson, J.-O. Phase stability and oxygen doping in the Cu–N–O system. *J. Cryst. Growth* **2010**, *312*, 5. [CrossRef]
47. Wilczopolska, M.; Nowakowska-Langier, K.; Okrasa, S.; Skowronski, L.; Minikayev, R.; Strzelecki, G.W.; Chodun, R.; Zdunek, K. Synthesis of copper nitride layers by the pulsed magnetron sputtering method carried out under various operating conditions. *Materials* **2021**, *14*, 2694. [CrossRef]
48. Sanjana, T.; Sunil, M.A.; Shaik, H.; Kumar, K.N. Studies on DC sputtered cuprous oxide thin films for solar cell absorber layers. *Mater. Chem. Phys.* **2022**, *281*, 125922. [CrossRef]
49. Zervos, M.; Othonos, A.; Pavloudis, T.; Giaremis, S.; Kioseoglou, J.; Mavridou, K.; Katsikini, M.; Pinakidou, F.; Paloura, E.C. Impact of Oxygen on the Properties of Cu$_3$N and Cu$_{3-x}$N$_{1-x}$O$_x$. *J. Phys. Chem. C* **2021**, *125*, 3680–3688. [CrossRef]
50. Rahmati, A.; Ghoohestani, M.; Badehian, H.; Baizaee, M.A. initio study of the structural, elastic, electronic and optical properties of Cu$_3$N. *Mater. Res.* **2014**, *17*, 303–310. [CrossRef]
51. Xiao, J.; Qi, M.; Gong, C.; Wang, Z.; Jiang, A.; Ma, J.; Cheng, Y. Crystal structure and optical properties of silver-doped copper nitride films (Cu$_3$N: Ag) prepared by magnetron sputtering. *J. Phys. D Appl. Phys.* **2018**, *51*, 55305. [CrossRef]
52. Borsa, D.; Grachev, S.; Presura, C.; Boerma, D. Growth and properties of Cu$_3$N films and Cu$_3$N/γ'-Fe$_4$N bilayers. *Appl. Phys. Lett.* **2002**, *80*, 1823–1825. [CrossRef]

Disclaimer/Publisher's Note: The statements, opinions and data contained in all publications are solely those of the individual author(s) and contributor(s) and not of MDPI and/or the editor(s). MDPI and/or the editor(s) disclaim responsibility for any injury to people or property resulting from any ideas, methods, instructions or products referred to in the content.

Article

Al$_2$O$_3$ Ceramic/Nanocellulose-Coated Non-Woven Separator for Lithium-Metal Batteries

Dong-Min Shin [1,2], Hyunsu Son [1,†], Ko Un Park [3], Junyoung Choi [1], Jungdon Suk [1,4,*], Eun Seck Kang [3], Dong-Won Kim [2] and Do Youb Kim [1,*]

1. Energy Materials Research Center, Korea Research Institute of Chemical Technology, 141 Gajeong-ro, Yuseong-gu, Daejeon 34114, Republic of Korea
2. Department of Chemical Engineering, Hanyang University, Seoul 04763, Republic of Korea
3. Advanced Materials Parts TP, Materials & Devices Advanced Research Center, Chief Technology Office, LG Electronics, W1, LG Science Park, 10, Magokjungang 10-ro, Gangseo-gu, Seoul 07796, Republic of Korea
4. Department of Chemical Convergence Materials, University of Science and Technology, 217 Gaejong-ro, Yuseong-gu, Daejeon 34113, Republic of Korea
* Correspondence: jdsuk@krict.re.kr (J.S.); dykim@krict.re.kr (D.Y.K.)
† Current Address: Advanced Pouch Cell Development Team 2, Mobility & IT Battery Development, LG Energy Solution, 188, Munji-ro, Yuseong-gu, Daejeon 34122, Republic of Korea.

Citation: Shin, D.-M.; Son, H.; Park, K.U.; Choi, J.; Suk, J.; Kang, E.S.; Kim, D.-W.; Kim, D.Y. Al$_2$O$_3$ Ceramic/Nanocellulose-Coated Non-Woven Separator for Lithium-Metal Batteries. *Coatings* **2023**, *13*, 916. https://doi.org/10.3390/coatings13050916

Academic Editor: Ning Sun

Received: 24 April 2023
Revised: 8 May 2023
Accepted: 11 May 2023
Published: 13 May 2023

Copyright: © 2023 by the authors. Licensee MDPI, Basel, Switzerland. This article is an open access article distributed under the terms and conditions of the Creative Commons Attribution (CC BY) license (https:// creativecommons.org/licenses/by/ 4.0/).

Abstract: Separators play an essential role in lithium (Li)-based secondary batteries by preventing direct contact between the two electrodes and providing conduction pathways for Li-ions in the battery cells. However, conventional polyolefin separators exhibit insufficient electrolyte wettability and thermal stability, and in particular, they are vulnerable to Li dendritic growth, which is a significant weakness in Li-metal batteries (LMBs). To improve the safety and electrochemical performance of LMBs, Al$_2$O$_3$ nanoparticles and nanocellulose (NC)-coated non-woven poly(vinylidene fluoride)/polyacrylonitrile separators were fabricated using a simple, water-based blade coating method. The Al$_2$O$_3$/NC-coated separator possessed a reasonably porous structure and a significant number of hydroxyl groups (-OH), which enhanced electrolyte uptake (394.8%) and ionic conductivity (1.493 mS/cm). The coated separator also exhibited reduced thermal shrinkage and alleviated uncontrollable Li dendritic growth compared with a bare separator. Consequently, Li-metal battery cells with a LiNi$_{0.8}$Co$_{0.1}$Mn$_{0.1}$O$_2$ cathode and an Al$_2$O$_3$/NC-coated separator using either liquid or solid polymer electrolytes exhibited improved rate capability, cycle stability, and safety compared with a cell with a bare separator. The present study demonstrates that combining appropriate materials in coatings on separator surfaces can enhance the safety and electrochemical performance of LMBs.

Keywords: nanocellulose; Al$_2$O$_3$; lithium-metal battery; safety; stability

1. Introduction

Lithium (Li)-ion batteries (LIBs) are regarded as excellent power sources for electronic devices, electric vehicles (EVs), and large-scale energy storage systems (ESSs) owing to their high energy density, good cycle performance, and small self-discharge capacity [1,2]. Typically, LIBs consist of four main components: the cathode, anode, separator, and electrolyte. Among them, the separator plays a decisive role in isolating the cathode and anode to prevent physical contact and providing channels for Li-ion travel inside the battery cell [3–5]. At present, microporous polyolefin separators, including polyethylene (PE) and polypropylene (PP), are widely used in LIBs owing to their good mechanical properties and excellent chemical stability. However, polyolefin separators have some fundamental limitations that must be resolved for general LIB applications [6–8]. Commonly, non-polar polyolefin separators exhibit relatively poor wetting with polar organic liquid electrolytes, which eventually leads to an increase in the internal resistance of LIBs [9–11]. Moreover, the relatively low melting temperature of commercial polyolefin separators (130 °C for PE and 165 °C for PP) leads to shrinkage of the separators in a high-temperature environment, which can cause a fire or even an explosion owing to direct contact between the cathode and anode in a battery cell [11,12]. Hence, their insufficient electrolyte wettability and poor thermal stability render them unable to meet the needs of high-performance LIBs.

To overcome these drawbacks, a large number of researchers have contributed to the development of novel LIB separators. Among various approaches, the use of an inorganic ceramic particle coating on the separator surface has attracted considerable attention owing to its simple operating process, relatively easy thickness control of the separator, and low cost [13–15]. In addition, inorganic ceramic particles such as SiO_2, Al_2O_3, and TiO_2 can not only improve the absorption of the liquid electrolyte, but also enhance the thermal stability of the separators [6,7,16–18]. For example, Shi et al. fabricated Al_2O_3 powder-coated PE separators [6], and Zheng et al. reported SiO_2 nanoparticle-coated PE separators [18]. These coated separators showed improved thermal stability, good wettability, and higher uptake of liquid electrolyte than bare PE separators. In addition, LIB cells using the coated separators exhibited improved electrochemical performance compared with cells using bare PE separators.

An alternative material to enhance the properties of separators is nanocellulose (NC), which is a cellulosic material with one dimension in the nanometer range, and is considered a promising material owing to its high mechanical properties and outstanding electrolyte wettability [19,20]. In addition, cellulose is the most abundant natural polymeric material on earth, is renewable and biodegradable with thermal dimensional stability. Despite these benefits, the relatively high production cost of NC is a major drawback. However, cellulose shows a good affinity with polar liquid electrolytes owing to its many hydrophilic components, such as hydroxyl groups (-OH) and has good mechanical properties with abundant hydrogen bonds between the cellulose fibers [9,20,21]. Accordingly, researchers have attempted to fabricate functional separators using NC to enhance the safety and electrochemical performance of LIB cells [22–26]. However, most studies have focused on constructing porous separators using NC as the base material. Recently, Pan et al. reported that an NC-modified PE separator could stabilize Li-metal anodes by regulating the Li-ion flux and mitigating Li dendrite growth, eventually improving the cycle performance of LIB cells [20]. The results demonstrated that NC is an excellent coating material for high-performance separators; nevertheless, studies on NC-coated separators have rarely been reported.

We speculated that if an appropriate amount of NC were incorporated into the ceramic particle coating layer, it could enhance the electrolyte uptake without pore clogging. Accordingly, a simple, water-based and cost-effective blade coating method was applied to fabricate Al_2O_3/NC composite-coated poly(vinylidene fluoride)/polyacrylonitrile (PVdF/PAN) non-woven separators. To the best of our knowledge, this is the first report on separators coated with ceramic particles and NC. Introducing NC in a coating layer can improve the hydrophilicity of the separator surface and the wettability in electrolytes. In addition, NC acts as a spacer between the Al_2O_3 particles, which provides the coated separator with a more suitable pore structure for electrolyte penetration. Consequently, the coated separator showed higher electrolyte uptake, up to 394%, and enhanced ionic conductivity. The coated separator also exhibited improved thermal stability and mitigated Li dendrite growth. As a result, NCM811//liquid electrolyte (LE)//Li LIB cells using the Al_2O_3/NC-coated separator exhibited enhanced rating capability and cycle stability compared with cells using a bare PVdF/PAN separator. Furthermore, a solid polymer electrolyte (SPE) was also applied to the coated separators to fabricate Li polymer battery (LPB) cells. The coated separator also enhanced the electrochemical performance of the NCM811//SPE//Li LPB cells at room temperature.

2. Experimental Section

2.1. Materials

A poly(vinylidene fluoride)/polyacrylonitrile non-woven separator (PVdF/PAN, thickness = 30 μm) was purchased from Amogreentech Co. Ltd. and used as a base separator. Aqueous nanocellulose (NC) solution (1 wt%) was obtained from LG Electronics (Seoul, Republic of Korea). Aluminum oxide (Al_2O_3, average particle size = 480 nm; AES-11, Sumitomo Chemical Co., Tokyo, Japan), sodium carboxymethyl cellulose (CMC, WS-C, Dai-ichi Kogyo Seiyaku Co., Ltd., Tokyo, Japan), disodium laureth sulfosuccinate (28 wt% ASCO® DLSS, AK Chemtech Co., Ltd., Seoul, Republic of Korea), poly(ethylene glycol) dimethyl ether (PEGDME, average Mw = 500, Sigma-Aldrich, St. Louis, MO, USA), and fluoroethylene carbonate (FEC, Sigma-Aldrich, 99%) were purchased and used as received. Bisphenol A ethoxylate diacrylate (BPh-A, average Mw = 688, Hannog Chemical, Gunsan, Republic of Korea) and *t*-butyl peroxipivalate (Seki Arkema Co., Haman-gun, Republic of Korea) were used as received. Lithium bis(trifluoromethanesulfonyl) imide (LiTFSI, Sigma-Aldrich, 99.95%) was dried in a vacuum oven at 120 °C before use. In all experiments, deionized (DI) water with a resistivity of 18.2 MΩ, which was prepared using a Milli-Q ultrapure water system (Millipore, Burlington, VT, USA), was used.

2.2. Preparation of Al_2O_3/NC-Coated Separators

Coating slurries with various Al_2O_3 and NC weight ratios were prepared by mixing aqueous stock solutions of Al_2O_3 (52 wt%) and NC (1 wt%) in various ratios. Aqueous solutions of 2 wt% CMC (as a binder) and 1 wt% DLSS (as a surfactant) were then added and mixed homogeneously. The coating slurries were then cast onto a bare PVdF/PAN separator using a doctor blade, and the coated separators were dried in a convection oven at 60 °C for 4 h. The Al_2O_3/NC-coated separator is denoted by Al_2O_3/NC-X, where X represents the mass ratio of NC (out of 10) in the Al_2O_3/NC composite. Table S1 in the Supporting Information (SI) summarizes the weight of Al_2O_3, the NC stock solutions for preparing coating solutions, and the resulting weight ratio between Al_2O_3 and NC.

2.3. Characterization

The surface morphologies of the samples were observed using a field-emission scanning electron microscope (FE-SEM, JSM-6700F, JEOL, Tokyo, Japan) at an accelerating voltage of 10 kV. The water contact angle was evaluated using a contact angle tester (SEO300A, Surface and Electrooptics Co., Suwon, Republic of Korea). The Gurley values for the samples were obtained using a Gurley densometer (Model 4110 N, Gurley Precision Instruments, Troy, MI, USA) by measuring the time required for 100 cc of air to pass through under a pressure of 0.02 MPa. The thermal shrinkage of the separators was calculated by measuring their dimensions before and after heating at different temperatures for 15 min according to the following equation:

$$\text{Thermal shrinkage} = (D - D_0)/D_0 \times 100\%, \quad (1)$$

where D_0 and D represent the areas of the separator before and after heating, respectively. The electrolyte uptake of the different separators was measured by immersing the separators in an electrolyte (1 M $LiPF_6$ in ethylene carbonate (EC)/diethyl carbonate (DEC) (1:1 v/v)). The electrolyte-soaked separators were weighed after removing redundant electrolyte using wipes. The electrolyte uptake can be calculated using Equation (2):

$$\text{Uptake} = (W - W_0)/W_0 \times 100\%, \quad (2)$$

where W and W_0 are the weights of the wet separator with the electrolyte and the dry separator, respectively.

2.4. Electrochemical Measurements

To evaluate the electrochemical performances of the samples, CR2032 coin-type cells were assembled in an Ar-filled glove box (H_2O < 1 ppm, MBraun, Velbert, Germany). The ionic conductivity was measured via electrochemical impedance spectroscopy (EIS) using a potentiostat with a built-in EIS (VMP3, Biologic, Seyssinet-Pariset, France) at room temperature. Cells for the EIS measurements were fabricated using two stainless steel (SUS) electrodes (0.125 mm thick, Hohsen Co., Osaka, Japan) with various separator samples. The scanning frequency ranged from 0.1 Hz to 65 kHz with a small perturbation voltage of 10 mV AC amplitude. The ionic conductivity was calculated using the following equation:

$$\sigma = d/(R_b \times S), \quad (3)$$

where σ, d, R_b, and S represent the ionic conductivity, thickness of the separator, bulk resistance, and electrode area, respectively. The electrochemical stability window was measured by linear sweep voltammetry (LSV) using a potentiostat (VMP3, Biologic) with a working electrode of SUS and a counter electrode of Li at a scan rate of 1.0 mV/s from open circuit voltage (OCV) to 6.0 V (vs. Li/Li$^+$).

Li plating/stripping tests and LIB performance were evaluated using Li//Li symmetric cells and LiNi$_{0.8}$Co$_{0.1}$Mn$_{0.1}$O$_2$(NCM811)//Li half-cells, respectively, with a liquid electrolyte (LE, 1.0 M LiPF$_6$ in EC/DEC (1:1 v/v) with 5 wt% of FEC) as well as a solid polymer electrolyte (SPE). The NCM811 (POSCO Chem., Pohang, Repblic of Korea) cathode was fabricated by casting a slurry containing NCM811 powder, Super P, and PVdF at a mass ratio of 65.5:1.5:2 in *N*-methyl-2-pyrrolidone (NMP) on an Al-foil current collector, which was then dried in an oven at 100 °C for 24 h. The loading of the NCM811 active material was fixed at ~7.0 mg/cm^2. To fabricate the SPE-loaded cells, an SPE precursor solution was prepared. The Li salt (LiTFSI) was mixed with a plasticizer, PEGDME, for 24 h. Then, BPh-A (as a crosslinking agent) and FEC (5 wt%, as an electrolyte additive) were added to the mixed solution. The ratio of PEGDME to BPh-A was fixed at 8:2 by weight, and the concentration of the Li salt was adjusted to obtain an [EO]/[Li$^+$] molar ratio of 20. *t*-BPP at 1 wt% with respect to BPh-A was mixed with the precursor solution as an initiator before use. Test cells loaded with SPE were assembled with the SPE precursor solution-soaked separator, and the cells were then placed in an oven at 90 °C for 30 min for crosslinking [27,28]. The LIB test was performed in the voltage range of 3.0–4.2 V (vs. Li/Li$^+$).

3. Results and Discussion

The Al$_2$O$_3$/NC composite-coated separators with various mass ratios of Al$_2$O$_3$ and NC (Al$_2$O$_3$/NC-X, where X represents the mass ratio of NC out of 10 of the Al$_2$O$_3$/NC composite) were fabricated by water-based coating slurries on a PVdF/PAN non-woven base separator using a doctor blade (see the Section 2 for details). Briefly, coating slurries were prepared by mixing Al$_2$O$_3$ particles, NC, sodium carboxymethyl cellulose (CMC) as a binder and disodium laureth sulfosuccinate (DLSS) as a surfactant in DI water. DLSS is an anionic surfactant and it is well-known that it is electrochemically stable up to 4.4 V vs. Li/Li$^+$ [29]. It not only improves dispersion stability of the coating materials in a slurry but also enhances the wettability of coating slurry on a separator surface [29,30]. In our experiments also, DLSS played a key role in enhancing the quality of coating layer. As shown in Figure S1 in the SI, a coating layer without DLSS was not stable enough and easily detached from a base separator surface even during a brief time in water. On the contrary, a coating slurry with DLSS could produce a conformal and stable coating layer due to well-dispersed coating materials and enhanced wettability on the base separator surface.

Figure 1 shows the top-view scanning electron microscope (SEM) images of the bare PVdF/PAN and Al$_2$O$_3$/NC-X separators. As shown in Figure 1a, a bare PVdF/PAN separator consists of numerous nanofibers with diameters of approximately 200–500 nm and a large number of voids between individual nanofibers. The surfaces of the coated separators were completely covered with coating materials (e.g., Al_2O_3 and NC). Figure 1b shows the SEM image of a separator coated with only Al_2O_3 (Al_2O_3/NC-O), where the base separator was concealed by an Al_2O_3 coating layer with pores smaller than those of the base separator. When the mass ratio of NC was relatively low (not greater than three), there were no obvious changes in the surface morphology compared with Al_2O_3/NC-0, even though NC was visible (Figure S2a in the SI and Figure 1c). The red arrows in Figure 1c indicate NC particles between the Al_2O_3 particles. As the NC mass ratio increased, a larger number of NC particles were observed on the surface of the coated separators, and the pores between the Al_2O_3 particles were blocked (Figure S2b in the SI and Figure 1d). A further increase in the mass ratio of NC eventually clogged most of the pores (Figure S2c in the SI), thereby hindering the migration of Li ions through the separator. The thickness of the coating layer was approximately 5 µm for all the coated separators (Figure S3 in the SI). XRD patterns obtained from the coated separators exhibited sharp peaks for α-Al_2O_3, and broad peaks for PAN and PVdF (Figure S4 in the SI). However, since the amount of NC was relatively small and had a relatively lower degree of crystallinity than the Al_2O_3 particles, peaks for NC were not observed. Figure S5 in the SI shows energy dispersive X-ray spectroscopy mapping images of coated separators.

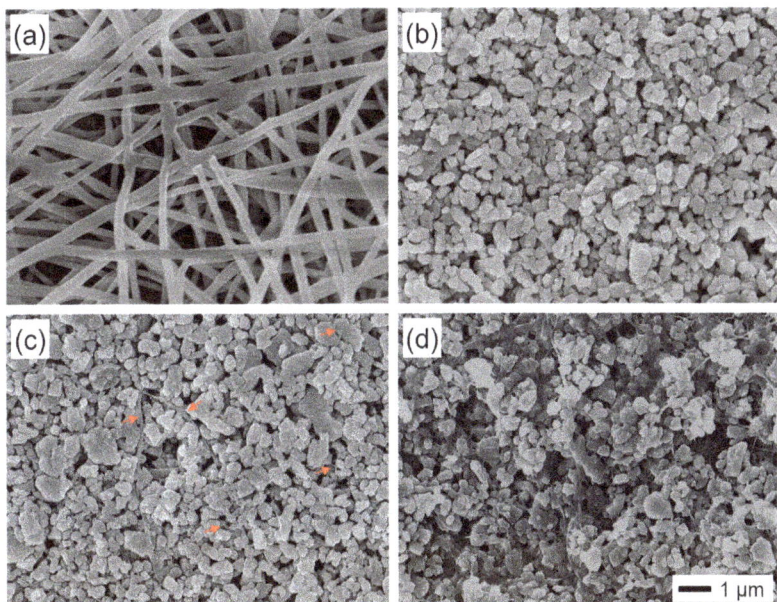

Figure 1. Top-view scanning electron microscope (SEM) images of (**a**) bare PVdF/PAN, (**b**) Al_2O_3/NC-0, (**c**) Al_2O_3/NC-3, and (**d**) Al_2O_3/NC-7 separators. Red arrows in (**c**) indicate NCs. All images have an identical magnification.

The wetting behavior is an important property of separators. Figure 2a shows the contact angles of deionized water on the separator samples. The bare PVdF/PAN separator exhibited a relatively high contact angle of 85° because of the hydrophobic nature of the bare PVdF/PAN separator surface. When the base separator was coated with Al_2O_3, the contact angle decreased substantially owing to the hydrophilic character of the Al_2O_3 particles, even though it still exhibited a contact angle of 20°. In contrast, the Al_2O_3/NC-3 and Al_2O_3/NC-7 separators were completely wet with deionized water and exhibited an astonishingly low contact angle of 0°. This could be attributed to the addition of NC with abundant hydrophilic polar -OH groups on its surface. A small contact angle tends to result in better wettability of the electrolyte, which can affect the electrolyte uptake, ionic conductivity of the separator, and eventually, the electrochemical performance of the battery [25,31]. The electrolyte wetting behavior of the separator samples was also evaluated using several electrolytes with various polarity indices (Figure S6 in the SI). All the separator samples, including the base separator, exhibited good wettability to the electrolyte solvents.

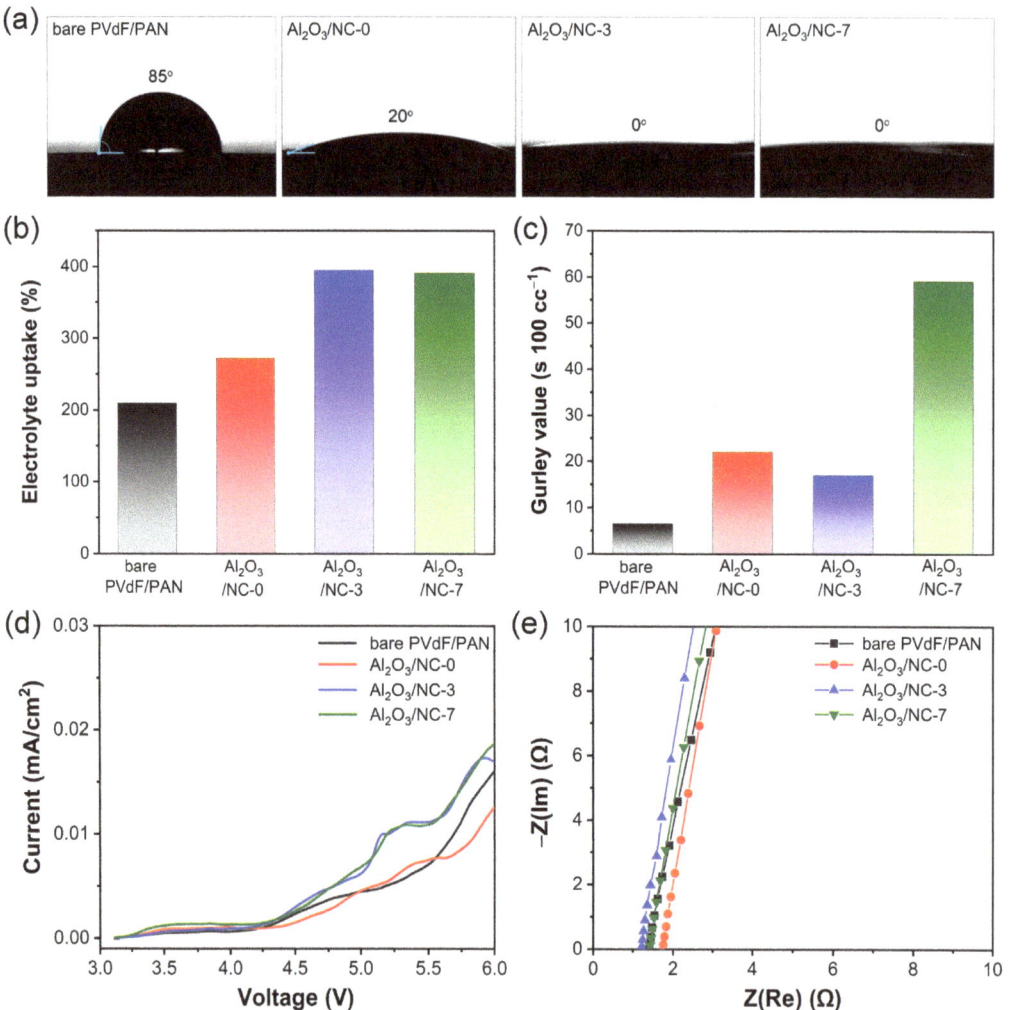

Figure 2. (**a**) Water contact angles on the separator samples, (**b**) electrolyte uptake, and (**c**) Gurley value of the separator samples. (**d**) Linear sweep voltammetry curves of SUS//Li cells and (**e**) Nyquist plots of SUS//SUS cells using the separator samples.

The electrolyte uptake of the separators, which represents the ability of a separator to retain an electrolyte solution, was also evaluated (Figure 2b). The electrolyte uptake of the bare PVdF/PAN was 209.5% and that of the Al_2O_3/NC-0 was 271.6%. The superior wetting ability of the separators coated with Al_2O_3 and NC enabled enhanced electrolyte uptake of 394.8% and 391.2% for Al_2O_3/NC-3 and Al_2O_3/NC-7, respectively.

The air permeability of the separator samples was determined by the Gurley method, which measures the time required for 100 cc of air to pass through the separator samples

(Figure 2c and Figure S7 in the SI). The bare PVdF/PAN had a large number of submicron pores and thus showed a relatively low Gurley value of 6.5 s. Al_2O_3/NC-0 exhibited an increased Gurley value of 22 s because the bare PVdF/PAN surface was covered with Al_2O_3 particles. However, when a small amount of NC was added to the coating layer, the Gurley values of the separators were slightly decreased; the values for Al_2O_3/NC-1 and Al_2O_3/NC-3 were 18 s and 17 s, respectively. This can be explained by the small amount of NC acting as a spacer, and the close packing of the Al_2O_3 particles was interrupted by the NC. The reduced amount of Al_2O_3 in the coating layer could be one of the reasons for the decreased Gurley values of the samples. As the NC mass ratio was further increased, the Gurley values of the separator samples rebounded because a relatively large amount of NC blocked the pores of the coating layers. Al_2O_3/NC-5, Al_2O_3/NC-7, and Al_2O_3/NC-9 exhibited Gurley values of 19 s, 59 s, and 231 s, respectively.

Figure 2d shows the linear sweep voltammetry (LSV) curves of the SUS//Li cells using the separator samples. Although the curves gradually increased from approximately 4.3 V, the current was relatively low and resembled that of the bare PVdF/PAN separator. This result implies that the electrochemical stability of the coated separators is comparable with that of the bare PVdF/PAN separator and is suitable for battery applications. Figure 2e shows Nyquist plots of the SUS/SUS cells obtained using the separator samples. The ionic conductivities of the separator samples were determined using the bulk resistances indicated by the high-frequency x-axis intercepts. The ionic conductivity of bare PVdF/PAN was 1.07 mS/cm and that of Al_2O_3/NC-0 was 1.04 mS/cm. However, the Al_2O_3/NC-3 and Al_2O_3/NC-7 separators exhibited enhanced ionic conductivities of 1.49 and 1.28 mS/cm, respectively. These values were approximately 40% and 20% higher than those for of the bare PVdF/PAN separator. Generally, the ionic conductivity of a separator is related to its electrolyte uptake ability and pore structure, which provides channels for the transportation of ions [31–33]. Therefore, the superior ionic conductivity of Al_2O_3/NC-3 was attributed to a relatively low Gurley value (17 s) and high electrolyte uptake (394.8%). The higher ionic conductivity of the Al_2O_3/NC-3 separator not only improved the discharge capacity but also promoted the transport of lithium ions [34,35]. Table 1 summarizes the properties of the separator samples.

Table 1. Comparison of properties of a bare PVdF/PAN and Al_2O_3/NC-X-coated separators.

Sample	Thickness (µm)	Electrolyte Uptake (%)	Gurley Number (s)	Ionic Conductivity (mS/cm)
bare PVdF/PAN	30	209.5	6.5	1.07
Al_2O_3@NC-0	35	271.6	22	1.04
Al_2O_3@NC-3	35	394.8	17	1.49
Al_2O_3@NC-7	35	391.2	59	1.28

The excellent thermal stability of separators is essential for battery safety as it can avoid explosions caused by short circuits at elevated temperatures. To verify the thermal stability of the separator samples, their morphological changes and shrinkage were examined after treatment at a series of temperatures for 15 min (Figure 3). The differences in the dimensional stability of the separator samples after treatment at elevated temperatures were obvious (Figure 3a). The shrinkage of the separator samples was quantified using the normalized cross-sectional area (Figure 3b). In the case of the bare PVdF/PAN separator, a 4% thermal shrinkage was observed at a relatively low temperature of 100 °C. It further shrank as temperature increased and maintained only 46% of its original size after being treated at 250 °C. Meanwhile, owing to the excellent thermal stability of the Al_2O_3 particles and NC, all the Al_2O_3/NC-X samples displayed significantly improved thermal durability compared with the bare PVdF/PAN separator. The Al_2O_3/NC-0 separator exhibited the best thermal stability; it maintained 95% of its original size after being treated at 250 °C. Because the thermal stability of NC is inferior to that of Al_2O_3, the thermal shrinkage of Al_2O_3/NC-X increased as the mass ratio of NC increased. After being heated at 150 °C,

Al$_2$O$_3$/NC-3 and Al$_2$O$_3$/NC-7 maintained 100% and 95% of their original sizes, respectively. Even after being heated at 250 °C, Al$_2$O$_3$/NC-3 and Al$_2$O$_3$/NC-7 still retained 79% and 72% of their original sizes, respectively. These values are far higher than those of the bare PVdF/PAN separator (46%). The morphological changes and shrinkages of all the separator samples after treatment at elevated temperatures are shown in Figures S8 and S9 in the SI, respectively. The results show that the Al$_2$O$_3$/NC coating layer significantly enhanced the thermal stability and effectively reduced the thermal shrinkage of the separators, which could improve the safety of LIBs using coated separators.

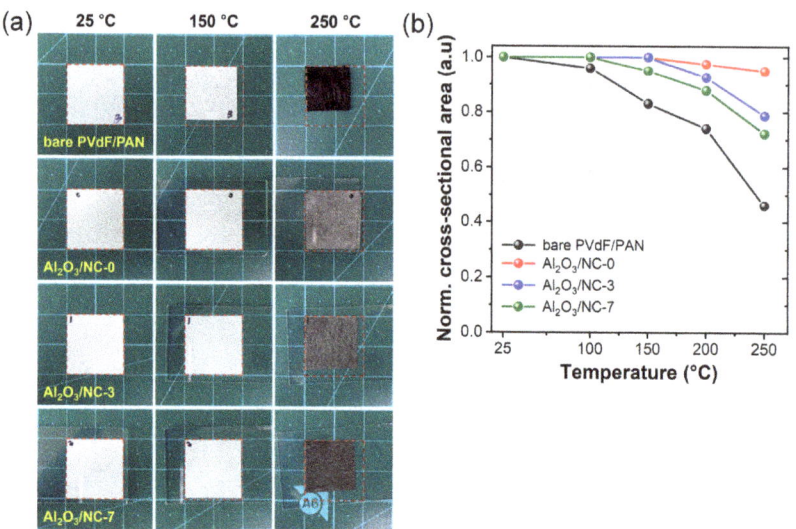

Figure 3. (**a**) Photographs showing thermal shrinkage of a bare PVdF/PAN and Al$_2$O$_3$/NC-X-coated separators at ambient and elevated temperatures. (**b**) Comparison of normalized cross-sectional area of separator samples at various temperatures.

To investigate the cyclic stability of the separators, Li plating/stripping tests were performed using Li//LE//Li symmetric cells with different separators. Figure 4a shows the voltage profiles of the cells with a capacity of 1.0 mAh/cm^2 at a current density of 1.0 mA/cm^2. In a cell using a bare PVdF/PAN separator, the profile exhibited a gradual increase in the voltage hysteresis and a sudden drop at approximately 160 h. This phenomenon could be attributed to an internal short circuit in the cell owing to the non-homogeneous Li-ion flux and continuous growth of sharp Li dendrites on the surface of the Li electrode during the test [35,36]. In contrast, the cells using the coated separators showed more stable voltage responses with lower overpotentials and longer cycle lives than the cell using a bare PVdF/PAN separator. The cells using Al$_2$O$_3$/NC-0, Al$_2$O$_3$/NC-3, and Al$_2$O$_3$/NC-7 separators also exhibited a gradual increase in the voltage hysteresis and short-circuited at approximately 240 h, 280 h, and 240 h, respectively. The improved cycle stability of the cells using the coated separators could be attributed to the enhanced electrolyte uptake and the coating layer could alleviating the growth of Li dendrites by regulating the Li-ion flux. For better comparison, Figure S10 in the SI shows the enlarged voltage profiles of these cells. Although the overpotentials gradually increased as the test progressed in all cells, cells using the coated separators displayed much lower overpotentials than that of the cell using a bare PVdF/PAN separator. Notably, the cell using the Al$_2$O$_3$/NC-3 separator exhibited the lowest overpotential and longest cycle life among the samples. The superior electrolyte uptake and high ionic conductivity of the Al$_2$O$_3$/NC-3 separator could facilitate the efficient movement of Li-ions in the cell and contribute to mitigating electrochemical polarization during cycling.

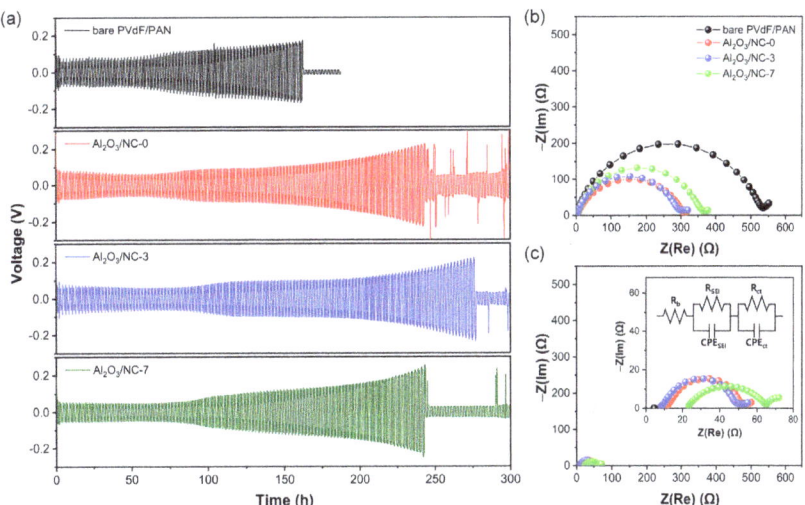

Figure 4. (a) Voltage profiles of Li plating/stripping in Li/LE/Li symmetric cells using different separators with a capacity of 1.0 mAh/cm² at a current density of 1.0 mA/cm². Nyquist plots of the symmetric cells using different separators (b) before and (c) after 100th plating/stripping cycle.

Figure 4b,c show the Nyquist plots of the cells using different separators before and after the 100th Li plating/stripping cycle, respectively. Generally, R_b in the high-frequency region of the alternating current (AC) impedance spectrum represents the bulk resistance, and the interfacial resistance of the middle- and low-frequency regions in the semi-cycle includes the solid electrolyte interface resistance (R_{SEI}) and charge transfer resistance (R_{ct}) [33,36,37]. According to the fitted results, the interfacial resistance of the bare PVdF/PAN, Al$_2$O$_3$/NC-0, Al$_2$O$_3$/NC-3, and Al$_2$O$_3$/NC-7 separators before cycling was 537.4, 306.8, 301.5 and 369.9 Ω, respectively. The Al$_2$O$_3$/NC-3 separator showed the lowest interfacial resistance compared with the other separators, which can be attributed to the superior properties of the Al$_2$O$_3$/NC-3 separator, such as better affinity with a liquid electrolyte and high electrolyte uptake [7,14,38,39]. Furthermore, the cell using the Al$_2$O$_3$/NC-3 separator retained a lower resistance after 100 cycles (Figure 4c); this is consistent with the lowest overpotentials at the 100th cycle in the voltage profiles of the Li plating/stripping test (Figure 4a and Figure S10 in the SI). These results suggest that the Al$_2$O$_3$/NC-3 separator possesses better interfacial compatibility with the electrolyte and lithium electrode, which, in turn, affects the high C-rate and stable cycle performance of the battery cell.

In addition, a functional separator with a higher mechanical strength can alleviate Li dendrite growth [40,41]. The growth of Li dendrite can cause a rapid decline in cell performance due to the continuous decomposition of an electrolyte and the formation of a thick SEI layer on the Li anode [42,43]. Additionally, it can lead to cell explosion due to short-circuiting. Therefore, investigating the morphological changes in Li is crucial in the development of Li metal-based batteries [44–50]. To verify the ability of the coated separator to suppress the dendritic growth of Li, the morphology of Li was examined after Li plating/stripping cycles using SEM. Figure 5a,b show the top-view SEM images of Li after the 10th Li plating/stripping cycle in Li/LE/Li symmetric cells using bare PVdF/PAN and Al$_2$O$_3$/NC-3 separators, respectively. In the cell using the bare PVdF/PAN separator, Li deposits were composed of numerous Li dendrites and masses, rendering a rough and porous morphology (Figure 5a). By contrast, Li deposits in the cell using the Al$_2$O$_3$/NC-3 separator showed a relatively dense and smooth surface. These SEM results prove that the Al$_2$O$_3$/NC coating layer uniformly distributed the Li ions through the pores of the coating layer and also inhibited the extensive growth of Li dendrites owing to its relatively high mechanical strength [51–53].

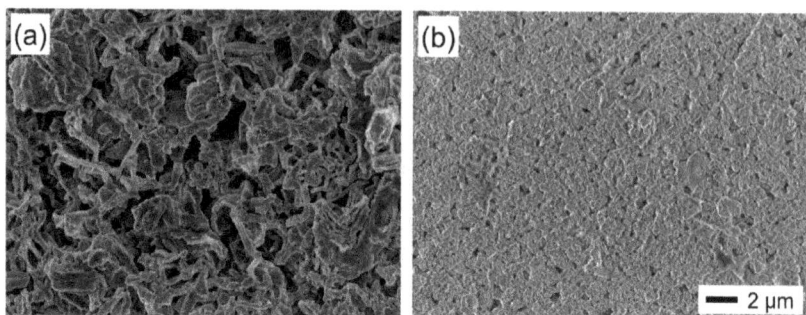

Figure 5. Top-view SEM images of lithium anodes in Li/LE/Li symmetric cells using (**a**) bare PVdF/PAN and (**b**) Al$_2$O$_3$@NC-3 separators after 10 plating/stripping cycles with a capacity of 1.0 mAh/cm^2 at a current density of 1.0 mA/cm^2. The images have an identical magnification.

Encouraged by these results, NCM811//LE//Li cells using different separators were assembled to investigate the effect of the coating layer on the electrochemical performance of the LIB cells. As shown in Figure 6a, all the cells displayed similar voltage profiles and specific capacities during the 1st cycle at a current density of 0.1 C-rate. Discharge capac-ities of 199.6, 194.1, 198.2, and 195.1 mAh/g were noted for the cells using the bare PVdF/PAN, Al$_2$O$_3$/NC-0, Al$_2$O$_3$/NC-3, and Al$_2$O$_3$/NC-7 separator, respectively. When the current density was gradually increased to 2.0 C-rate, the cell using the Al$_2$O$_3$/NC-3 sepa-rator displayed the best rate capability among all the cells. The cell using the Al$_2$O$_3$/NC-3 separator delivered a discharge capacity of 146.4 mAh/g at 2.0 C-rate, which was higher than those for the cells using the bare PVdF/PAN (138.1 mAh/g), Al$_2$O$_3$/NC-0 (140.6 mAh/g), and Al$_2$O$_3$/NC-7 (121.1 mAh/g) (Figure 6b and Figure S11 in the SI). The higher rate capability of the cell using Al$_2$O$_3$/NC-3 can be attributed to the improved electrolyte uptake and ionic conductivity of the separator, owing to the Al$_2$O$_3$ and NC coating layers.

Figure 6c shows the cycle performance of the cells using different separators at 0.5 C-rate. Based on these results, the highest capacity retention was achieved by the cell using the Al$_2$O$_3$/NC-3 separator (65.9%) after 120 charge/discharge cycles, as compared with the cells using the bare PVdF/PAN (57.1%), Al$_2$O$_3$/NC-0 (61.5%), and Al$_2$O$_3$/NC-7 (42.8%). Furthermore, electrochemical impedance spectroscopy (EIS) was performed on the cells using the bare PVdF/PAN and Al$_2$O$_3$/NC-3 separators; it was found that the cell using the Al$_2$O$_3$/NC-3 separator exhibited a lower resistance after the 1st and 100th cycles (Figure S12 in the SI). The lower resistance of the cell was beneficial for reducing the electrochemical polarization of the cell [54], which could contribute toward the improved cycle stability of the cell using the Al$_2$O$_3$/NC-3 separator. In addition, its ability to suppress Li dendrite growth and stabilize the Li deposit of the Al$_2$O$_3$/NC-3 separator, as shown in Figure 5, could also be a reason for the improved cycle stability of the cell.

To further confirm the thermal safety of the cells, the open-circuit voltages (OCVs) of the cells were monitored using bare PVdF/PAN and Al$_2$O$_3$/NC-3 separators before and after heat treatment at 120 °C. As shown in Figure S13 of the SI, the OCV of the cell using a bare PVdF/PAN separator decreased abruptly after heating at 120 °C; this was likely caused by a short circuit induced by the thermal shrinkage of the separator. By contrast, the cell using the Al$_2$O$_3$/NC-3 separator maintained its OCV at approximately 3.4 V (vs. Li/Li$^+$), even after being heated at 120 °C, without an abrupt decrease in the OCV.

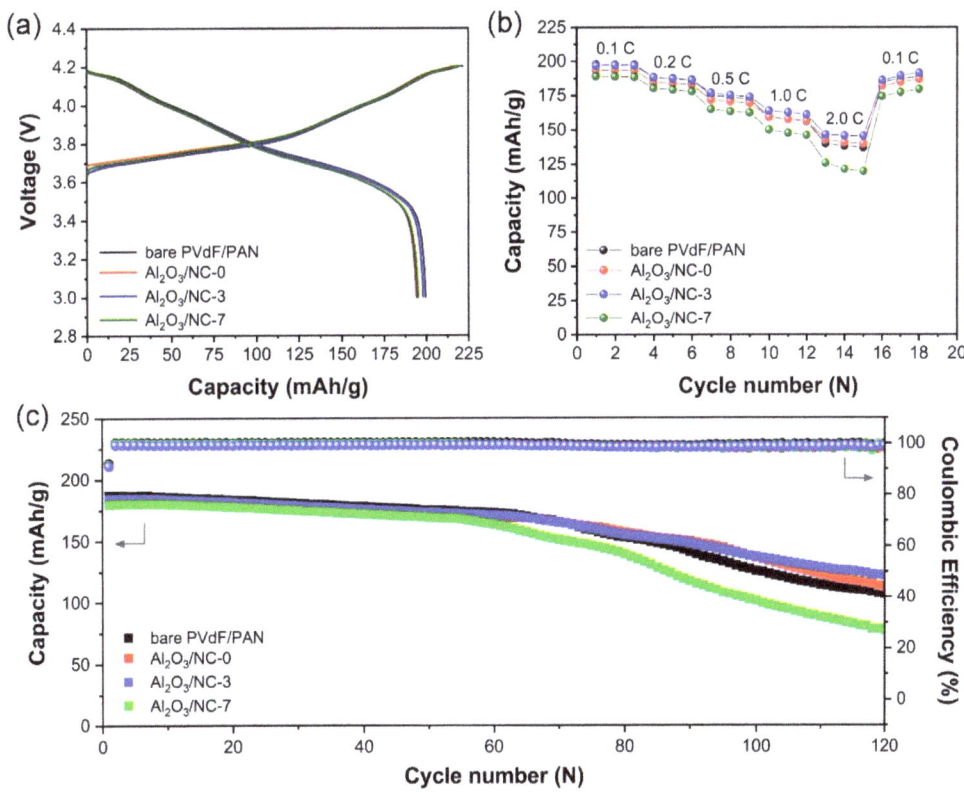

Figure 6. Electrochemical performance of NCM811/LE/Li half-cells using different separators; (a) voltage profiles at 0.1 C-rate, (b) C-rate performance, and (c) cycle stability at 0.5 C-rate.

Furthermore, we evaluated the applicability of the coated separators to LPBs by adopting an SPE. Figure S14 of the SI presents the voltage profiles during the Li plating/stripping test for the Li//SPE//Li symmetric cells using different separators at room temperature. As the ionic conductivity of the SPE used in this study was lower than that of the LE [27], the voltage profiles displayed higher overpotentials compared with those for the LE cases, even at lower current densities. Among these, the cell using Al_2O_3/NC-3 showed the best performance, with a stable voltage profile and the lowest overpotential across the experimental period. This result indicates that the Li ions could be sufficiently transferred through the SPE system within the separators. This result also implies that the porous structure of the Al_2O_3/NC-3 separator could induce a uniform Li distribution across the lithium electrode [52,55].

Moreover, NCM811//SPE//Li LPB cells with different separators were assembled, and their electrochemical performance was evaluated at room temperature. The cells exhibited inherently inferior electrochemical performances than the cells using LE, owing to the lower ionic conductivity of the SPE. However, the cells using Al_2O_3/NC-3 showed electrochemical performances equal to or greater than those of the other cells (Figure 7a,b and Figure S15 in the SI). Especially for the cycling test at 0.2 C-rate, the cell using Al_2O_3/NC-3 exhibited the highest discharge capacity of 123.5 mAh/g and a capacity retention of 77.8% at the 100th cycle. These values exceeded those for the cells using the bare PVdF/PAN (98.0 mAh/g, 65.1%), Al_2O_3/NC-0 (107.6 mAh/g, 73.4%), and Al_2O_3/NC-7 (52.0 mAh/g, 34.1%). These results also indicate that coated separators can be applied to LPBs to enhance their electrochemical performance. Finally, the electrochemical performance of our

LMB cells was compared with those of previously reported cells using a NCM811 cathode (Table S2 in the SI).

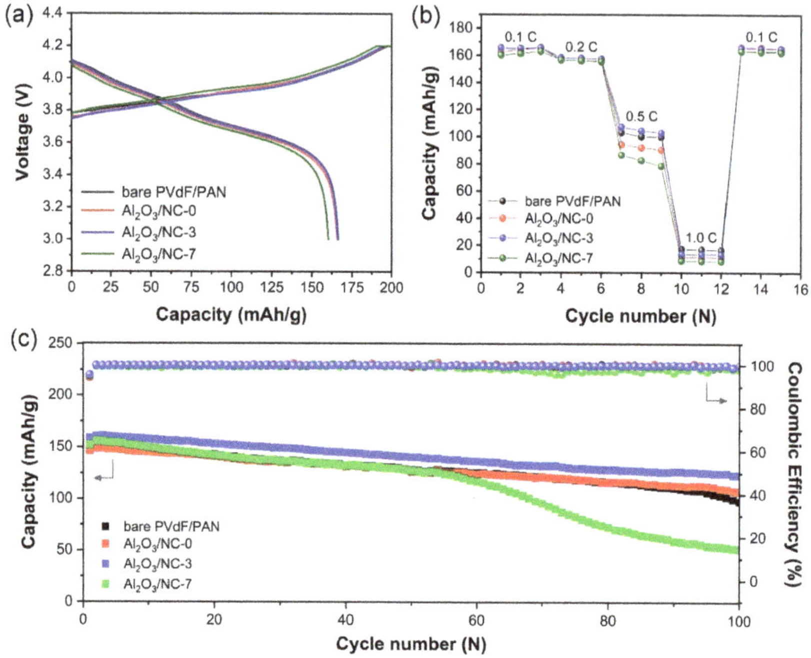

Figure 7. Electrochemical performance of NCM811/SPE/Li half-cells using different separators: (a) voltage profiles at 0.1 C-rate, (b) C-rate performance, and (c) cycle stability at 0.2 C-rate.

4. Conclusions

In summary, Al_2O_3/NC-coated PVdF/PAN separators were prepared for the first time by a simple aqueous blade coating method. The effects of the weight ratio between Al_2O_3 and NC in the coating layer on thermal stability and electrochemical performance of the separators were systematically investigated. Thanks to the inclusion of hydrophilic NC in the coating layer, the optimized Al_2O_3/NC-3-coated separator exhibited a good porous structure on the surface as well as improved electrolyte wettability, electrolyte uptake, and ionic conductivity. The separator also showed improved thermal durability with reduced thermal shrinkage at elevated temperatures in comparison to a bare separator. In addition, it was found that when the Al_2O_3/NC-3-coated separator was used, uncontrollable Li dendrite growth could be effectively alleviated by delocalized Li-ions and mechanical suppression. As a consequence, NCM811//LE//Li LIB cells using the Al_2O_3/NC-3 separator showed enhanced rate capability and superior cyclic performance than a cell using a bare separator. Furthermore, an SPE was also applied to the coated separators to evaluate their electrochemical performance in LPB cells. At room temperature, the NCM811//SPE//Li cell using the Al_2O_3/NC-3 separator also exhibited improved electrochemical performance compared with the cell with a bare separator. These findings suggest that combination of appropriate materials in the coating on a separator surface can enhance the safety and electrochemical performance of LMBs. Overall, Al_2O_3/NC-coated separators have the potential to provide large-scale Li-based secondary batteries with higher safety and electrochemical performance for application in EVs and ESS.

Supplementary Materials: The following supporting information can be downloaded at: https://www.mdpi.com/article/10.3390/coatings13050916/s1. Table S1: Weight of Al_2O_3 and NC stock solutions for preparing coating solutions and resulting weight ratio between Al_2O_3 and NC, Figure S1: Photographs of Al_2O_3/NC coated separators using coating slurries with or without DLSS, Figure S2: Top-view SEM images of (a) Al_2O_3/NC-1, (b) Al_2O_3/NC-5, and (c) Al_2O_3/NC-9 coated separators, Figure S3: (left) Cross-sectional SEM image of a Al_2O_3/NC-3 coated separator, and (right) photograph images showing separator samples and their thicknesses, Figure S4: XRD patterns of Al_2O_3/NC-0 and Al_2O_3/NC-3 separators, Figure S5: SEM-EDS mapping images of C, O, and Al for Al_2O_3/NC-0, Al_2O_3/NC-3, and Al_2O_3/NC-7 separators, Figure S6: Photographs showing wetting behavior of various solvents on different separators, Figure S7: Gurley value of the separator samples, Figure S8: Photographs showing thermal shrinkage of bare PVdF/PAN and Al_2O_3/NC coated separators at ambient and elevated temperatures, Figure S9: Comparison of normalized cross-sectional area of all separator samples, Figure S10: Enlarged voltage profiles of Li plating/stripping in Li/LE/Li symmetric cells using different separators with a capacity of 1.0 mAh/cm^2 at a current density of 1.0 mA/cm^2 shown in Figure 4, Figure S11: Charge-discharge voltage profiles of NCM811/LE/Li cells using different separators; (a) bare PVdF/PAN, (b) Al_2O_3@NC-0, (c) Al_2O_3@NC-3 and (d) Al_2O_3@NC-7 separators, Figure S12: Nyquist plots of NCM811/LE/Li half cells using bare PVdF/PAN and Al_2O_3/NC-3 coated separators (a) after 1 cycle and (b) 100 cycles at 0.5 C-rate, Figure S13: Plot of open circuit voltages of the NCM811/LE/Li cells using a bare PVdF/PAN and Al_2O_3/NC-3 separators at room temperature before and after heating at 120 °C, Figure S14: Voltage profiles of Li plating/stripping in Li/SPE/Li symmetric cells using different separators with a capacity of 0.4 mAh/cm^2 at a current density of 0.4 mA/cm^2, Figure S15: Charge-discharge voltage profiles of NCM811/SPE/Li cells using different separators; (a) bare PVdF/PAN, (b) Al_2O_3@NC-0, (c) Al_2O_3@NC-3 and (d) Al_2O_3@NC-7 separators, Table S2:Comparison of electrochemical performance of LMB cells using NCM811 cathode [56–69].

Author Contributions: Conceptualization, D.-M.S. and D.Y.K.; methodology, D.-M.S., H.S., K.U.P. and J.C.; validation, E.S.K., D.-W.K. and J.S.; formal analysis, D.-M.S. and K.U.P.; investigation, D.-M.S. and D.Y.K.; data curation, D.-M.S.; writing—original draft preparation, D.-M.S.; writing—review and editing, D.Y.K.; supervision, D.Y.K.; project administration, E.S.K. and J.S.; funding acquisition, E.S.K. and J.S. All authors have read and agreed to the published version of the manuscript.

Funding: This research was supported by Nano·Material Technology Development Program through the National Research Foundation of Korea (NRF) funded by Ministry of Science and ICT (NRF-2020M3H4A3081876 and NRF-2020M3H4A3082530).

Institutional Review Board Statement: Not applicable.

Informed Consent Statement: Not applicable.

Data Availability Statement: The data presented in this study are available on request from the corresponding author.

Conflicts of Interest: The authors declare no conflict of interest.

References

1. Kim, T.; Song, W.; Son, D.Y.; Ono, L.K.; Zi, Y. Lithium-ion batteries: Outlook on present, future, and hybridized technologies. *J. Mater. Chem. A* **2019**, *7*, 2942–2964. [CrossRef]
2. Zubi, G.; Dufo-López, R.; Carvalho, M.; Pasaoglu, G. The lithium-ion battery: State of the art and future perspectives. *Renew. Sustain. Energy Rev.* **2018**, *89*, 292–308. [CrossRef]
3. Yuan, B.; Wen, K.; Chen, D.; Liu, Y.; Dong, Y.; Feng, C.; Han, Y.; Han, J.; Zhang, Y.; Xia, C.; et al. Composite Separators for Robust High Rate Lithium Ion Batteries. *Adv. Func. Mater.* **2021**, *31*, 2101420. [CrossRef]
4. Lu, W.; Yuan, Z.; Zhao, Y.; Zhang, H.; Zhang, H.; Li, X. Porous membranes in secondary battery technologies. *Chem. Soc. Rev.* **2017**, *46*, 2199–2236. [CrossRef]
5. Xiang, Y.; Li, J.; Lei, J.; Liu, D.; Xie, Z.; Qu, D.; Li, K.; Deng, T.; Tang, H. Advanced Separators for Lithium-Ion and Lithium-Sulfur batteries: A Review of Recent Progress. *ChemSusChem* **2016**, *9*, 3023–3039. [CrossRef] [PubMed]
6. Shi, C.; Zhang, P.; Chen, L.; Yang, P.; Zhao, J. Effect of a thin ceramic-coating layer on thermal and electrochemical properties of polyethylene separator for lithium-ion batteries. *J. Power Sources* **2014**, *270*, 547–553. [CrossRef]
7. Yang, P.; Zhang, P.; Shi, C.; Chen, L.; Dai, J.; Zhao, J. The functional separator coated with core-shell structured silica-poly(methyl methacrylate) sub-microspheres for lithium-ion batteries. *J. Membr. Sci.* **2015**, *474*, 148–155. [CrossRef]

8. Shi, J.L.; Fang, L.F.; Li, H.; Zhang, H.; Zhu, B.K.; Zhu, L.P. Improved thermal and electrochemical performances of PMMA modified PE separator skeleton prepared via dopamine-initiated ATRP for lithium ion batteries. *J. Membr. Sci.* **2013**, *437*, 160–168. [CrossRef]
9. Zhu, C.; Zhang, J.; Xu, J.; Yin, X.; Wu, J.; Chen, S.; Zhu, Z.; Wang, L.; Wang, H. Facile fabrication of cellulose/polyphenylene sulfide composite separator for lithium-ion batteries. *Carbohydr. Polym.* **2020**, *248*, 116753. [CrossRef]
10. Kim, K.J.; Kwon, Y.K.; Yim, T.; Choi, W. Functional separator with lower resistance toward lithium ion transport for enhancing the electrochemical performance of lithium ion batteries. *J. Ind. Eng. Chem.* **2019**, *71*, 228–233. [CrossRef]
11. Sun, G.; Guo, J.; Niu, H.; Chen, N.; Zhang, M.; Tian, G.; Qi, S.; Wu, D. The design of a multifunctional separator regulating the lithium ion flux for advanced lithium-ion batteries. *RSC Adv.* **2019**, *9*, 40084–40091. [CrossRef]
12. Wang, Z.; Chen, J.; Ye, B.; Pang, P.; Ma, Z.; Chen, H.; Nan, J. A pore-controllable polyamine (PAI) layer-coated polyolefin (PE) separator for pouch lithium-ion batteries with enhanced safety. *J. Solid State Electrochem.* **2020**, *24*, 843–853. [CrossRef]
13. Lee, Y.; Lee, H.; Lee, T.; Ryou, M.H.; Lee, Y.M. Synergistic thermal stabilization of ceramic/co-polyimide coated polypropylene separators for lithium-ion batteries. *J. Power Sources* **2015**, *294*, 537–544. [CrossRef]
14. Li, W.; Li, X.; Yuan, A.; Xie, X.; Xia, B. Al_2O_3/poly(ethylene terephthalate) composite separator for high-safety lithium-ion batteries. *Ionics* **2016**, *22*, 2143–2149. [CrossRef]
15. Ding, L.; Yan, N.; Zhang, S.; Xu, R.; Wu, T.; Yang, F.; Cao, Y.; Xiang, M. Low-Cost Mass Manufacturing Technique for the Shutdown-Functionalized Lithium-Ion Battery Separator Based on Al_2O_3 Coating Online Construction during the β-iPP Cavitation Process. *ACS Appl. Mater. Interfaces* **2022**, *14*, 6714–6728. [CrossRef] [PubMed]
16. Parikh, D.; Jafta, C.J.; Thapaliya, B.P.; Sharma, J.; Meyer, H.M., III; Silkowski, C.; Li, J. Al_2O_3/TiO_2 coated separators: Roll-to-roll processing and implications for improved battery safety and performance. *J. Power Sources* **2012**, *507*, 230259. [CrossRef]
17. Shekarian, E.; Nasr, M.R.J.; Mohammadi, T.; Bakhtiari, O.; Javanbakht, M. Enhanced Wettability and Electrolyte Uptake of Coated Commercial Polypropylene Separators with Inorganic Nanopowders for Application in Lithium-ion Battery. *J. Nanostruct.* **2019**, *9*, 736–750. [CrossRef]
18. Zheng, H.; Wang, Z.; Shi, L.; Zhao, Y.; Yuan, S. Enhanced thermal stability and lithium ion conductivity of polyethylene separator by coating colloidal SiO_2 nanoparticles with porous shell. *J. Colloid Interface Sci.* **2019**, *554*, 29–38. [CrossRef]
19. Xu, Q.; Kong, Q.; Liu, Z.; Wang, X.; Liu, R.; Zhang, J.; Yue, L.; Duan, Y.; Cui, G. Cellulose/Polysulfonamide Composite Membrane as a High Performance Lithium-Ion Battery Separator. *ACS Sustain. Chem. Eng.* **2014**, *2*, 194–199. [CrossRef]
20. Pan, R.; Xu, X.; Sun, R.; Wang, Z.; Lindh, J.; Edstrom, K.; Stromme, M.; Nyholm, L. Nanocellulose Modified Polyethylene Separators for Lithium Metal Batteries. *Small* **2018**, *14*, e1704371. [CrossRef]
21. Sheng, J.; Tong, S.; He, Z.; Yang, R. Recent developments of cellulose materials for lithium-ion battery separators. *Cellulose* **2017**, *24*, 4103–4122. [CrossRef]
22. Chun, S.J.; Choi, E.S.; Lee, E.H.; Kim, J.H.; Lee, S.Y.; Lee, S.Y. Eco-friendly cellulose nanofiber paper-derived separator membranes featuring tunable nanoporous network channels for lithium-ion batteries. *J. Mater. Chem.* **2012**, *22*, 16618–16626. [CrossRef]
23. Pan, R.; Wang, Z.; Sun, R.; Lindh, J.; Edstron, K.; Stromme, M.; Nyholm, L. Polydopamine-based redox-active separators for lithium-ion batteries. *J. Mater.* **2019**, *5*, 204–213. [CrossRef]
24. Wang, Z.; Pan, R.; Sun, R.; Edstrom, K.; Stromme, M.; Nyholm, L. Nanocellulose Structured Paper-Based Lithium Metal Batteries. *ACS Appl. Energy Mater.* **2018**, *1*, 4341–4350. [CrossRef]
25. Huang, Q.; Zhao, C. Bacterial cellulose nanofiber membrane for use as lithium-ion battery separator. *IOP Conf. Ser. Earth Environ. Sci.* **2021**, *647*, 012069. [CrossRef]
26. Kim, J.H.; Gu, M.; Lee, D.H.; Kim, J.H.; Oh, Y.S.; Min, S.H.; Kim, B.S.; Lee, S.Y. Functionalized Nanocellulose-Integrated Heterolayered Nanomats toward Smart Battery Separators. *Nano Lett.* **2016**, *16*, 5533–5541. [CrossRef]
27. Choi, W.; Kang, Y.; Kim, I.J.; Seong, B.G.; Yu, W.R.; Kim, D.W. Stable Cycling of a 4 V Class Lithium Polymer Battery Enabled by In Situ Cross-Linked Ethylene Oxide/Propylene Oxide Copolymer Electrolytes with Controlled Molecular Structures. *ACS Appl. Mater. Interfaces* **2021**, *13*, 35664–35676. [CrossRef]
28. Kim, H.; Kim, D.Y.; Suk, J.; Kang, Y.; Lee, J.B.; Kim, H.J.; Kim, D.W. Stable cycling via absolute intercalation in graphite-based lithium-ion battery incorporated by solidified ether-based polymer electrolyte. *Mater. Adv.* **2021**, *2*, 3898–3905. [CrossRef]
29. Jeon, H.; Yeon, D.; Lee, T.; Park, J.; Ryou, M.H.; Lee, Y.M. A water-based Al_2O_3 ceramic coating for polyethylene-based microporous separator for lithium-ion batteries. *J. Power Sources* **2016**, *315*, 161–168. [CrossRef]
30. Kennedy, S.; Kim, J.T.; Lee, Y.M.; Rhiri, I.; Ryou, S.Y. Upgrading the Properties of Ceramic-Coated Separator for Lithium Secondary Batteries by Changing the Mixing Order of the Water-Based Ceramic Slurry Components. *Batteries* **2022**, *8*, 64. [CrossRef]
31. Liu, X.; Wei, S.; Ning, R.; Sun, Z.; Li, X. Preparation and Characteristics of Lithium Battery Separator Based on Cellulose Modification by Water-Soluble Polyimide Impregnated. *IOP Conf. Ser. Earth Environ. Sci.* **2021**, *687*, 012119. [CrossRef]
32. Gou, J.; Liu, W.; Tang, A. A novel method to prepare a highly porous separator based on nanocellulose with multi-scale pore structures and its application for rechargeable lithium ion batteries. *J. Membr. Sci.* **2021**, *639*, 119750. [CrossRef]
33. Xu, R.; Sheng, L.; Gong, H.; Kong, Y.; Yang, Y.; Li, M.; Bai, Y.; Song, S.; Liu, G.; Wang, T.; et al. High-Performance Al_2O_3/PAALi Composite Separator Prepared by Water-Based Slurry for High-Power Density Lithium-Based Battery. *Adv. Eng. Mater.* **2020**, *23*, 2001009. [CrossRef]
34. Yang, C.; Tong, H.; Luo, C.; Yuan, S.; Chen, G.; Yang, Y. Boehmite particle coating modified microporous polyethylene membrane: A promising separator for lithium ion batteries. *J. Power Sources* **2017**, *348*, 80–86. [CrossRef]

35. Ryu, J.; Han, D.Y.; Hong, D.; Park, S. A polymeric separator membrane with chemoresistance and high Li-ion flux high-energy-density lithium metal batteries. *Energy Storage Mater.* **2022**, *45*, 941–951. [CrossRef]
36. Li, J.; Wang, Q.; Wang, Z.; Cao, Y.; Zhu, J.; Lou, Y.; Zhao, Y.; Shi, L.; Yuan, S. Evaporation and in-situ gelation induced porous hybrid film without template enhancing the performance of lithium ion battery separator. *J. Colloid Interface Sci.* **2021**, *595*, 142–150. [CrossRef]
37. Wang, Y.; Yin, C.; Song, Z.; Wang, Q.; Lan, Y.; Luo, J.; Bo, L.; Yue, Z.; Sun, F.; Li, X. Application of PVDF Organic Particles Coating on Polyethylene Separator for Lithium Ion Batteries. *Materials* **2019**, *12*, 3125. [CrossRef]
38. Thiangtham, S.; Saito, N.; Manuspiya, H. Asymmetric Porous and Highly Hydrophilic Sulfonated Cellulose/Biomembrane Functioning as a Separator in a Lithium-Ion Battery. *ACS Appl. Energy Mater.* **2022**, *5*, 6206–6218. [CrossRef]
39. Zhang, T.; Qu, H.; Sun, K. Development of polydopamine coated electrospun PAN/PMMA nanofibrous membrane as composite separator for lithium-ion batteries. *Mater. Lett.* **2019**, *245*, 10–13. [CrossRef]
40. Lei, O.K.; Zhang, Q.; Wu, X.Y.; Wei, X.; Zhang, J.; Wang, K.X.; Chen, J.S. Towards ultra-stable lithium metal batteries: Interfacial ionic flux regulated through LiAl LDH-modified polypropylene separator. *Chem. Eng. J.* **2020**, *395*, 125187. [CrossRef]
41. Li, D.; Wang, H.; Luo, L.; Zhu, J.; Li, J.; Liu, P.; Yu, Y.; Jiang, M. Electrospun Separator Based on Sulfonated Polyoxadiazole with Outstanding Thermal Stability and Electrochemical Properties for Lithium-Ion Batteries. *ACS Appl. Energy Mater.* **2021**, *4*, 879–887. [CrossRef]
42. Xu, W.; Wang, J.; Ding, F.; Chen, X.; Nasybulin, E.; Zhang, Y.; Zhang, J.G. Lithium metal anodes for rechargeable batteries. *Energy Environ. Sci.* **2014**, *7*, 513–537. [CrossRef]
43. Zhang, J.G.; Xu, W.; Xiao, J.; Cao, X.; Liu, J. Lithium Metal Anodes with Nonaqueous Electrolytes. *Chem. Rev.* **2020**, *120*, 13312–13348. [CrossRef]
44. Sun, F.; He, X.; Jiang, X.; Osenberg, M.; Li, J.; Zhou, D.; Dong, K.; Hilger, A.; Zhu, X.; Gao, R.; et al. Advancing knowledge of electrochemically generated lithium microstructure and performance decay of lithium ion battery by synchrotron X-ray tomography. *Mater. Today* **2019**, *27*, 21–32. [CrossRef]
45. Sun, F.; Zielke, L.; Markötter, H.; Hilger, A.; Zhou, D.; Moroni, R.; Zengerle, R.; Thiele, S.; Banhart, J.; Manke, I. Morphological Evolution of Electrochemically Plated/Stripped Lithium Microstructures Investigated by Synchrotron X-ray Phase Contrast Tomography. *ACS Nano* **2016**, *10*, 7990–7997. [CrossRef]
46. Sun, F.; Moroni, R.; Dong, K.; Markötter, H.; Zhou, D.; Hilger, A.; Zielke, L.; Zengerle, R.; Thiele, S.; Banhart, J.; et al. Study of the Mechanisms of Internal Short Circuit in a Li/Li Cell by Synchrotron X-ray Phase Contrast Tomography. *ACS Energy Lett.* **2017**, *2*, 94–104. [CrossRef]
47. Sun, F.; Osenberg, M.; Dong, K.; Zhou, D.; Hilger, A.; Jafta, C.J.; Risse, S.; Lu, Y.; Markötter, H.; Manke, I. Correlating Morphological Evolution of Li Electrodes with Degrading Electrochemical Performance of Li/LiCoO$_2$ and Li/S Battery Systems: Investigated by Synchrotron X-ray Phase Contrast Tomography. *ACS Energy Lett.* **2018**, *3*, 356–365. [CrossRef]
48. Sun, F.; Yang, C.; Manke, I.; Chen, L.; Dong, S. Li-based anode: Is dendrite-free sufficient? *Mater. Today* **2020**, *38*, 7–9. [CrossRef]
49. Sun, F.; Gao, R.; Zhou, D.; Osenberg, M.; Dong, K.; Kardjilov, N.; Hilger, A.; Markötter, H.; Bieker, P.M.; Liu, X.; et al. Revealing Hidden Facts of Li Anode in Cycled Lithium−Oxygen Batteries through X-ray and Neutron Tomography. *ACS Energy Lett.* **2019**, *4*, 306–316. [CrossRef]
50. Sun, F.; Zhou, D.; He, X.; Osenberg, M.; Dong, K.; Chen, L.; Mei, S.; Hilger, A.; Markötter, H.; Lu, Y.; et al. Morphological Reversibility of Modified Li-Based Anodes for Next-Generation Batteries. *ACS Energy Lett.* **2020**, *5*, 152–161. [CrossRef]
51. Huang, Z.; Chen, Y.; Han, Q.; Su, M.; Liu, Y.; Wang, S.; Wang, H. Vapor-induced phase inversion of poly (m-phenylene isophthalamide) modified polyethylene separator for high-performance lithium-ion batteries. *Chem. Eng. J.* **2022**, *429*, 132429. [CrossRef]
52. Choi, J.; Yang, K.; Bae, H.S.; Phiri, I.; Ahn, H.J.; Won, J.C.; Lee, Y.M.; Kim, Y.H.; Ryou, M.H. Highly Stable Porous Polyimide Sponge as a Separator for Lithium-Metal Secondary Batteries. *Nanomaterials* **2020**, *10*, 1976. [CrossRef] [PubMed]
53. Choi, S.; Mugobera, S.; Ko, J.M.; Lee, K.S. Lithium Dendrite-suppressing separator with high thermal stability by rod-like ZnO coating for lithium batteries. *Colloids Surf. A Physicochem. Eng. Asp.* **2021**, *631*, 127722. [CrossRef]
54. Yin, Y.; Zhao, W.; Wang, A.; Yue, H.; Cao, Z.; Shi, Z.; Li, S.; Yang, S. Cation-Selective Dual-Functional Separator as an Effective Polysulfide Barrier and a Li Dendrite Inhibitor for Lithium-Sulfur Batteries. *ACS Appl. Energy Mater.* **2020**, *3*, 11855–11862. [CrossRef]
55. Han, D.H.; Zhang, M.; Lu, P.X.; Wan, Y.L.; Chen, Q.L.; Niu, H.Y.; Yu, Z.W. A multifunctional separator with Mg(OH)$_2$ nanoflake coatings for safe lithium-metal batteries. *J. Energy Chem.* **2021**, *52*, 75–83. [CrossRef]
56. Yang, L.-Y.; Cao, J.-H.; Liang, W.-H.; Wang, Y.-K.; Wu, D.-Y. Effects of the Separator MOF-Al$_2$O$_3$ Coating on Battery Rate Performance and Solid−Electrolyte Interphase Formation. *ACS Appl. Mater. Interfaces* **2022**, *14*, 13722. [CrossRef]
57. Cheng, C.; Liu, H.; Ouyang, C.; Hu, N.; Zha, G.; Hou, H. A high-temperature stable composite polyurethane separator coated Al$_2$O$_3$ particles for lithium ion battery. *Compos. Commun.* **2022**, *33*, 101217. [CrossRef]
58. Liang, T.; Cao, J.-H.; Liang, W.-H.; Li, Q.; He, L.; Wu, D.-Y. Asymmetrically coated LAGP/PP/PVDF–HFP composite separator film and its effect on the improvement of NCM battery performance. *RSC Adv.* **2019**, *9*, 41151. [CrossRef]
59. Rao, Q.-S.; Liao, S.-Y.; Huang, X.-W.; Li, Y.-Z.; Liu, Y.-D.; Min, Y.-G. Assembly of MXene/PP Separator and Its Enhancement for Ni-Rich LiNi$_{0.8}$Co$_{0.1}$Mn$_{0.1}$O$_2$ Electrochemical Performance. *Polymers* **2020**, *12*, 2192. [CrossRef]

60. Li, X.; Liu, K.; Yan, Y.; Yu, J.; Dong, N.; Liu, B.; Tian, G.; Qi, S.; Wu, D. Thermostable and nonflammable polyimide/zirconia compound separator for lithium-ion batteries with superior electrochemical and safe properties. *J. Colloid Interface Sci.* **2022**, *625*, 936. [CrossRef]
61. Wu, D.; Dong, N.; Wang, R.; Qi, S.; Liu, B.; Wu, D. In situ construction of High-safety and Non-flammable polyimide "Ceramic" Lithium-ion battery separator via SiO_2 Nano-Encapsulation. *Chem. Eng. J.* **2021**, *420*, 129992. [CrossRef]
62. Han, D.; Wang, X.; Zhou, Y.-N.; Zhang, J.; Liu, Z.; Xiao, Z.; Zhou, J.; Wang, Z.; Zheng, J.; Jia, Z.; et al. A Graphene-Coated Thermal Conductive Separator to Eliminate the Dendrite-Induced Local Hotspots for Stable Lithium Cycling. *Adv. Energy Mater.* **2022**, *12*, 2201190. [CrossRef]
63. Yu, J.; Dong, N.; Liu, B.; Tian, G.; Qi, S.; Wu, D. A newly-developed heat-resistance polyimide microsphere coating to enhance the thermal stability of commercial polyolefin separators for advanced lithium-ion battery. *Chem. Eng. J.* **2022**, *442*, 136314. [CrossRef]
64. Sun, S.; Wang, J.; Chen, X.; Ma, Q.; Wang, Y.; Yang, K.; Yao, X.; Yang, Z.; Liu, J.; Xu, H.; et al. Thermally Stable and Dendrite-Resistant Separators toward Highly Robust Lithium Metal Batteries. *Adv. Energy Mater.* **2022**, *12*, 2202206. [CrossRef]
65. Yao, H.; Yan, L.; Shen, J.; Wang, T.; Chen, P.; Cong, X.; Zhang, S.; Jiang, H.; Zhao, X. Controllably regulating ion transport in lithium metal batteries via pore effect of metal–organic framework-based separators. *Appl. Surf. Sci.* **2022**, *589*, 152885. [CrossRef]
66. Zou, Z.; Hu, Z.; Pu, H. Lithium-ion battery separators based-on nanolayer co-extrusion prepared polypropylene nanobelts reinforced cellulose. *J. Membr. Sci.* **2023**, *666*, 121120. [CrossRef]
67. Hu, W.; Fu, W.; Jhulki, S.; Chen, L.; Narla, A.; Sun, Z.; Wang, F.; Magasinski, A.; Yushin, G. Heat-resistant Al_2O_3 nanowire-polyetherimide separator for safer and faster lithium-ion batteries. *J. Mater. Sci. Technol.* **2023**, *142*, 112. [CrossRef]
68. Baek, M.; Yoo, J.; Kim, Y.; Seo, H.; Lee, S.-M.; Jo, H.; Woo, S.-G.; Kim, J.-H. Surface Modification of Polyethylene Separator for Li-Ion Batteries via Imine Formation. *Int. J. Energy Res.* **2023**, *2023*, 4624762. [CrossRef]
69. Kim, Y.; Jang, Y.-J.; Seo, H.; Lee, J.-N.; Woo, S.-G.; Kim, J.-H. Poly(ether ether ketone)-Induced Surface Modification of Polyethylene Separators for Li-Ion Batteries. *Energies* **2023**, *16*, 627. [CrossRef]

Disclaimer/Publisher's Note: The statements, opinions and data contained in all publications are solely those of the individual author(s) and contributor(s) and not of MDPI and/or the editor(s). MDPI and/or the editor(s) disclaim responsibility for any injury to people or property resulting from any ideas, methods, instructions or products referred to in the content.

Article

Hard Carbons Derived from Phenyl Hyper-Crosslinked Polymers for Lithium-Ion Batteries

Ziyang Guo [1,2], Xiaodong Tian [3], Yan Song [3,4,*], Tao Yang [3], Zihui Ma [3,4], Xiangjie Gong [3,4] and Chao Wang [1,2,*]

1. College of Materials Science and Engineering, North University of China, Taiyuan 030051, China
2. Advance Energy Materials and Systems Institute, North University of China, Taiyuan 264000, China
3. CAS Key Laboratory for Carbon Materials, Institute of Coal Chemistry, Chinese Academy of Sciences, Taiyuan 030001, China
4. Center of Materials Science and Optoelectronics Engineering, University of Chinese Academy of Sciences, Beijing 100049, China
* Correspondence: yansong1026@126.com (Y.S.); wangchao_nuc@126.com (C.W.); Tel.: +86-351-2021135 (Y.S.)

Abstract: Hyper-crosslinked polymers are attracting extensive attention owing to their ease of design and synthesis. Based on the flexibility of its molecular design, a hyper-crosslinked polymer with a π-conjugated structure and its derived carbon were synthesized by the Friedel–Crafts reaction. The polymer and its derived hard carbon material were characterized by FTIR, ^{13}C NMR, Raman, BET, and other characterization tools. The electrochemical properties of both materials as anode electrodes of lithium-ion batteries were investigated. Benefiting from the highly cross-linked skeleton and conjugated structure, the as-prepared carbon materials still had high specific surface area (583 m^2 g^{-1}) and porosity (0.378 cm^3 g^{-1}) values. The hard carbon (CHCPB) anode possessed the powerful reversible capacity of 699 mAh g^{-1} at 0.1A g^{-1}, and it had an excellent rate of performance of 165 mAh g^{-1} at the large current density of 5.0 A g^{-1}. Long-cycle performance for 2000 charge/discharge cycles displayed that the capacity was kept at 148 mAh g^{-1} under 2 A g^{-1}. This work contributes to a better understanding of the properties of hard carbon materials derived from hyper-crosslinked polymers and how this class of materials can be further exploited in various applications.

Keywords: hyper-crosslinked polymers; π-conjugated; hard carbon; anode materials; lithium-ion batteries

Citation: Guo, Z.; Tian, X.; Song, Y.; Yang, T.; Ma, Z.; Gong, X.; Wang, C. Hard Carbons Derived from Phenyl Hyper-Crosslinked Polymers for Lithium-Ion Batteries. *Coatings* **2023**, *13*, 421. https://doi.org/10.3390/coatings13020421

Academic Editor: Octavian Buiu

Received: 2 December 2022
Revised: 9 February 2023
Accepted: 10 February 2023
Published: 13 February 2023

Copyright: © 2023 by the authors. Licensee MDPI, Basel, Switzerland. This article is an open access article distributed under the terms and conditions of the Creative Commons Attribution (CC BY) license (https://creativecommons.org/licenses/by/4.0/).

1. Introduction

Lithium-ion batteries (LIBs) are widely used as smart portable electronic devices and powered vehicles because of their high energy/power density and long cycle life [1–3]. As a key component of LIBs, the anode material plays an important role in improving the overall performance. Graphite, a commercial anode material, cannot meet the increasing demand of energy and power density due to its limited theoretical capacity (372 mAh g^{-1}) and poor rate performance [4–6]. Therefore, it is challenging to develop next-generation advanced anode materials with high reversible capacity, excellent rate performance, and long cycle life.

Hard carbon is the preferred anode material compared with graphite and soft carbon due to its short-range graphitization domains, rich pores, and abundant edges and defects. The unique structures facilitate the transport of Li$^+$ ions and provide them with abundant active sites [7,8]. According to research, nanotextures and synthesis conditions of the precursors can modulate the microstructures of the obtained hard carbon and thus affect its electrochemical performance [9–12]. There are various precursor substances of the preparation of hard carbon, such as coal tar pitch (CTP), biomass, resin, and polymer [13–15]. Fujimoto et al. prepared hard carbon with different oxygen contents under three different pretreatment conditions using CTP as the precursor [13]. Fromm et al. prepared hard carbon from bamboo precursors to investigate the effect of heat treatment temperatures on the microstructure and electrochemical properties of the as-obtained hard carbon [14]. Gao and co-workers successfully prepared resin-based hard carbon by pyrolyzing and carbonizing the phenolic resin precursor at 1100 °C [15]. However, there will be an inevitable limitation for CTP produced from fossil resources for future applications from the perspective of sustainable development. The heterogeneity of biomass composition, the high cost of resin, and the low carbon yield all limit their use as the promising precursors for hard carbon materials.

Hyper-crosslinked polymers (HCPs), synthesized by the Friedel–Crafts reaction [16], can provide a more ideal method for hard carbon owing to the characteristics of extensive monomers resources and high carbon yield. Noteworthily, the framework structure and the surface properties of HCPs are flexible and adjustable, which can be modified by changing the geometric shape of the monomer, the type of functional group and chain length of crosslinker, or by doping hetero atoms. This means that high performance carbon electrode materials with oriented structure are expected to be made by rational design and preparation. Recent research has shown that HCP-derived carbon materials by various strategies exhibit great potential in the field of electrochemical energy storage [17–22]. For instance, porous carbon nanotubes (HCPTs) with different morphologies and sizes were obtained by adjusting the initial monomer concentration and molecular sizes of HCPs, which were prepared by a self-templated, surfactant-free strategy. The authors found that the obtained PCNTs showed excellent electrochemical performance for supercapacitors [23]. A porous carbon material carbonized from hyper-crosslinked porous polymers gave a specific capacity of 1221 mAh g^{-1} at 0.1 A g^{-1} and a satisfactory capacity of supercapacitors [19]. Guan et al. successfully fabricated honeycomblike porous carbons with specific pore structures by adjusting the dosage of chloroform, which showed an ultrahigh surface area of 3473.0 m^2 g^{-1} and abundant mesopores with sizes of 2–5 nm [24]. A microporous carbon (MICHP) derived from hyper-crosslinked polystyrene was obtained through a facile molecular crosslinker design strategy, which exhibited a surface area of 387 m^2 g^{-1} [25].

The current work profitably synthesizes a hyper-crosslinked polymer (HCP) with a π-conjugated structure by the Friedel–Crafts reaction of benzene and 1,4-dimethoxybenzene (DMB). Then, a hard carbon material with a layer-stacked structure is obtained by high-temperature treatment of the HCP precursor. The hard carbon material inherits the rigid skeleton of the HCP precursor and has a similar specific surface area and pore volume. When employed as an anode of LIBs, it shows a satisfying rate capacity of 699 mAh g^{-1} at 0.1 A g^{-1} and 165 mAh g^{-1} at the large current density of 5.0 A g^{-1}. After cycling 2000 times at 2.0 A g^{-1}, the reversible capacity can reach 148 mAh g^{-1}. This work provides a new promising strategy and fundamental understanding of carbon-based electrodes derived from HCPs for rechargeable batteries.

2. Materials and Methods
2.1. Materials

1,4-Dimethoxybenzene (DMB, 99%), nitrobenzene, and $FeCl_3$ (anhydrous) were purchased from MACKMIN Co., Ltd. (Shanghai, China). Benzene and methanol were purchased from Kermel Co., Ltd. (Tianjin, China). Conductive carbon black, polyvinylidene fluoride (PVDF), and 1 M $LiPF_6$ in a mixed solvent of vinyl carbonate (EC), and diethyl carbonate (DEC) (1:1, vol/vol) with 5.0% fluoroethylene carbonate (FEC) was obtained from Suzhou DuoDuo Chemical Technology Co., Ltd. (Suzhou, China). The commercial hard carbon (CHC) was obtained from KUREHA Chemical Co., Ltd. (Shanghai, China).

2.2. Synthetic Procedures for Hyper-Crosslinked Polymers and Hard Carbon Materials

Benzene (0.13 g, 1.67 mmol) and DMB (0.69 g, 4.99 mmol) were added into 15 mL of nitrobenzene. Then, anhydrous $FeCl_3$ (2.44 g, 15.02 mmol) was added to the above-mixed solution at room temperature. After that, the reaction system was stirred for 5 h at 45 °C, 5 h at 80 °C, and 24 h at 120 °C to obtain a porous polymer. After cooling to room temperature, the crude product was washed with hydrochloric acid and methanol in turn until the filtrate became colorless. Then, the target product was extracted with methanol for 48 h in a Soxhlet extractor and dried under vacuum at 60 °C for 24 h. The final product was named HCPB.

The HCPB was thermally treated at 1200 °C for 60 min under an argon flow in a tube furnace to realize carbonization after manual grinding. The resulting hard carbons were denoted as CHCPB (Scheme 1).

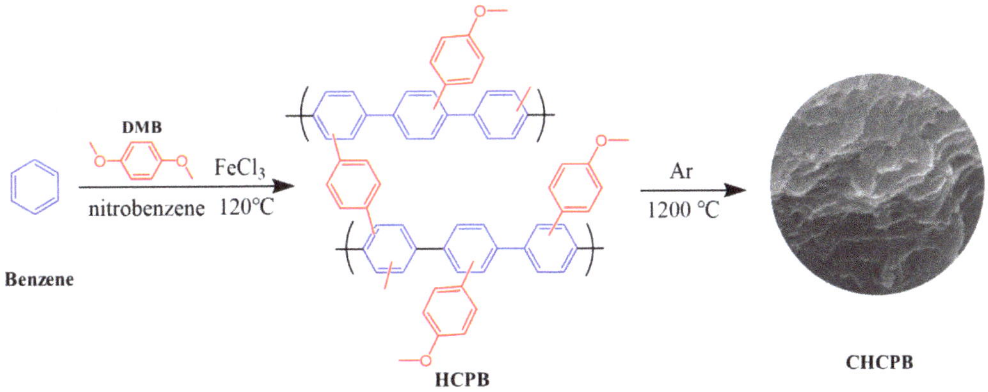

Scheme 1. Schematic diagram of the synthesis of CHCPB.

2.3. Structural Characterization

The composition structure of products was studied by Fourier transform-infrared (FTIR, Bruker, Bremen, Germany) spectra on a Bruker VERTEX 80v. Solid-state ^{13}C CP/MAS NMR spectra were performed on a WB 600 MHz Bruker Avance III spectrometer (Bruker, Bremen, Germany). The surface areas were calculated from nitrogen adsorption data by Brunauer–Emmett–Teller (BET) analysis. Pore size distributions were calculated by DFT methods via the adsorption branch. The morphology of the products was investigated using scanning electron microscopy (SEM, JSM-7001F, Jeol, Tokyo, Japan) and transmission electron microscopy (TEM, JEM-2100F, Jeol, Tokyo, Japan). The structure of products was determined by X-ray diffraction (XRD, D8 ADVANCE A25, Bruker, Bremen, Germany) and Raman spectrometry (Horiba LabRAM HR Evolution, Horiba, Kyoto, Japan). The X-ray photoelectron spectroscopy (XPS, AXIS ULTRA DLD, Kratos, Manchester, UK) was used to study the surface chemical composition of obtained samples. The elemental analysis (EA) measurements were performed on a Vario EL CUBE elemental analyzer (Elementar, Frankfurt, Germany). The thickness and size of the nanosheets were obtained on a Bruker Dimension Icon atomic force microscope (AFM, Bruker, Bremen, Germany) in tapping mode.

2.4. Electrochemical Measurement

The electrochemical features of hard carbon materials were explored by assembling a CR2016 coin type cell in the glovebox. A 1 M amount of LiPF$_6$ in the mixed solvent of vinyl carbonate (EC) and diethyl carbonate (DEC) (1:1, vol/vol) with 5.0% fluoroethylene carbonate (FEC) was used as the electrolyte. Celgard 2400 membranes were used as the separator, and metal lithium slices acted as the counter electrode. The working electrode was made from a mixture of active materials, conductive carbon black, and polyvinylidene fluoride solution (2% PVDF) at the weight ratio of 6:3:1, with N-methylpyrrolidone (NMP) as the solvent. The obtained slurry was loaded onto the copper foil and dried at 80 °C overnight. After that, the whole copper foil was punched to the circular pole piece with the diameter of 10 mm, and the mass loading was obtained in the range of 0.45–0.57 mg cm^{-2}.

The galvanostatic charge and discharge measurements were carried on a LAND BT2000 battery system in the voltage range of 0.01–3 V. Electrochemical impedance spectroscopy (EIS) and cyclic voltammetry (CV) were tested on a 660 E electrochemical working station.

3. Results

3.1. The Structure and Morphology Characterization of the Materials

The yield of HCPs was ~90% owing to the presence of $FeCl_3$ catalyst. The Fourier transform infrared (FT-IR) spectroscopy of the HCPB is exhibited in Figure S1. Stretching vibrations at 2932 and 2828 cm^{-1} and bending vibrations at 1427 cm^{-1} of C–H, as well as skeleton vibration at 1616 and 1460 cm^{-1} of the benzene ring were observed. The small peak at 1140 cm^{-1} could be attributed to the C–O bonds of the methoxy groups derived from unreacted DMB. More detailed characterization of HCPB was observed by ^{13}C CP/MAS NMR spectroscopy (Figure S2). The resonance peaks at 124, 114, and 102 ppm belonged to the carbon in the benzene ring attached to the aromatic ring and non-substituted aromatic carbon, respectively. The resonance peaks at 55 and 153 ppm were ascribed to the unreacted methyl in the crosslinking agent and the aromatic carbon connected to the methoxy group. The above results indicated that DMB formed highly cross-linked networks with benzene rings as bridges [26].

The Raman spectra of the two samples are shown in Figure 1a. The peaks at 1190, 1435, and 1636 cm^{-1} were assigned to the stretching vibration of C–H in the plane and the vibration of aromatic rings and benzene rings, respectively. In addition, Raman peaks at 1337 and 1591 cm^{-1} were attributed to the D and G bands, which are typical spectral features of conjugated carbon frameworks [20,27]. After high temperature treatment, the spectrum of CHCPB exhibited only two typical D and G peaks, indicating the very stable conjugated skeleton. The ratio of the D-band (defect-induced mode) to the G-band (graphitic mode) denoted the degree of graphitization of material [28]. For CHCPB, the I_D/I_G value was 1.05, which suggested the existence of CHCPB defects. Both XPS and EA measurements revealed the presence of O elements in CHCPB (Figure S3 and Table S1). It was evident that the oxygen functional groups could not only increase the accessibility of electrode materials in electrolyte but also enhanced the storage capacity of Li^+ ions [29].

The N_2 adsorption–desorption isotherms were obtained to analyze the changes of the pore structure and specific surface area. Type I isotherms were found for the two samples. As shown in Figure 1c, they all had a sharp adsorption capacity at relatively low pressure. The adsorption almost reached saturation within a very small pressure variation range, indicating the microporous characteristic. The adsorption curve rose slowly in the pressure range of 0.3–1.0, and a clear hysteresis loop was observed. This indicated the presence of mesopores and few macropores in HCPB. However, the desorption curve did not cooperate with the adsorption curve in the low-pressure region, perhaps because the deformation of the polymer skeleton may have led to narrowed micropore pores, and the adsorbed molecules were unable to desorbed completely. The pore size distributions of CHCPB and HCPB were mainly concentrated in 0.7 nm and 0.5–1.2 nm, respectively (Figure 1d), which was consistent with the above analysis. The specific surface area and pore structure parameters of the samples are summarized in Table 1. CHCPB had a similar specific surface area as HCPB. However, the micropore pore volume increased from 0.148 to 0.226 $cm^3\ g^{-1}$, while the mesopore pore volume decreased from 0.197 to 0.070 $cm^3\ g^{-1}$. The possible reason was the slight shrinkage of the skeleton of the material after carbonization, which increased the micropores and reduced the mesopores and macropores. Overall, the skeleton of CHCPB still maintained the original rigid skeleton.

X-ray diffraction (XRD) was used to investigate the phase structure of the materials (Figure 1b). It can be clearly seen that the curve of HCPB only showed a broad peak at 24°, indicating its amorphous nature. However, CHCPB had a sharp peak at 26° and a small peak at 42°, corresponding to the (002) and (100) planes of graphite [30], respectively, which suggested that the heat treatment strengthened the π–π conjugated and increased the order degree. CHCPB tended to form layer-stacking structures due to the strong interlayer π–π interactions. Furthermore, two small peaks could be observed at 33.6° and 36.2°, which may have originated from traces of Fe_2O_3 formed by the residual catalyst.

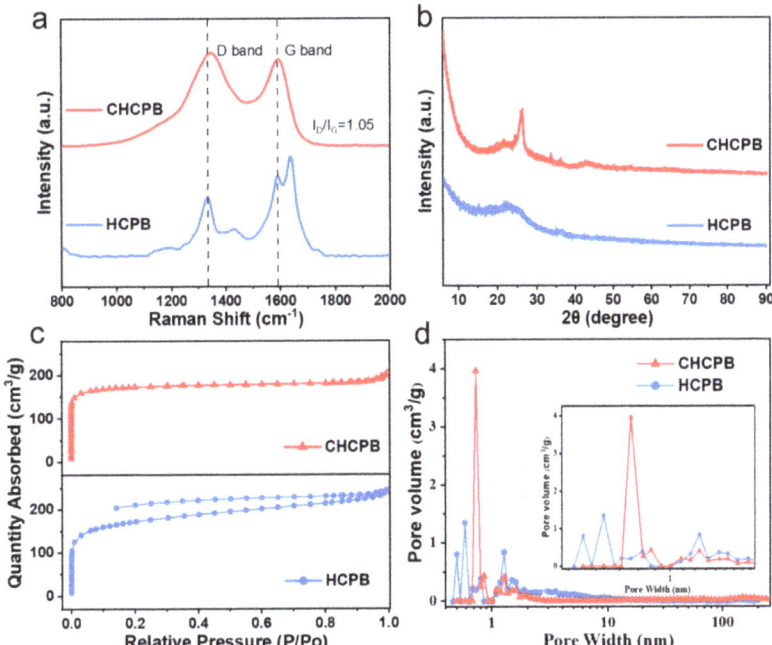

Figure 1. (a) Raman spectra and (b) XRD patterns of HCPB and CHCPB; (c) N_2 adsorption-desorption isothermal curves; and the pore size distribution curves (d) of HCPB and CHCPB.

Table 1. Porosity parameters of polymer and hard carbon.

Sample	S_{BET} (m² g⁻¹)	V_{micro} (cm³ g⁻¹)	V_{meso} (cm³ g⁻¹)	V_{total} (cm³ g⁻¹)
HCPB	593	0.15	0.20	0.38
CHCPB	583	0.23	0.07	0.32

The morphologies of materials were observed by SEM (Figure 2a,b). It could be seen that HCPB displayed a clear network structure. However, the CHCPB presented a layer-stacking structure, which was caused by strong π–π interactions among molecules. In addition, these two samples were also examined by TEM, and the disordered microcrystal structure of the CHCPB and HCPB could be distinctly observed (Figure 2c,d). The selected area electron diffraction (SAED) image of HCPB exhibited diffuse diffraction rings, further confirming the amorphous structure. On the contrary, CHCPB exhibited a partial short-range ordered graphite domain structure and sharper diffraction rings, which proved the microcrystal structure and layer-by-layer stacking nature.

To further prove the existence of the layered structure, CHCPB was stripped into a colloidal dispersion by the ultrasonic-assisted solvent exfoliation method, exhibiting a clear Tyndall effect. The AFM image further confirmed the precise information of CHCPB nanosheets (Figure 2e,f). The thickness of the nanosheets varied from 2 to 4 nm, which corresponded to the few-layered structure of CHCPB. Notably, the transverse dimensions of the nanosheets changed from 550 to 2200 nm, resulting in an aspect ratio of height and thus confirming their two-dimensional (2D) stratification characteristics, which was in agreement with the SEM and TEM results.

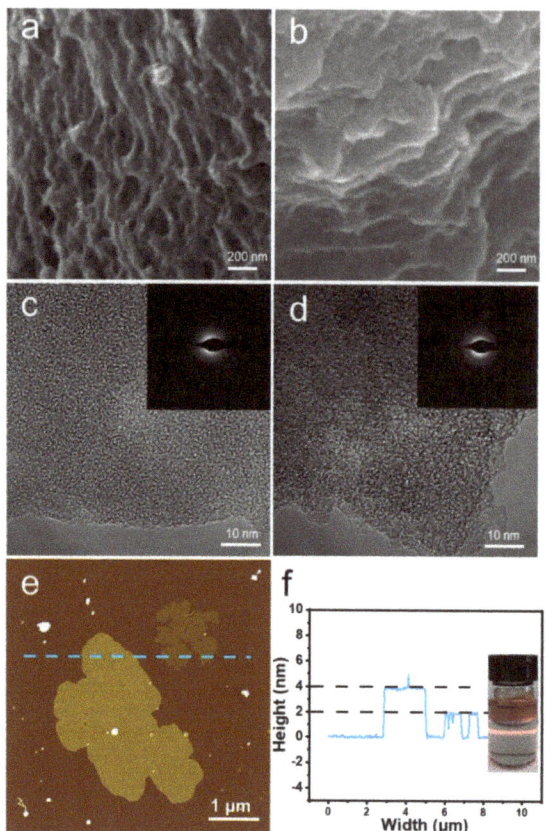

Figure 2. SEM images of (**a**) HCPB and (**b**) CHCPB; HRTEM images of (**c**) HCPB and (**d**) CHCPB; Inset: the SAED images; (**e**) AFM image of CHCPB nanosheet; (**f**) height profile along the line; Inset: optical image of CHCPB nanosheet in DMF solution.

3.2. Li Ion Storage Performances

Figure 3a,b show the CV curves of two electrode materials between 0.01 and 3 V at a scan rate of 0.1 mV s^{-1}. In the first anodic sweep, oxidation peaks appeared at 1.1 V, corresponding to the de-lithiation process of the HCPB electrode. From the first cycle, an oxidation peak at 1.9 V and a reduction peak at 1.7 V were formed, corresponding to the lithiation and de-lithiation of Li$^+$ within π-conjugated HCPB [27]. The anodic peak at 2.35 V was related to the process of forming CuO, and the copper probably derived from the current collector made of commercial copper foil [17,31]. For CHCPB, a broad peak at 1.15 V assigned to de-intercalation of Li$^+$ from the CHCPB electrode was also observed. Two obvious reduction peaks located at around 0.8 and 1.3 V during the first cathodic sweep sharply declined in the next two cycles, revealing the massive irreversible capacity loss ascribed to the development of solid electrolyte interphase (SEI) film on the electrode surface and other irreversible redox reactions between functional groups of HCPB/CHCPB with electrolyte [32]. The stable formation of SEI was beneficial to prevent further dissolution of electrolyte, thus enhancing the stabilization of the batteries [33]. The CV curves of the second and third circles were nearly overlapped in the subsequent cycles, which revealed the excellent redox activity of CHCPB [28]. Galvanostatic charge–discharge (GCD) tests of HCPB and CHCPB were further carried out at 0.1 A g^{-1} (Figure 3c,d). CHCPB delivered a higher initial charge/discharge capacity (790.6/2260.8 mAh g^{-1}) than that of HCPB (474.4/2042.0 mAh g^{-1}). This may have been due to the defects and layered structure of CHCPB after high-temperature carbonization, which provided abundant storage sites for Li$^+$ ions [34]. A small platform close to 1.0 V at the first discharge curve was related to the formation of SEI film and the irreversible lithium-ion insertion reaction. The potential platform of discharge curves at around 1.0 V disappeared, and the charge–discharge curves hardly changed during the following cycles, suggesting that the CHCPB electrode was highly reversible.

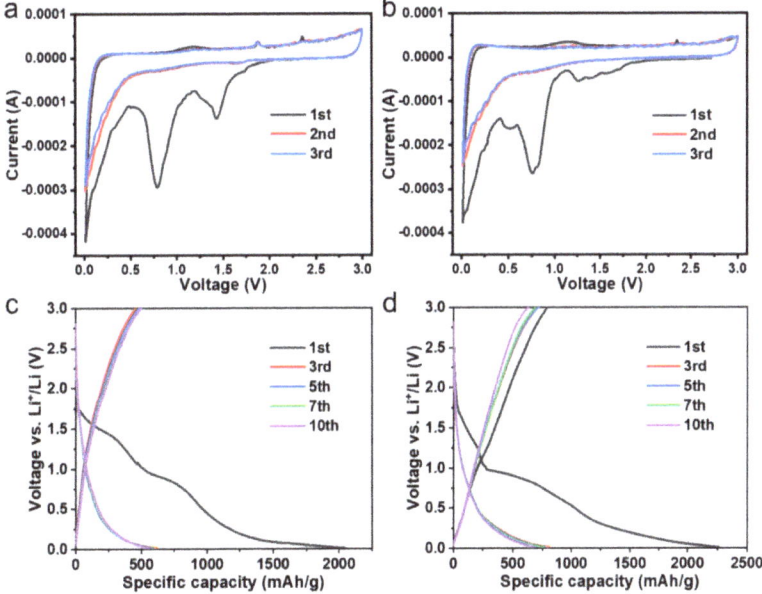

Figure 3. Cyclic voltammetry curves of (**a**) HCPB and (**b**) CHCPB at 0.1 mV s^{-1}; GCD curves of (**c**) HCPB and (**d**) CHCPB at a current density of 0.1 A g^{-1}.

The presence of the π-conjugated structure allowed for the layer-by-layer arrangement of molecules, resulting in stable charged and discharged states, and enabling the CHCPB-based LIBs to achieve fast ionic transport, high-rate capability, and good cycling stability. The rate properties were measured at the current density of 0.1, 0.2, 0.3, 0.5, 1, and 2 A g^{-1}, respectively (Figure 4a). The discharge capacities of CHCPB were 699, 579, 527, 428, 338, and 261 mAh g^{-1}, respectively, much better than that of HCPB and most of the reported polymer-derived hard carbons (Table S2). Moreover, at the current density of 5 A g^{-1}, the discharge capacity of CHCPB was still as high as 165 mAh g^{-1}. When the current density recovered to 0.1 A g^{-1}, CHCPB reached a higher capacity (704 mAh g^{-1}) than that of its initial cycle, which was assigned to the activation of electrodes during the process of charging/discharging [20,28]. In stark contrast, the capacities of CHC only exhibited 246 mAh g^{-1} at 0.1 A g^{-1} and 121 mAh g^{-1} at 5 A g^{-1} (Figure S4). The excellent rate performance of the CHCPB electrode profited from the enhanced ionic conductivity of the stacked layered structure, which was of great significance for fast charging/discharging. Figure S5 presents the 1st, 2nd, 100th, 200th, 500th, 1000th, and 2000th GCD curves of HCPB and CHCPB at 2 A g^{-1}. The initial charge/discharge capacities were 331.7/1490.3 mAh g^{-1} for CHCPB and 249.9/1461.1 mAh g^{-1} for HCPB, and their coulombic efficiency (CE) values were 22.26% and 17.10%, respectively. The low CE may have resulted from the formation of the SEI layer [35]. Figure 4c displays the long-term cycling test results at 2 A g^{-1}. In the initial 50 cycles, the discharge capacity rapidly decreased, which may be owing to the continuous formation and decomposition of SEI film and the volume change of electrodes during the Li$^+$ insertion and extraction [36]. At the 200th cycle, the cells achieved charge/discharge capacities that increased to 313.4/315.5 mAh g^{-1} with a CE of 99.33% for CHCPB and 295.3/297.8 mAh g^{-1} with a CE of 99.16% for HCPB, which were caused by electrode activation. The capacity of the two electrodes gradually decayed in the subsequent cycles, denoting that the electrode activation process was completed. The specific discharge capacity of CHCPB decreased more slowly and was maintained at 148.0 mAh g^{-1} after 2000 cycles, which was higher than that of HCPB (92.7 mAh g^{-1}) and CHC (63.0 mAh g^{-1}). The relatively stable cycling performance of CHCPB might be attributed to its rigid skeleton. The above results indicated that CHCPB could serve as a potential anode material for high performance LIBs.

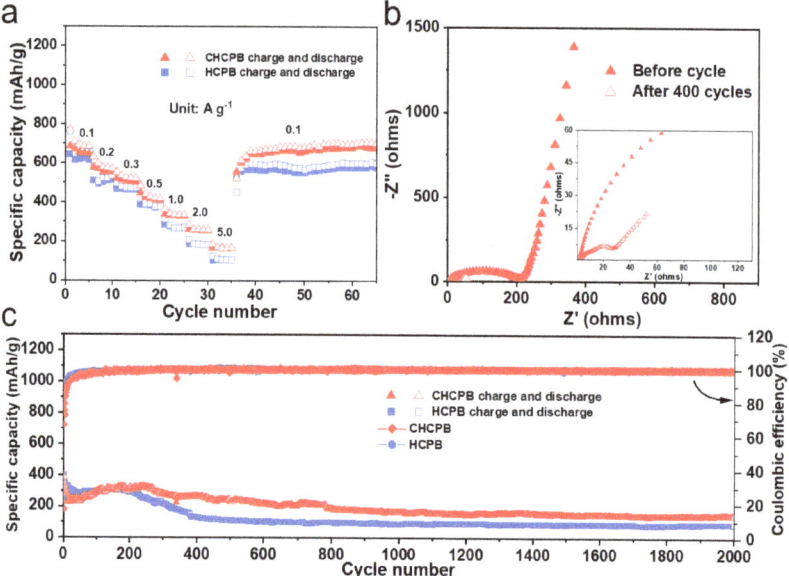

Figure 4. (a) Rate properties from 0.1 to 5 A g^{-1} and back to 0.1 A g^{-1} for HCPB and CHCPB; (b) EIS of CHCPB electrodes before and after the cycle; (c) long-term running properties of HCPB and CHCPB at 2 A g^{-1}.

Electrochemical impedance spectroscopy (EIS) was used to examine the interfacial properties of HCPB and CHCPB electrodes, and the fitting circuit model for the two electrodes can be found in Figure S6. The truncation of the Z_{real} axis in the high-frequency region belonged to the electrolyte resistance (R_s). The semi-circle at the high-frequency region referred to resistance of the SEI film (R_f). A semi-circle in the middle-frequency region was the charge transfer impedance (R_{ct}) between the electrode and electrolyte. The slope line in the low-frequency region was the Warburg impedance (Z_W) related to the diffusion of lithium ions in the electrode material [37]. As illustrated in Figure S7, the HCPB electrode showed 373 Ω of the R_{ct} before the cycle. After the 400th cycle, a substantial increase in charge transport resistance was observed, which indicated the poor kinetic properties of charge carrier insertion/extraction. The above result can be ascribed to the presence of the network structure, whose crosslinked nodes blocked transmission channels, thus limiting the release of lithium ions. For the CHCPB electrode, the R_{ct} values before and after 400 cycles were 154 and 15 Ω, respectively (Figure 4b), which were due to the formation of stable SEI film during charging and discharging. The data further demonstrated the higher electrical conductivity of CHCPB compared with HCPB. Moreover, the stable SEI film formed during cycling also decreased the interfacial resistance of the electrode, which was beneficial to the charge transfer and Li$^+$ transport.

The storage mechanism of the Li$^+$ ion was analyzed by testing the CV curves at different scan rates (Figure 5). The faster the scanning rate, the higher the peak current of the CV curve. The relationship between peak current (i) and sweep speed (v) was given by the following equation:

$$i = av^b \tag{1}$$

$$\log i = b \log v + \log a \tag{2}$$

where a and b are adjustable constants. When the b value approached 0.5, the storage process was diffusion-controlled (insertion/extraction). When the b value was close to 1.0, a capacitive process (surface adsorption/desorption behavior) dominated [38]. The calculated b value was 0.72 for HCPB and 0.85 for CHCPB, which meant that the lithium storage process in the two electrodes was primarily controlled by the capacitive process. The results verified that the surface capacitive behavior of the two electrodes dominates the storage capacity, being put down to the high specific surface area and the short ion diffusion distance [20].

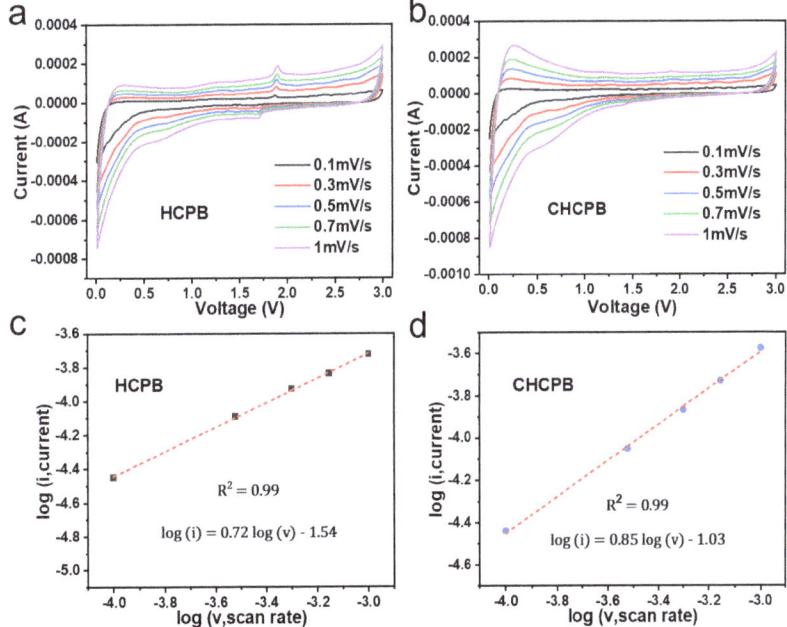

Figure 5. (**a**,**b**) CV curves of HCPB and CHCPB with scan rates from 0.1 to 1.0 mV s^{-1}; (**c**,**d**) log (*i*)-log (*b*) graphs of HCPB and CHCPB.

The superior electrochemical properties of the CHCPB electrode could be attributed to its unique characteristics: (1) the high-temperature heat treatment and its π-conjugated structure enhanced the conductivity and promoted the electron transport; (2) the abundant micropores of the CHCPB electrode provided high specific surface area and a number of defects, thus providing abundant storage sites for lithium ions and improved capacity; (3) the stacked layered structure afforded fast channels for ion transport and thus better rate performance. In addition, the rigid backbone knitted by the aromatic structural units led to good cycling stability at high current densities of CHCPB.

4. Conclusions

In this work, the synthetic HCPB was carbonized at 1200 °C under argon atmosphere to obtain a hard carbon material (CHCPB). The layered structure of CHCPB improved the stability and enriched the storage sites for Li$^+$ ions. The CHCPB electrode had a high reversible capacity of 699 mAh g^{-1} at 0.1 A g^{-1}. At the same time, the CHCPB electrode also showed a great rate of performance. The capacities were 699, 579, 527, 428, 338, 261, 165 and 704 mAh g^{-1} at 0.1, 0.2, 0.3, 0.5, 1.0, 2.0, 5.0, and 0.1 A g^{-1}, respectively for Li | | CHCPB batteries. The CHCPB electrode displayed outstanding electrochemical performance because of the larger specific surface area, more defects, faster ion transport, and a more stable structure. Based on this study, we can also endow the precursors with different pore structures and functionalized crosslinking networks through a flexible molecular design to meet different application requirements.

Supplementary Materials: The following supporting information can be downloaded at: https://www.mdpi.com/article/10.3390/coatings13020421/s1, Figure S1: FTIR spectra of HCPB; Figure S2: ^{13}C CP/MAS NMR spectra of HCPB; Figure S3: The XPS spectrum of CHCPB; Figure S4: (a) The rate property and (b) long-term running property at 2 A g^{-1} of commercial hard carbon (CHC); Figure S5: Galvanostatic charge–discharge curves of (a) HCPB and (b) CHCPB at 2 A g^{-1}. Figure S6: The EIS fitting circuit model of HCPB and CHCPB electrodes; Figure S7: EIS of the HCPB electrode before (a) and after cycling (b); Table S1. The elemental composition determined by EA. Table S2. Comparison of the electrochemical performances with some reported polymers-derived carbon materials.

Author Contributions: Investigation, Z.G.; writing—original draft preparation, Z.G.; writing—review and editing, Y.S., X.T. and T.Y.; supervision, Y.S. and C.W.; conceptualization, X.T. and T.Y.; formal analysis, Z.M. and X.G.; project administration, Y.S.; funding acquisition, Y.S., X.T. and T.Y. All authors have read and agreed to the published version of the manuscript.

Funding: This work was supported by the National Natural Science Foundation of China (Nos. 52072383, U21A2061, and No. 22209197), Natural Science Foundation of Shanxi Province (202203021211002), Shanxi Province Science Foundation for Youths (No. SQ2019001), and the Innovation Fund of Institute of Coal Chemistry, Chinese Academy of Sciences (No. SCJC-XCL-2022-08).

Institutional Review Board Statement: Not applicable.

Informed Consent Statement: Not applicable.

Data Availability Statement: All data that support the findings of this study are included within the article.

Conflicts of Interest: The authors declare no conflict of interest.

References

1. Zhang, H.; Liu, Q.; Fang, Y.; Teng, C.; Liu, X.; Fang, P.; Tong, Y.; Lu, X. Boosting Zn-Ion Energy Storage Capability of Hierarchically Porous Carbon by Promoting Chemical Adsorption. *Adv. Mater.* **2019**, *31*, 1904948. [CrossRef]
2. Zhang, C.; Ma, Y.; Zhang, X.; Abdolhosseinzadeh, S.; Sheng, H.; Lan, W.; Pakdel, A.; Heier, J.; Nüesch, F. Two-Dimensional Transition Metal Carbides and Nitrides (MXenes): Synthesis, Properties, and Electrochemical Energy Storage Applications. *Energy Environ. Mater.* **2020**, *3*, 29–55. [CrossRef]
3. Lu, Y.; Wang, J.; Zeng, S.; Zhou, L.; Xu, W.; Zheng, D.; Liu, J.; Zeng, Y.; Lu, X. An ultrathin defect-rich Co$_3$O$_4$ nanosheet cathode for high-energy and durable aqueous zinc ion batteries. *J. Mater. Chem. A* **2019**, *7*, 21678–21683. [CrossRef]
4. Zhang, C.; Park, S.-H.; O'Brien, S.E.; Seral-Ascaso, A.; Liang, M.; Hanlon, D.; Krishnan, D.; Crossley, A.; McEvoy, N.; Coleman, J.N.; et al. Liquid exfoliation of interlayer spacing-tunable 2D vanadium oxide nanosheets: High capacity and rate handling Li-ion battery cathodes. *Nano Energy* **2017**, *39*, 151–161. [CrossRef]
5. Tang, H.; Li, W.; Pan, L.; Cullen, C.P.; Liu, Y.; Pakdel, A.; Long, D.; Yang, J.; McEvoy, N.; Duesberg, G.S.; et al. In Situ Formed Protective Barrier Enabled by Sulfur@Titanium Carbide (MXene) Ink for Achieving High-Capacity, Long Lifetime Li-S Batteries. *Adv. Sci.* **2018**, *5*, 1800502. [CrossRef]
6. Abdolhosseinzadeh, S.; Schneider, R.; Verma, A.; Heier, J.; Nüesch, F.; Zhang, C. Turning Trash into Treasure: Additive Free MXene Sediment Inks for Screen-Printed Micro-Supercapacitors. *Adv. Mater.* **2020**, *32*, 2000716. [CrossRef]
7. Zhao, L.-F.; Hu, Z.; Lai, W.-H.; Tao, Y.; Peng, J.; Miao, Z.-C.; Wang, Y.-X.; Chou, S.-L.; Liu, H.-K.; Dou, S.-X. Hard Carbon Anodes: Fundamental Understanding and Commercial Perspectives for Na-Ion Batteries beyond Li-Ion and K-Ion Counterparts. *Adv. Energy Mater.* **2021**, *11*, 2002704. [CrossRef]
8. Zhang, X.; Yang, Y.; Zhou, Z. Towards practical lithium-metal anodes. *Chem. Soc. Rev.* **2020**, *49*, 3040–3071. [CrossRef]

9. Dahbi, M.; Kiso, M.; Kubota, K.; Horiba, T.; Chafik, T.; Hida, K.; Matsuyama, T.; Komaba, S. Synthesis of hard carbon from argan shells for Na-ion batteries. *J. Mater. Chem. A* **2017**, *5*, 9917–9928. [CrossRef]
10. Long, W.; Fang, B.; Ignaszak, A.; Wu, Z.; Wang, Y.-J.; Wilkinson, D. Biomass-derived nanostructured carbons and their composites as anode materials for lithium ion batteries. *Chem. Soc. Rev.* **2017**, *46*, 7176–7190. [CrossRef] [PubMed]
11. Beda, A.; Taberna, P.-L.; Simon, P.; Matei Ghimbeu, C. Hard carbons derived from green phenolic resins for Na-ion batteries. *Carbon* **2018**, *139*, 248–257. [CrossRef]
12. Saurel, D.; Orayech, B.; Xiao, B.; Carriazo, D.; Li, X.; Rojo, T. From Charge Storage Mechanism to Performance: A Roadmap toward High Specific Energy Sodium-Ion Batteries through Carbon Anode Optimization. *Adv. Energy Mater.* **2018**, *8*, 1703268. [CrossRef]
13. Fujimoto, H.; Tokumitsu, K.; Mabuchi, A.; Chinnasamy, N.; Kasuh, T. The anode performance of the hard carbon for the lithium ion battery derived from the oxygen-containing aromatic precursors. *J. Power Sources* **2010**, *195*, 7452–7456. [CrossRef]
14. Fromm, O.; Heckmann, A.; Rodehorst, U.C.; Frerichs, J.; Becker, D.; Winter, M.; Placke, T. Carbons from biomass precursors as anode materials for lithium ion batteries: New insights into carbonization and graphitization behavior and into their correlation to electrochemical performance. *Carbon* **2018**, *128*, 147–163. [CrossRef]
15. Ni, J.; Huang, Y.; Gao, L. A high-performance hard carbon for Li-ion batteries and supercapacitors application. *J. Power Sources* **2013**, *223*, 306–311. [CrossRef]
16. Xu, S.; Luo, Y.; Tan, B. Recent Development of Hypercrosslinked Microporous Organic Polymers. *Macromol. Rapid Commun.* **2013**, *34*, 471–484. [CrossRef]
17. Zhang, Q.; Dai, Q.; Yan, C.; Su, C.; Li, A. Nitrogen-doped porous carbon nanoparticle derived from nitrogen containing conjugated microporous polymer as high performance lithium battery anode. *J. Alloys Compd.* **2017**, *714*, 204–212. [CrossRef]
18. Yang, X.; Wei, C.; Sun, C.; Li, X.; Chen, Y. High performance anode of lithium-ion batteries derived from an advanced carbonaceous porous network. *J. Alloys Compd.* **2017**, *693*, 777–781. [CrossRef]
19. Zhang, C.; Kong, R.; Wang, X.; Xu, Y.; Wang, F.; Ren, W.; Wang, Y.; Su, F.; Jiang, J.-X. Porous carbons derived from hypercrosslinked porous polymers for gas adsorption and energy storage. *Carbon* **2017**, *114*, 608–618. [CrossRef]
20. Li, C.; Kong, D.; Wang, B.; Du, H.; Zhao, J.; Dong, Y.; Xie, Y. Conjugated microporous polymer derived N, O and S co-doped sheet-like carbon materials as anode materials for high-performance lithium-ion batteries. *J. Taiwan Inst. Chem. Eng.* **2022**, *134*, 104293. [CrossRef]
21. Zhang, Q.; Xu, Z.; Bai, Y.; Zhang, Y. Low temperature synthesis and superior lithium storage properties of fluorine-rich tubular porous carbon. *J. Alloys Compd.* **2022**, *901*, 163657. [CrossRef]
22. Zhang, Q.; Zhang, Y.; Lian, F.; Xu, Z.; Wang, X. Synthesis and lithium storage properties of nitrogen-containing hard carbon from conjugated microporous polymers. *Ionics* **2022**, *28*, 2623–2633. [CrossRef]
23. Wang, X.; Mu, P.; Zhang, C.; Chen, Y.; Zeng, J.; Wang, F.; Jiang, J.X. Control Synthesis of Tubular Hyper-Cross-Linked Polymers for Highly Porous Carbon Nanotubes. *ACS Appl. Mater. Interfaces* **2017**, *9*, 20779–20786. [CrossRef] [PubMed]
24. Guan, T.; Zhao, J.; Zhang, G.; Wang, J.; Zhang, D.; Li, K. Template-Free Synthesis of Honeycomblike Porous Carbon Rich in Specific 2–5 nm Mesopores from a Pitch-Based Polymer for a High-Performance Supercapacitor. *ACS Sustain. Chem. Eng.* **2018**, *7*, 2116–2126. [CrossRef]
25. Xu, F.; Han, H.; Qiu, Y.; Zhang, E.; Repich, H.; Qu, C.; Yu, H.; Wang, H.; Kaskel, S. Facile regulation of carbon framework from the microporous to low-porous via molecular crosslinker design and enhanced Na storage. *Carbon* **2020**, *167*, 896–905. [CrossRef]
26. Wang, S.; Zhang, C.; Shu, Y.; Jiang, S.; Xia, Q.; Chen, L.; Jin, S.; Hussain, I.; Cooper, A.I.; Tan, B. Layered microporous polymers by solvent knitting method. *Sci. Adv.* **2017**, *3*, e1602610. [CrossRef] [PubMed]
27. Ye, X.-L.; Huang, Y.-Q.; Tang, X.-Y.; Xu, J.; Peng, C.; Tan, Y.-Z. Two-dimensional extended π-conjugated triphenylene-core covalent organic polymer. *J. Mater. Chem. A* **2019**, *7*, 3066–3071. [CrossRef]
28. Li, C.; Guo, X.; Du, H.; Zhao, J.; Liu, L.; Yuan, Q.; Fu, C. The synthesis of the D-A-type polymers containing benzo[1,2-b:6,5-b′]dithiophene-4,5-dione unit, their composites with carbon, and the lithium storage performances as electrode materials. *J. Solid State Electrochem.* **2021**, *25*, 1847–1859. [CrossRef]
29. Piedboeuf, M.-L.C.; Job, N.; Aqil, A.; Busby, Y.; Fierro, V.; Celzard, A.; Detrembleur, C.; Léonard, A.F. Understanding the Influence of Surface Oxygen Groups on the Electrochemical Behavior of Porous Carbons as Anodes for Lithium-Ion Batteries. *ACS Appl. Mater. Interfaces* **2020**, *12*, 36054–36065. [CrossRef]
30. Ni, B.; Li, Y.; Chen, T.; Lu, T.; Pan, L. Covalent organic frameworks converted N, B co-doped carbon spheres with excellent lithium ion storage performance at high current density. *J. Colloid Interface Sci.* **2019**, *542*, 213–221. [CrossRef] [PubMed]
31. Mao, Y.; Duan, H.; Xu, B.; Zhang, L.; Hu, Y.; Zhao, C.; Wang, Z.; Chen, L.; Yang, Y. Lithium storage in nitrogen-rich mesoporous carbon materials. *Energy Environ. Sci.* **2012**, *5*, 7950–7955. [CrossRef]
32. Yang, Y.; Yuan, J.; Huang, S.; Chen, Z.; Lu, C.; Yang, C.; Zhai, G.; Zhu, J.; Zhuang, X. Porphyrinic conjugated microporous polymer anode for Li-ion batteries. *J. Power Sources* **2022**, *531*, 231340. [CrossRef]
33. Yang, L.; Wei, W.; Ma, Y.; Xu, Y.; Chang, G. Intermolecular channel expansion induced by cation-π interactions to enhance lithium storage in a crosslinked π-conjugated organic anode. *J. Power Sources* **2020**, *449*, 227551. [CrossRef]
34. Zhang, Y.; Li, J.; Gong, Z.; Xie, J.; Lu, T.; Pan, L. Nitrogen and sulfur co-doped vanadium carbide MXene for highly reversible lithium-ion storage. *J. Colloid Interface Sci.* **2021**, *587*, 489–498. [CrossRef] [PubMed]

35. Song, Z.; Zhou, H. Towards sustainable and versatile energy storage devices: An overview of organic electrode materials. *Energy Environ. Sci.* **2013**, *6*, 2280–2301. [CrossRef]
36. Liu, N.; Lu, Z.; Zhao, J.; McDowell, M.T.; Lee, H.-W.; Zhao, W.; Cui, Y. A pomegranate-inspired nanoscale design for large-volume-change lithium battery anodes. *Nat. Nanotechnol.* **2014**, *9*, 187–192. [CrossRef] [PubMed]
37. Li, Z.; Zhong, W.; Cheng, A.; Li, Z.; Li, L.; Zhang, H. Novel hyper-crosslinked polymer anode for lithium-ion batteries with highly reversible capacity and long cycling stability. *Electrochim. Acta* **2018**, *281*, 162–169. [CrossRef]
38. Jiang, G.; Qiu, Y.; Lu, Q.; Zhuang, W.; Xu, X.; Kaskel, S.; Xu, F.; Wang, H. Mesoporous Thin-Wall Molybdenum Nitride for Fast and Stable Na/Li Storage. *ACS Appl. Mater. Interfaces* **2019**, *11*, 41188–41195. [CrossRef]

Disclaimer/Publisher's Note: The statements, opinions and data contained in all publications are solely those of the individual author(s) and contributor(s) and not of MDPI and/or the editor(s). MDPI and/or the editor(s) disclaim responsibility for any injury to people or property resulting from any ideas, methods, instructions or products referred to in the content.

Communication

Facile Synthesis of Polyacrylic Acid/Graphene Oxide Composite Hydrogel Electrolyte for High-Performance Flexible Supercapacitors

Yue Xin, Zhaoxin Yu, Razium Ali Soomro and Ning Sun *

Beijing Key Laboratory of Electrochemical Process and Technology for Materials, College of Materials Science & Engineering, Beijing University of Chemical Technology, Beijing 100029, China
* Correspondence: ningsun@mail.buct.edu.cn

Abstract: The development of hydrogel electrolytes plays a critical role in high-performance flexible supercapacitor devices. Herein, a composite hydrogel electrolyte of polyacrylic acid (PAA) and graphene oxide (GO) has been successfully prepared, where the oxygen-containing functional groups of GO may crosslink and form hydrogen bonds with carboxyl on the molecular chain of PAA, thereby significantly enhancing the mechanical properties of a PAA-based gel electrolyte. The tensile strength increases from 4.0 MPa for pristine PAA gel to 6.1 MPa for PAA/GO composite gel, with the elongation at break rising from 1556% to 1950%. Meanwhile, GO promotes the transportation of electrolyte ions, which are favorable for enhancing the ionic conductivity of the PAA hydrogel. As a result, the assembled supercapacitor based on PAA/GO composite hydrogel electrolyte shows enhanced capacitance retention of 64.3% at a large current density of 20 A g^{-1} and excellent cycling stability over 10,000 cycles at 5 A g^{-1}. Furthermore, the fabricated flexible supercapacitor devices could maintain outstanding electrochemical performance at various bending angles of 0–90°, indicating a promising prospect for the PAA/GO hydrogel electrolyte in flexible wearable fields.

Keywords: hydrogel electrolyte; supercapacitor; polyacrylic acid (PAA); graphene oxide (GO); flexible

Citation: Xin, Y.; Yu, Z.; Soomro, R.A.; Sun, N. Facile Synthesis of Polyacrylic Acid/Graphene Oxide Composite Hydrogel Electrolyte for High-Performance Flexible Supercapacitors. *Coatings* **2023**, *13*, 382. https://doi.org/10.3390/coatings13020382

Academic Editor: Mihai Anastasescu

Received: 10 January 2023
Revised: 27 January 2023
Accepted: 1 February 2023
Published: 7 February 2023

Copyright: © 2023 by the authors. Licensee MDPI, Basel, Switzerland. This article is an open access article distributed under the terms and conditions of the Creative Commons Attribution (CC BY) license (https://creativecommons.org/licenses/by/4.0/).

1. Introduction

A supercapacitor is a kind of energy storage device with the highlighted advantages of high power density and a long cycle life [1–4]. With the development and popularization of smart wearable devices, flexible energy storage devices have gradually become a research hotspot [5–7]. Flexible supercapacitors are composed of flexible electrodes and electrolytes. Among them, the flexible electrolyte can not only transport ions but also play the role of separator, thus simplifying the structure of supercapacitors. However, the present electrochemical performance of flexible supercapacitors is still far from satisfactory. The key to realizing excellent electrochemical performance for flexible supercapacitor devices is designing flexible electrodes and electrolytes with excellent electrochemical and mechanical properties [8]. Flexible electrolytes can avoid electrolyte leaks and keep the role of transporting ions under severe deformation. Compared to traditional energy storage devices using liquid electrolytes, devices based on solid electrolytes are safer, stable and variable in shape, which will bring bright prospects for wearable electronic devices in the future [9].

There are three types of common solid electrolytes: ceramic electrolytes [10], gel electrolytes [11] and polyelectrolytes [12]. Among these solid electrolytes, gel polymer electrolytes are considered the ideal ones that have the advantages of high ionic conductivity, low electronic conductivity, suitable mechanical properties, high stability and easy processability [13,14]. Polyacrylic acid hydrogel is a common gel electrolyte for supercapacitors with relatively high ionic conductivity, good hydrophilicity, nontoxicity and low price [15–20]. However, the mechanical property of polyacrylic acid-based gel usually cannot satisfy the requirement of flexible supercapacitors.

Various strategies, including copolymerization, double crosslinking networks, and doping, have been proposed to improve the mechanical strength of polyacrylic acid-based gel electrolytes without sacrificing ionic conductivity. For example, Liao et al. [18] prepared a hydrogel electrolyte with considerable ionic transportation and a noticeable stress response by copolymerizing PAA and octadecyl methacrylate in a vinyl-treated sponge. The polymer chains were grafted onto the 3D interconnected sponge and built a physical framework, thus exhibiting excellent mechanical properties. Huang et al. [21] prepared a new electrolyte by combining polyacrylic acid and vinyl hybrid silica nanoparticles, and the obtained double crosslinking hydrolyte electrolyte could be stretched over 3700% without any cracks. Additives have also been adopted to improve the conductivity and mechanical properties of polyacrylic acid gel. Guo et al. [22] reported Fe^{3+} crosslinked PAA hydrogel electrolyte by adding $FeCl_3$ into an AA monomer during the polymerization process. The ionic bond between Fe^{3+} and carboxylic acid ions, as well as the intra- and intermolecular hydrogen bonding, endow the supramolecular hydrogel with excellent mechanical properties (extensibility > 700% and stress > 400 kPa). Graphene oxide (GO) has been widely applied in advanced electrochemical devices [23–26]. The surface polar groups and conjugate structure of GO can efficiently promote the transportation of electrolyte ions; therefore, GO is regarded as an excellent ionic conductive additive to the polymer matrix [27–31].

In this study, PAA/GO composite hydrogel electrolytes were successfully prepared by incorporating GO during the polymerization of the AA monomer, as illustrated in Figure 1a. As the abundant surface groups on GO can attract protons and propagate through hydrogen-bonding networks, the transport distance of ions in the gel is greatly shortened with the improvement of ionic conductivity (Figure 1b). Moreover, the oxygen-containing functional groups of GO can crosslink and form hydrogen bonds with the COOH group on the PAA molecular chain, thus improving the mechanical properties of the gel electrolyte. Therefore, compared with pristine PAA gel, the obtained PAA/GO hydrogel electrolytes realize better electrochemical and mechanical properties. After being coupled with commercial active carbon electrodes, the assembled supercapacitor using an optimum PAA/0.5%GO gel electrolyte exhibits excellent electrochemical performance with a high capacitance retention of 64.3% at an enhanced current density of 20 A g^{-1} and outstanding cycling stability with no capacity decay over 10,000 cycles.

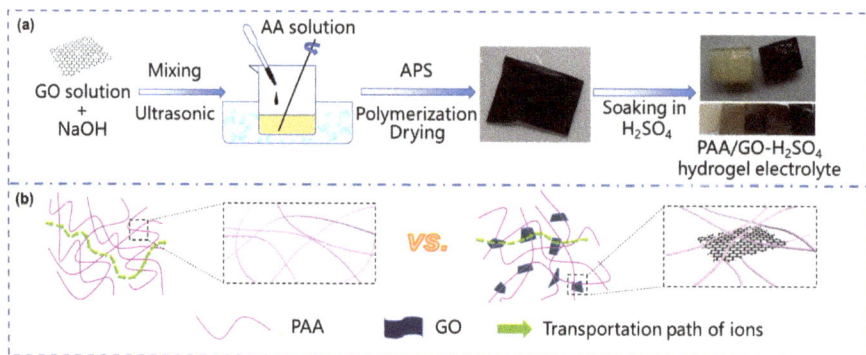

Figure 1. (a) Preparation process of PAA/GO composite gel electrolyte and (b) the schematic diagram for the role of GO in PAA/GO composite gel electrolyte.

2. Experimental

2.1. Materials Synthesis

Graphene oxide (GO) was prepared by the modified Hummer's method [32]. The preparation process of PAA/GO hydrogel electrolyte is shown in Figure 1. First, NaOH powder was slowly added to GO aqueous solution with a concentration of 10 mg mL^{-1} followed by stirring and ultrasonic treatment. 3 g of acrylic acid (AA) monomer was dispersed in 2 mL water and stirred for 10 min in an ice bath. GO/NaOH solution was added to the acrylic acid solution slowly to control the polymerization speed, and then 0.048 g ammonium persulfate (APS) as initiator was introduced and stirred for 15 min. The obtained AA/GO dispersion was injected into a mold and placed in an oven at a constant temperature of 40 °C for 40 h. After polymerization, the polyacrylic acid gel was cut into the desired shape and dried at 90 °C for 90 min. Finally, the PAA/GO gel electrolyte was prepared after being soaked in 3 M H_2SO_4 and further removing bubbles in the vacuum dryer. Gel electrolytes with different amounts of GO (0, 0.1 wt%, 0.5 wt%, 1 wt%, 2 wt%) were prepared and named PAA, PAA/0.1%GO, PAA/0.5%GO, PAA/1%GO, PAA/2%GO.

2.2. Materials Characterization

A scanning electron microscope (SEM, HITACHI S480) was used to observe the surface morphology of the gel electrolyte. Fourier transform infrared spectroscopy (FTIR) was measured by a Nicolet 6700 spectrometer. X-ray photoelectron spectroscopy (XPS) was tested using ESCALAB 250. The tensile properties of hydrogels were measured by the electronic universal testing machine CMT6103. The test conditions were as follows: 25 °C, sample with a size of 3 cm × 2 cm × 0.5 mm, and crosshead speed of 100 mm/min.

2.3. Electrochemical Measurements

All the electrochemical tests of the obtained gel electrolyte were based on the commercial activated carbon (YP50, Kuraray Co., Ltd.) electrode, which was prepared by mixing the activated carbon, acetylene black with polytetrafluoroethylene (PTFE) binder with a mass ratio of 85:10:5 in ethanol. The slurry was dried, rolled into uniform sheets, and cut into round slices with a diameter of 5 mm. After being dried at 120 °C for 6 h in a vacuum oven, the electrodes were obtained with an active material loading of 6 mg cm^{-2}. For the assembly of a solid-state supercapacitor, PAA/GO-H_2SO_4 hydrogel electrolytes were cut into a round slice with a diameter of 8 mm. YP50 was used as the working electrode and counter electrode, and Ag/AgCl was used as a reference electrode. The tests were carried out in a three-electrode system, and the operation potential range was set in the range of 0–0.9 V. For the assembly of a flexible supercapacitor, the gel electrolytes and electrodes were cut into 1.5 cm × 2 cm rectangle slices, with carbon cloth as the flexible current collector. The galvanostatic charge/discharge test was conducted on an Arbin Supercapacitor Test System (BT-G). The cyclic voltammetry (CV) curve and electrochemical impedance spectrum (EIS), with frequencies ranging from 10 mHz to 100 kHz, were measured on the CHI1100C (Chenhua) electrochemical workstation.

3. Results and Discussion

The appearance of the PAA hydrogel is light yellow and transparent (Figure 1), while that of PAA/GO composite gel gets black, and the color gradually deepens with the increase of GO content. The TEM images in Figure A1 show that GO sheets are 5–8 μm in lateral size. The uniform apparent color of PAA/GO gel suggests the homogeneous mixing of PAA and GO nanosheets. Figure 2a–d shows the scanning electron microscopy (SEM) images of PAA and PAA/GO gel with different GO contents. When the GO content is low, the surface of the PAA/GO gel is smooth and flat, which is favorable for electrolytes to fully contact the electrodes. However, when the content of GO increases to 2%, obvious wrinkles appear due to the aggregation of GO layers, which greatly affects the uniformity of gel electrolytes. As shown in Figure 2e,f, the PAA/GO gel can endure bending and torsion well, demonstrating its excellent mechanical strength.

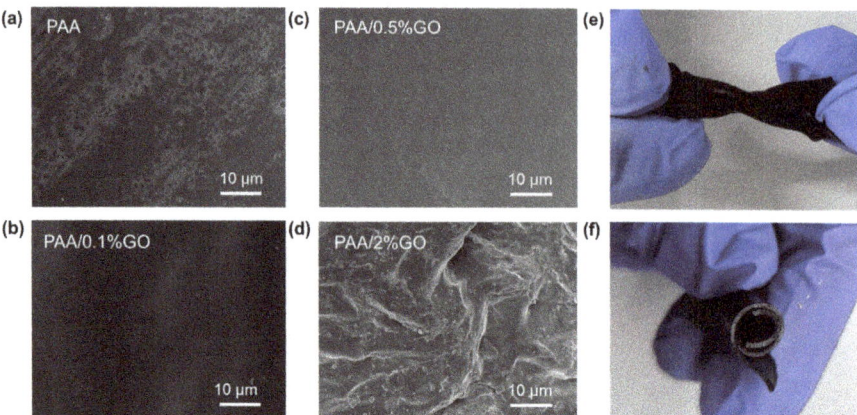

Figure 2. Scanning electron microscope (SEM) images of (**a**) PAA, (**b**) PAA/0.1%GO, (**c**) PAA/0.5%GO, and (**d**) PAA/2%GO. (**e**) Torsion properties and (**f**) rolling properties of composite PAA/0.5%GO electrolytes.

Since the oxygen-containing functional groups of GO can interact with the COOH of polyacrylic chains, GO acts as an effective crosslinking agent in PAA gel, significantly improving the tensile strength of PAA/GO composite gel. The hydrogel can produce large elastic deformation and quickly return to its original shape, as shown in Figure 3a. Figure 3b presents the stress-strain curve of PAA and PAA/0.5%GO, and the tensile strength increases from 4.0 MPa for PAA to 6.1 MPa for PAA/GO composite gel, with elongation at break enhancing from 1556% to 1950%, demonstrating that the addition of GO significantly improved the mechanical properties of PAA gel. The comparison with previously reported polymer electrolytes in Table A1 also verifies the prominent advantage of PAA/GO gel electrolyte.

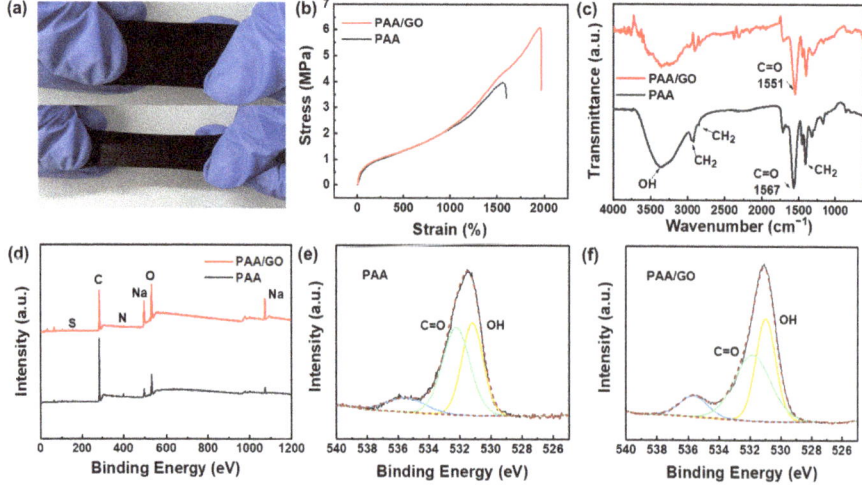

Figure 3. (**a**) Tensile performance of PAA/0.5%GO composite gel electrolyte. (**b**) Tensile stress-strain curve comparison of the PAA and PAA/0.5%GO composite gel electrolyte. (**c**) FTIR and (**d**) XPS full spectrum of PAA and PAA/0.5%GO. O1s peak fitting of (**e**) PAA and (**f**) PAA/GO.

Figure 3c shows the Fourier transform infrared spectroscopy (FTIR) of PAA and PAA/GO hydrogel, where the broad peak at 3360 cm^{-1} is related to the OH stretching vibration. The bending and stretching vibration of CH$_2$ on the polyacrylic acid molecular chain was observed at 1412 cm^{-1}, 2936 cm^{-1}, and 2847 cm^{-1} [33]. The typical bending vibration of C=O in PAA gel appears at 1567 cm^{-1}, while the peak of PAA/GO shifts to 1551 cm^{-1}, which may be ascribed to the influence of GO, which changes the microstructure of PAA gel [34].

XPS was performed to further analyze the elemental composition and content of the PAA gel and PAA/GO gel, as shown in Figure 3d. The PAA and PAA/GO gel are mainly composed of C, O and Na elements, with a minor amount of S and N, which results from the initiator APS. Due to the abundant oxygen-containing groups in GO, the PAA/GO gel displays a relatively higher O content than the pristine PAA gel [34]. Peak fitting of O1s was conducted to further determine the specific functional groups' types in the PAA and PAA/GO gels, as shown in Figure 3e,f. The O1s can be deconvoluted into OH, C=O and absorbed H$_2$O species located at 531 eV, 532.5 eV and 536 eV, respectively [35–37]. Compared with PAA, the PAA/GO gel shows a decreased proportion of C=O and enhanced OH functional groups due to the addition of GO, which not only facilitates crosslinking with the COOH of PAA but also helps the transportation of electrolyte ions, thus improving the conductivity of gel electrolytes.

Ionic conductivity plays an essential role in determining the electrochemical performance of solid electrolytes. Figure 4a shows the Nyquist plots of PAA/GO electrolytes with different GO contents, and the intersection points of the oblique lines with the X-axis are usually considered intrinsic resistance of the ion gel, as the ohmic resistance of the testing device is negligible. The ionic conductivity was calculated according to formula $\rho = L/RS$, where L is the thickness, R is the resistance of gel electrolyte and S is the geometric area of the electrode/electrolyte interface [38,39]. With the increase of GO content, the conductivity value of PAA/GO electrolyte steadily rises due to the transportation effect of oxygenic groups on ions, and the introduction of GO significantly shortens the ion transport path and thus reduces the internal resistance of gel electrolyte. The conductivity reaches the highest when the GO content is 0.5%, as displayed in Figure 4b. Further increasing the proportion of GO, the conductivity of PAA/GO electrolyte gradually decreases because GO nanosheet aggregation hinders ion transportation.

The electrochemical performance of the obtained PAA/GO gel electrolyte was evaluated by assembling supercapacitors with commercial active carbon as electrodes and the PAA/GO gel as electrolyte. Figure 4c,d compares the CV curves of PAA and PAA/0.5%GO electrolyte at various scan rates ranging from 10 mV s^{-1} to 200 mV s^{-1}. The curve shapes of PAA and PAA/0.5%GO are similar to a rectangle, and the pseudocapacitance peaks at about 0.4 V can mainly be attributed to the contribution from PAA [40]. With the increase in scan rate, the shape of the CV curve gradually becomes a flat shuttle. However, PAA/0.5%GO could still maintain a larger area than PAA, indicating that PAA/0.5%GO's better rate performance and maintained electrochemical double-layer capacitance behavior even at high scan rates. The capacitive contribution at different scan rates for the electrode using PAA/0.5%GO electrolyte was further quantitatively determined according to Dunn's method [41], as shown in Figure A2. With the increase in scan rates from 2 mV s^{-1} to 50 mV s^{-1}, the electrochemical double layer capacitance ratio rises from 71.4% to 92.1%. Figure 4e,f show the charge–discharge curves under different current densities of the assembled supercapacitors using PAA/0.5%GO gel electrolyte. The highly symmetrical charge–discharge curves with no obvious IR drop, even under a high current density of 20 A g^{-1}, suggest the small internal resistance and good ionic conductivity of the PAA/GO gel electrolyte. The specific capacitance of the supercapacitors using the PAA/GO gel electrolyte with different GO contents was calculated based on the charge–discharge curve, and similar specific capacitance values were observed at a small current density of 0.2 A g^{-1} for the PAA/GO gel electrolyte and the conventional PAA gel electrolyte, as shown in Figure 4g. The superiority of PAA/GO gel electrolyte becomes prominent with the increased current densities. The specific capacitance can maintain 127 F g^{-1} for PAA/0.5%GO when the current density rises to 20 A g^{-1}, much higher than 53 F g^{-1} for the PAA electrolyte. Compared to the capacitance value at 0.2 A g^{-1}, the capacitance retention for PAA/0.5%GO at an increased current density of 20 A g^{-1} is as high as 64.3%, further verifying the favorable effect of GO on the improvement of rapid ion transportation. Here, the comparatively good capacitance retention of PAA/0.5%GO than PAA/1%GO at a large current density of 20 A g^{-1} may be mainly ascribed to the higher conductivity of the gel electrolytes. In order to examine the long-term cycling stability of the gel electrolyte, the supercapacitor using PAA/0.5%GO gel electrolyte was galvanostatically charged-discharged at a current density of 5 A g^{-1} (Figure 4h). The capacity retention over 10,000 cycles is as high as 103.6%, demonstrating that the gel electrolyte has excellent cycle stability and great application potential. From the charge–discharge curves of the first ten cycles and the last ten cycles, it can be found that the charge–discharge behavior is consistent with isosceles triangle shapes and no obvious voltage drops, further proving the excellent electrochemical stability of PAA/GO gel electrolyte in the long-term charge–discharge cycling process.

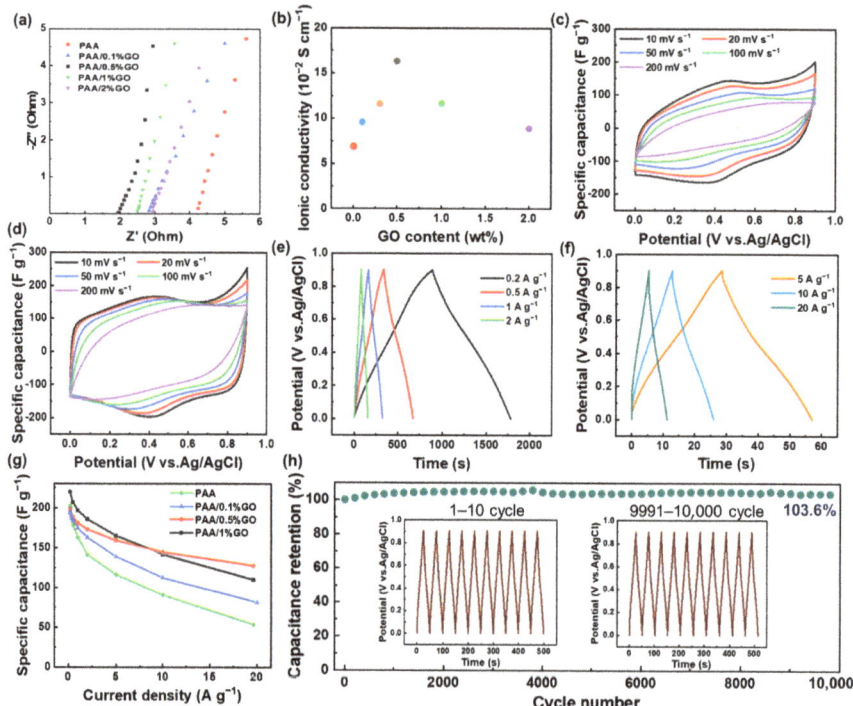

Figure 4. (a) Nyquist plots of PAA/GO electrolytes with different GO contents. (b) Ionic conductivity varies according to GO content. CV curves of the (c) PAA and (d) PAA/0.5%GO solid supercapacitor in 3 M H$_2$SO$_4$. (e,f) Galvanostatic charge–discharge curves at different current densities. (g) Specific capacitances as a function of current densities. (h) The cycling stability of PAA/0.5%GO at 5 A g^{-1} for 10,000 cycles.

To further investigate the practical application prospect of the PAA/GO gel electrolyte in flexible supercapacitors, carbon cloth was used as the current collector, with YP50F as electrode material, PAA/0.5%GO as the gel electrolyte and plastic as external packaging to assemble the packaged supercapacitor device. As shown in Figure 5a, the supercapacitor demonstrates excellent flexibility under various bending conditions (0°, 30°, 60° and 90°). Furthermore, three supercapacitors were connected in series and wrapped around the wrist to simulate the practical application of flexible wearable devices, and it was found that the little bulb could be well lit (Figure 5b), indicating the potential application prospects of the flexible supercapacitor with PAA/0.5%GO as the gel electrolyte.

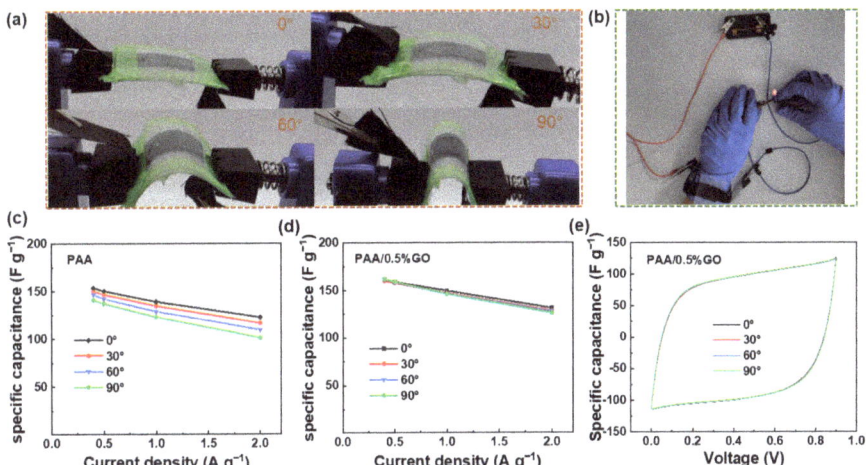

Figure 5. (**a**) Digital photographs of flexible supercapacitor tested at various bending angles of 0°, 30°, 60° and 90°. (**b**) Digital display of a small bulb lighted by three flexible supercapacitors connected in series. Rate performance for (**c**) PAA and (**d**) PAA/0.5%GO gel electrolyte at different bending angles. (**e**) CV curves at a scanning rate of 50 mV s^{-1} for PAA/0.5%GO gel electrolyte at different bending angles.

Figure 5c,d compares the bending performance of PAA and PAA/0.5%GO hydrogel electrolyte. For PAA gel, with the increase of bending angle, both capacitance and rate performance decline dramatically, which can be ascribed to insufficient contact between the electrode and the electrolyte under bending conditions that affect the diffusion of electrolyte ions. While for PAA/0.5%GO gel, little change can be observed in the rate curves at four different angles, further proving the excellent bending performance of PAA/0.5%GO hydrogel. Figure 5e shows the CV curve of PAA/0.5%GO at a scanning rate of 50 mV s^{-1} under various bending conditions. The coincident rectangular curves indicate that the flexible supercapacitor can still maintain double-layer capacitive behavior in the bending state with outstanding capacitance performance.

4. Conclusions

In summary, we successfully fabricated high-performance PAA/GO composite hydrogel electrolytes for flexible supercapacitors. The effective crosslinking between GO and the PAA polymer chain significantly enhanced the mechanical properties of the gel electrolyte, and the tensile strength increased from 4.0 MPa for PAA to 6.1 MPa for PAA/GO composite gel, with the elongation at break enhancing from 1556% to 1950%. Meanwhile, the addition of GO could improve the ionic conductivity of PAA gel and promote the transportation of electrolyte ions. As a result, the assembled electric double-layer capacitor using an optimum PAA/0.5%GO gel electrolyte exhibits greatly improved electrochemical performance with a high capacitance retention of 64.3% at an enhanced current density of 20 A g^{-1} and excellent cycling stability over 10,000 cycles at 5 A g^{-1}. Moreover, the fabricated flexible supercapacitor devices can maintain outstanding electrochemical performance at various bending conditions of 0–90°, demonstrating the promising practical application prospects of the PAA/GO electrolyte.

Author Contributions: Conceptualization, Y.X. and N.S.; methodology, Y.X.; validation, Y.X. and Z.Y.; formal analysis, Y.X.; investigation, Y.X.; resources, Y.X. and R.A.S.; data curation, Y.X. and Z.Y.; writing—original draft preparation, Y.X.; writing—review and editing, R.A.S. and N.S.; visualization, Y.X.; supervision, N.S.; project administration, N.S.; funding acquisition, N.S. All authors have read and agreed to the published version of the manuscript.

Funding: This work was financially supported by the Fundamental Research Funds for the Central Universities (buctrc202141).

Institutional Review Board Statement: Not applicable.

Informed Consent Statement: Not applicable.

Data Availability Statement: All data that support the findings of this study are included within the article.

Conflicts of Interest: The authors declare no conflict of interest.

Appendix A

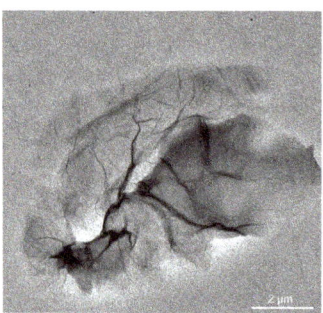

Figure A1. Transmission electron microscope (TEM) image of graphene oxide.

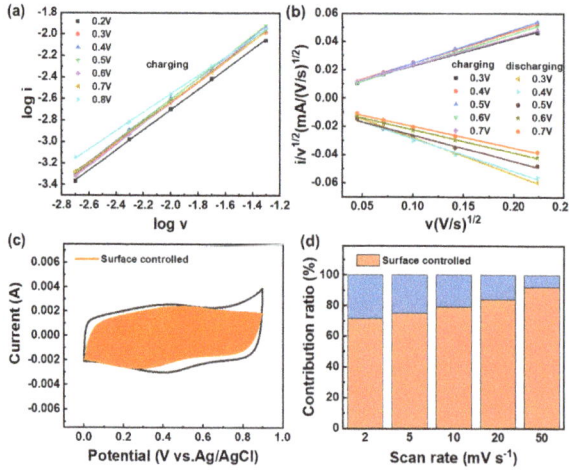

Figure A2. Evaluating the contribution from diffusion controlled and surface-controlled capacitance. (**a**) Plot of log i vs. log ν for determining the slope b; (**b**) Plot of square root of scan rate ($v^{1/2}$) vs. current/$v^{1/2}$ ($i/v^{1/2}$) at different potentials during charging and discharging; (**c**) Contribution from surface-controlled current towards charge storage at 10 mV s^{-1}; (**d**) Contribution ratio of surface-controlled capacitance at different scan rates.

Table A1. Comparison of stress, strain and ionic conductivity of PAA/0.5%GO gel electrolyte in this work with the polymer electrolytes reported in recent works.

	Stress (MPa)	Strain (%)	Ionic Conductivity (10^{-2} S cm^{-1})	Reference
S-PAA	2.2	1100%	1.5	[18]
VSNPs-PAA	0.12	3700%	0.75	[21]
Fe^{3+}/PAA	0.4	700%	9	[22]
Li-AG/PAM DN	1.1	2780%	4.1	[42]
BC-reinforced PAM	0.33	1300%	12.5	[43]
PK10 AGPE	2.22	1243%	21	[44]
B-PVA/KCl/GO	\	\	4.75	[45]
PAA/GO	6.1	1950%	16.81	This work

References

1. Zhang, J.; Luo, J.; Guo, Z.; Liu, Z.; Duan, C.; Dou, S.; Yuan, Q.; Liu, P.; Ji, K.; Zeng, C.; et al. Ultrafast Manufacturing of Ultrafine Structure to Achieve An Energy Density of Over 120 Wh kg^{-1} in Supercapacitors. *Adv. Energy Mater.* **2022**, *27*, 2203061. [CrossRef]
2. Muralee Gopi, C.V.V.; Vinodh, R.; Sambasivam, S.; Obaidat, I.M.; Kim, H.-J. Recent progress of advanced energy storage materials for flexible and wearable supercapacitor: From design and development to applications. *J. Energy Storage* **2020**, *27*, 101035. [CrossRef]
3. Xiao, J.; Han, J.; Zhang, C.; Ling, G.; Kang, F.; Yang, Q.H. Dimensionality, Function and Performance of Carbon Materials in Energy Storage Devices. *Adv. Energy Mater.* **2021**, *12*, 2100775. [CrossRef]
4. Yu, S.; Sun, N.; Hu, L.; Wang, L.; Zhu, Q.; Guan, Y.; Xu, B. Self-template and self-activation synthesis of nitrogen-doped hierarchical porous carbon for supercapacitors. *J. Power Source* **2018**, *405*, 132–141. [CrossRef]
5. Wang, Y.; Wu, X.; Han, Y.; Li, T. Flexible supercapacitor: Overview and outlooks. *J. Energy Storage* **2021**, *42*, 103053. [CrossRef]
6. Thomas, S.A.; Patra, A.; Al-Shehri, B.M.; Selvaraj, M.; Aravind, A.; Rout, C.S. MXene based hybrid materials for supercapacitors: Recent developments and future perspectives. *J. Energy Storage* **2022**, *55*, 105765. [CrossRef]
7. Swain, N.; Tripathy, A.; Thirumurugan, A.; Saravanakumar, B.; Schmidt-Mende, L.; Ramadoss, A. A brief review on stretchable, compressible, and deformable supercapacitor for smart devices. *Chem. Eng. J.* **2022**, *446*, 136876. [CrossRef]
8. Dubal, D.P.; Chodankar, N.R.; Kim, D.H.; Gomez-Romero, P. Towards flexible solid-state supercapacitors for smart and wearable electronics. *Chem. Soc. Rev.* **2018**, *47*, 2065–2129. [CrossRef]
9. Ma, G.; Dong, M.; Sun, K.; Feng, E.; Peng, H.; Lei, Z. A redox mediator doped gel polymer as an electrolyte and separator for a high performance solid state supercapacitor. *J. Mater. Chem. A* **2015**, *3*, 4035–4041. [CrossRef]
10. Menkin, S.; Lifshitz, M.; Haimovich, A.; Goor, M.; Blanga, R.; Greenbaum, S.G.; Goldbourt, A.; Golodnitsky, D. Evaluation of ion-transport in composite polymer-in-ceramic electrolytes. Case study of active and inert ceramics. *Electrochim. Acta* **2019**, *304*, 447–455. [CrossRef]
11. Liu, S.; Zhong, Y.; Zhang, X.; Pi, M.; Wang, X.; Zhu, R.; Cui, W.; Ran, R. Highly Deformable, Conductive Double-Network Hydrogel Electrolytes for Durable and Flexible Supercapacitors. *ACS Appl. Mater. Interfaces* **2022**, *14*, 15641–15652. [CrossRef]
12. Luangaramvej, P.; Dubas, S.T. Two-step polyaniline loading in polyelectrolyte complex membranes for improved pseudo-capacitor electrodes. *e-Polymers* **2021**, *21*, 194–199. [CrossRef]
13. Na, R.; Huo, P.; Zhang, X.; Zhang, S.; Du, Y.; Zhu, K.; Lu, Y.; Zhang, M.; Luan, J.; Wang, G. A flexible solid-state supercapacitor based on a poly(aryl ether ketone)–poly(ethylene glycol) copolymer solid polymer electrolyte for high temperature applications. *RSC Adv.* **2016**, *6*, 65186–65195. [CrossRef]
14. Alipoori, S.; Mazinani, S.; Aboutalebi, S.H.; Sharif, F. Review of PVA-based gel polymer electrolytes in flexible solid-state supercapacitors: Opportunities and challenges. *J. Energy Storage* **2020**, *27*, 101072. [CrossRef]
15. Chan, C.Y.; Wang, Z.; Jia, H.; Ng, P.F.; Chow, L.; Fei, B. Recent advances of hydrogel electrolytes in flexible energy storage devices. *J. Mater. Chem. A* **2021**, *9*, 2043–2069. [CrossRef]
16. Hu, M.; Wang, J.; Liu, J.; Zhang, J.; Ma, X.; Huang, Y. An intrinsically compressible and stretchable all-in-one configured supercapacitor. *ChemComm* **2018**, *54*, 6200–6203. [CrossRef]
17. Zhang, X.; Pei, Z.; Wang, C.; Yuan, Z.; Wei, L.; Pan, Y.; Mahmood, A.; Shao, Q.; Chen, Y. Flexible Zinc-Ion Hybrid Fiber Capacitors with Ultrahigh Energy Density and Long Cycling Life for Wearable Electronics. *Small* **2019**, *15*, 1903817. [CrossRef]
18. Liao, H.; Zhou, F.; Zhang, Z.; Yang, J. A self-healable and mechanical toughness flexible supercapacitor based on polyacrylic acid hydrogel electrolyte. *Chem. Eng. J.* **2019**, *357*, 428–434. [CrossRef]
19. Abubshait, H.A.; Saad, M.; Iqbal, S.; Abubshait, S.A.; Bahadur, A.; Raheel, M.; Alshammari, F.H.; Alwadai, N.; Alrbyawi, H.; Abourehab, M.A.S.; et al. Co-doped zinc oxide nanoparticles embedded in Polyvinylalcohol Hydrogel as solar light derived photocatalyst disinfection and removal of coloured pollutants. *J. Mol. Struct.* **2023**, *1271*, 134100. [CrossRef]

20. Iqbal, S.; Javed, M.; Qamar, M.A.; Bahadur, A.; Fayyaz, M.; Akbar, A.; Alsaab, H.O.; Awwad, N.S.; Ibrahium, H.A. Synthesis of Cu-ZnO/Polyacrylic Acid Hydrogel as Visible-Light-Driven Photocatalyst for Organic Pollutant Degradation. *ChemistrySelect* **2022**, *7*, e202103694. [CrossRef]
21. Huang, Y.; Zhong, M.; Huang, Y.; Zhu, M.; Pei, Z.; Wang, Z.; Xue, Q.; Xie, X.; Zhi, C. A self-healable and highly stretchable supercapacitor based on a dual crosslinked polyelectrolyte. *Nat. Commun.* **2015**, *6*, 10310. [CrossRef] [PubMed]
22. Guo, Y.; Zhou, X.; Tang, Q.; Bao, H.; Wang, G.; Saha, P. A self-healable and easily recyclable supramolecular hydrogel electrolyte for flexible supercapacitors. *J. Mater. Chem. A* **2016**, *4*, 8769–8776. [CrossRef]
23. Cao, Y.-C.; Xu, C.; Wu, X.; Wang, X.; Xing, L.; Scott, K. A poly (ethylene oxide)/graphene oxide electrolyte membrane for low temperature polymer fuel cells. *J. Power Source* **2011**, *196*, 8377–8382. [CrossRef]
24. Huang, Y.-F.; Wu, P.-F.; Zhang, M.-Q.; Ruan, W.-H.; Giannelis, E.P. Boron cross-linked graphene oxide/polyvinyl alcohol nanocomposite gel electrolyte for flexible solid-state electric double layer capacitor with high performance. *Electrochim. Acta* **2014**, *132*, 103–111. [CrossRef]
25. Xu, B.; Wang, H.; Zhu, Q.; Sun, N.; Anasori, B.; Hu, L.; Wang, F.; Guan, Y.; Gogotsi, Y. Reduced graphene oxide as a multi-functional conductive binder for supercapacitor electrodes. *Energy Storage Mater.* **2018**, *12*, 128–136. [CrossRef]
26. Sun, N.; Guan, Y.; Liu, Y.-T.; Zhu, Q.; Shen, J.; Liu, H.; Zhou, S.; Xu, B. Facile synthesis of free-standing, flexible hard carbon anode for high-performance sodium ion batteries using graphene as a multi-functional binder. *Carbon* **2018**, *137*, 475–483. [CrossRef]
27. Stankovich, S.; Dikin, D.A.; Piner, R.D.; Kohlhaas, K.A.; Kleinhammes, A.; Jia, Y.; Wu, Y.; Nguyen, S.T.; Ruoff, R.S. Synthesis of graphene-based nanosheets via chemical reduction of exfoliated graphite oxide. *Carbon* **2007**, *45*, 1558–1565. [CrossRef]
28. Zhu, Y.; Murali, S.; Cai, W.; Li, X.; Suk, J.W.; Potts, J.R.; Ruoff, R.S. Graphene and graphene oxide: Synthesis, properties, and applications. *Adv. Mater.* **2010**, *22*, 3906–3924. [CrossRef]
29. Huang, Y.; Zhu, M.; Huang, Y.; Pei, Z.; Li, H.; Wang, Z.; Xue, Q.; Zhi, C. Multifunctional Energy Storage and Conversion Devices. *Adv. Mater.* **2016**, *28*, 8344–8364. [CrossRef]
30. Pan, C.; Liu, L.; Gai, G. Recent Progress of Graphene-Containing Polymer Hydrogels: Preparations, Properties, and Applications. *Macromol. Mater. Eng.* **2017**, *302*, 1700184. [CrossRef]
31. Tian, Y.; Yu, Z.; Cao, L.; Zhang, X.L.; Sun, C.; Wang, D.-W. Graphene oxide: An emerging electromaterial for energy storage and conversion. *J. Energy Chem.* **2021**, *55*, 323–344. [CrossRef]
32. Marcano, D.C.; Kosynkin, D.V.; Berlin, J.M.; Sinitskii, A.; Sun, Z.; Slesarev, A.; Alemany, L.B.; Lu, W.; Tour, J.M. Improved Synthesis of Graphene Oxide. *ACS Nano* **2010**, *4*, 4806–4814. [CrossRef]
33. Han, Q.; Chen, L.; Li, W.; Zhou, Z.; Fang, Z.; Xu, Z.; Qian, X. Self-assembled three-dimensional double network graphene oxide/polyacrylic acid hybrid aerogel for removal of Cu(2+) from aqueous solution. *Environ. Sci. Pollut. Res.* **2018**, *25*, 34438–34447. [CrossRef]
34. Al-Gaashani, R.; Najjar, A.; Zakaria, Y.; Mansour, S.; Atieh, M.A. XPS and structural studies of high quality graphene oxide and reduced graphene oxide prepared by different chemical oxidation methods. *Ceram. Int.* **2019**, *45*, 14439–14448. [CrossRef]
35. Piloto, C.; Shafiei, M.; Khan, H.; Gupta, B.; Tesfamichael, T.; Motta, N. Sensing performance of reduced graphene oxide-Fe doped WO3 hybrids to NO2 and humidity at room temperature. *Appl. Surf. Sci.* **2018**, *434*, 126–133. [CrossRef]
36. Kong, W.; Yue, Q.; Li, Q.; Gao, B. Adsorption of Cd(2+) on GO/PAA hydrogel and preliminary recycle to GO/PAA-CdS as efficient photocatalyst. *Sci. Total Environ.* **2019**, *668*, 1165–1174. [CrossRef]
37. Feng, Z.; Feng, C.; Chen, N.; Lu, W.; Wang, S. Preparation of composite hydrogel with high mechanical strength and reusability for removal of Cu(II) and Pb(II) from water. *Sep. Purif. Technol.* **2022**, *300*, 121894. [CrossRef]
38. Wang, S.H.; Hou, S.S.; Kuo, P.L.; Teng, H. Poly(ethylene oxide)-co-poly(propylene oxide)-based gel electrolyte with high ionic conductivity and mechanical integrity for lithium-ion batteries. *ACS Appl. Mater. Interfaces* **2013**, *5*, 8477–8485. [CrossRef]
39. Yang, X.; Zhang, F.; Zhang, L.; Zhang, T.; Huang, Y.; Chen, Y. A High-Performance Graphene Oxide-Doped Ion Gel as Gel Polymer Electrolyte for All-Solid-State Supercapacitor Applications. *Adv. Funct. Mater.* **2013**, *23*, 3353–3360. [CrossRef]
40. Fan, X.; Lu, Y.; Xu, H.; Kong, X.; Wang, J. Reversible redox reaction on the oxygen-containing functional groups of an electrochemically modified graphite electrode for the pseudo-capacitance. *J. Mater. Chem.* **2011**, *21*, 18753. [CrossRef]
41. Islam, T.; Hasan, M.M.; Shah, S.S.; Karim, M.R.; Al-Mubaddel, F.S.; Zahir, M.H.; Dar, M.A.; Hossain, M.D.; Aziz, M.A.; Ahammad, A.J.S. High yield activated porous coal carbon nanosheets from Boropukuria coal mine as supercapacitor material: Investigation of the charge storing mechanism at the interfacial region. *J Energy Storage* **2020**, *32*, 101908. [CrossRef]
42. Lin, T.; Shi, M.; Huang, F.; Peng, J.; Bai, Q.; Li, J.; Zhai, M. One-Pot Synthesis of a Double-Network Hydrogel Electrolyte with Extraordinarily Excellent Mechanical Properties for a Highly Compressible and Bendable Flexible Supercapacitor. *ACS Appl. Mater. Interfaces* **2018**, *10*, 29684–29693. [CrossRef] [PubMed]
43. Li, X.; Yuan, L.; Liu, R.; He, H.; Hao, J.; Lu, Y.; Wang, Y.; Liang, G.; Yuan, G.; Guo, Z. Engineering Textile Electrode and Bacterial Cellulose Nanofiber Reinforced Hydrogel Electrolyte to Enable High-Performance Flexible All-Solid-State Supercapacitors. *Adv. Energy Mater.* **2021**, *11*, 2003010. [CrossRef]

44. Hu, X.; Fan, L.; Qin, G.; Shen, Z.; Chen, J.; Wang, M.; Yang, J.; Chen, Q. Flexible and low temperature resistant double network alkaline gel polymer electrolyte with dual-role KOH for supercapacitor. *J. Power Source* **2019**, *414*, 201–209. [CrossRef]
45. Peng, H.; Lv, Y.; Wei, G.; Zhou, J.; Gao, X.; Sun, K.; Ma, G.; Lei, Z. A flexible and self-healing hydrogel electrolyte for smart supercapacitor. *J. Power Source* **2019**, *431*, 210–219. [CrossRef]

Disclaimer/Publisher's Note: The statements, opinions and data contained in all publications are solely those of the individual author(s) and contributor(s) and not of MDPI and/or the editor(s). MDPI and/or the editor(s) disclaim responsibility for any injury to people or property resulting from any ideas, methods, instructions or products referred to in the content.

Article

Impact of Iron Pyrite Nanoparticles Sizes in Photovoltaic Performance

Refka Sai [1,2,*] and Rasha A. Abumousa [3]

[1] Laboratory of Semiconductors, Nanostructures and Advanced Technologies, Borj Cedria Science and Technology Park, BP 95, Hammam-Lif 2050, Tunisia
[2] Departement de Physique, Faculté des Sciences de Bizerte, Université de Carthage, Tunis 2036, Tunisia
[3] Department of Mathematics and Sciences, College of Humanities and Sciences, Prince Sultan University, Riyadh 11586, Saudi Arabia
* Correspondence: refkasai@hotmail.com

Abstract: With rising energy demand and depleted traditional fuels, solar cells offer a sustainable and clean option. In recent years, and due to its acceptable band gap, high absorption coefficient, and inexpensive cost, iron pyrite (FeS_2) is a popular material for solar cells. Earth abundance and nontoxicity further boost its photovoltaic possibilities. The current study examined the influence of sulfurization at 350–400 °C on iron pyrite layers fabricated using spray pyrolysis. The morphology and size from TEM confirmed the XRD results of synthesizing a pyrite FeS_2 with an average particle size of 10–23 nm at 350–400 °C, respectively. The direct band gap calculated by DFT as a function of temperature was found to be consistent with the experimental findings, 0.87 eV (0.87) and 0.90 eV (0.95) at 350 °C and 400 °C, respectively. We found high-performing photovoltaic cells on ITO/ZnO/FeS_2/MoO_3/Au/Ag, obtained with an excellent quality of nanoparticles and nanostructures of FeS_2 pyrite, which improved with the method of preparation and growth parameters.

Keywords: iron pyrite; band energy; organic photovoltaic cells

1. Introduction

In the last decades, transition metals experienced a renewal of interest due to their excellent electrical, transport, magnetic, and optical properties [1–3]. In general, FeS_2 pyrite is an ideal material for the fabrication of solar cells and photovoltaic devices [4–9] due to its high absorption coefficient ($\alpha > 10^5$ cm^{-1} for $h\nu > 1.3 - 1.4$ eV) [5], its small band gap (about 0.95 eV) [10], high photocurrent quantum efficiency (> 90%) [6], and low material cost [4–11].

FeS_2 pyrite was one of the first crystal structures that resulted from Bragg, in 1914 [12], with his XRD system. It has a simple cubic structure similar to that of rock salt.

FeS_2 pyrite is a good option for thin film photovoltaic. Considering its potential and current importance [4–9], many experimental [4–9,13–40] and theoretical works [41–60] have been interested to FeS_2. Schlegel et al. [33] determined the transition and reflectivity spectrum of a single crystal of FeS_2. They showed that FeS_2 pyrite has an empty 3d e_g band at 300 K, a completely filled 3 d t_{2g}, and an indirect band gap equal to 0.95 eV. Kou et al. [25] found a band gap at 297 K of about 0.84 eV. Karguppikar et al. [40] reported that pyrite can be an indirect semiconductor from its conductivity properties, Hall effect data, and optical gap of 0.92 eV. Sun et al. [37] determined an FeS_2 pyrite thin film by sulfurizing oxide precursor films. From their UV-vis absorbance spectroscopy and X-ray photoelectron spectroscopy (XPS), they showed a direct band gap of about 0.75 eV, an indirect band gap of about 1.19 eV, and a high absorption efficiency ($\alpha > 10^5$ cm^{-1}). Yu et al. [39] used the chemical bath deposition (CBD) method. They reported that the band gap of FeS_2 can be increased from 0.86 to 1.31 eV when doped by Mn.

Many other preparation methods, such as spray pyrolysis, metal organic chemical vapor deposition (MOCVD), and ion beam sputtering have been declared for the synthesis of nanocrystals, nanowires, and crystallites of FeS_2.

Mostly all experiments found a band gap between 0.84 and 1.03 eV. For theoretical study, Bullet [42] used the first principle local density approximation (LDA) calculation to investigate the optical properties of iron pyrite. He found an indirect band gap of 0.4 eV for marcasite and 0.7 eV for pyrite. This value is smaller by 0.25 eV when compared to the experimental indirect band gap (0.95 eV). Zhoa et al. [59] performed the self-consistent linear combination of atomic orbital (LCAO) formalism to determine the electronic properties of iron pyrite. Their smallest theoretical direct band gap was about 0.64 eV, and they found an indirect band gap of 0.59 eV. Additionally, Opahlele et al. [53,54] determined the electronic properties of FeS_2 utilizing an LDA potential parameterized by the Perdew-Zunger (LDA-PZ). Their calculated band gap was about 0.85 eV. Muscat et al. [52] employed the periodic LCAO method with the CRYSTAL 98 package and pseudopotential technique with CASTEP Software package.

Wadia et al. [61,62] showed a complete research study a few years ago and investigated 23 potential materials for photovoltaics and found that FeS_2 pyrite was the best one, beating all materials in terms of cost. It was confirmed that the extraction cost of silicon was 57 times more than that of FeS_2 (USD 1.7 for silicon compared to USD 0.03 for FeS_2). Additionally, the silicon energy output for extractions was 12 times bigger than that of FeS_2 (24 KWh kg^{-1} vs. 2 KWh kg^{-1} for 24 KWh kg^{-1}). Rahman et al. [62] showed that FeS_2 is much more cost-effective than silicon if they are produced with taxation and the same regulations in the same country. All these beneficial and interesting features make FeS_2 an excellent candidate for photovoltaic performance.

We included the prepared structure to show the real effect of the structure on band gap for devices that are not photovoltaic. The biggest dilemma for iron pyrite is attributed to the structure of pyrite, for which we complete studied to include the impact of the nanoparticles of iron pyrite on photovoltaic performance. This work aimed to improvise a new progress in the use of iron pyrite in photovoltaics.

2. Experimental

2.1. Materials and Method

We used spray pyrolysis for the fabrication of our sample. Amorphous iron oxide films were placed on normal glass substrates by spray pyrolysis. Then, the found films were heated under a sulphur atmosphere. We started by cleaning the glass substrates. The process was as follows: first, we put the films in an acidic solution for 3 h. After that, they were kept in a detergent solution and washed with distilled water. Then, they were kept in a solution of methanol and washed utilizing an ultrasonic cleaner for 15 min. Finally, the substrate was cleaned with distilled water and dried under a stream of nitrogen. In the second procedure, we chose to spray $FeCl_3 \cdot 6H_2O$ (0.05M) for 5 min in an aqueous-based solution onto glass substrates. After we dissolved $FeCl_3 \cdot 6H_2O$ in deionized water (ionization reaction: $2H_2O \rightarrow H_3O^+ + OH^-$), a dark amorphous iron oxide layer was obtained. The jet flow rate and the distance nozzle-substrate were about 7 mL/min and 45 cm. Carrier gas was used by compressed air. We obtained dark layers. Then, it was heated under a sulphur atmosphere ($\sim 10^{-6}$ torr) at two sulfurization temperature (350 °C–400 °C). However, we succeeded in obtaining FeS_2 pyrite via spray pyrolysis.

2.2. Characterization

The crystal structure of the FeS_2 pyrite was analyzed by powder X-ray diffraction (XRD) using a Siemens D500 diffractometer (Siemens Bruker, Germany) (CuKα radiation λ = 1.54201 Å). The parameter lattice and crystal structure were obtained using the Reitveld method by utilizing the PDXL program. Raman spectra were determined to further study the phase evolution with increasing temperature. The morphology and size of nanocrystals were recorded using transmission electron microscopy (TEM) using JEOL

2010 (200 KV) microscopy. Optical absorption was investigated using a SHIMADZU 3100s spectrophotometer (SHIMADZU, Columbia, MD, USA).

2.2.1. X-ray Diffraction:

XRD patterns of the FeS$_2$ pyrite sample are shown in Figure 1. Typical diffraction peaks at 2θ = 28.71°, 33.43°, 37.25°, 40°, 57.79°, 59.98°, 61.89°, and 64.31° attributed respectively to plan (111), (200), (210), (211), (220), (311), (222), (230), and (321), corresponding with the norm diffraction data of the FeS$_2$ (JCPDS card n°028-0076; space group Pa3). No other impurities such as marcasite, pyrrhotite, or greigite compounds were detected in the XRD patterns, confirming the high purity of the obtained sample. Powder XRD patterns appeared in a cube of crystalline in a pyrite structure, which the disulfide ions localized in octahedral coordinated with Fe metal ions within a space group symmetry of T_h^6 (Pa3). The significant effect of temperature can be observed on the position of sulfur (S). The sulfur position changed when the temperature increased.

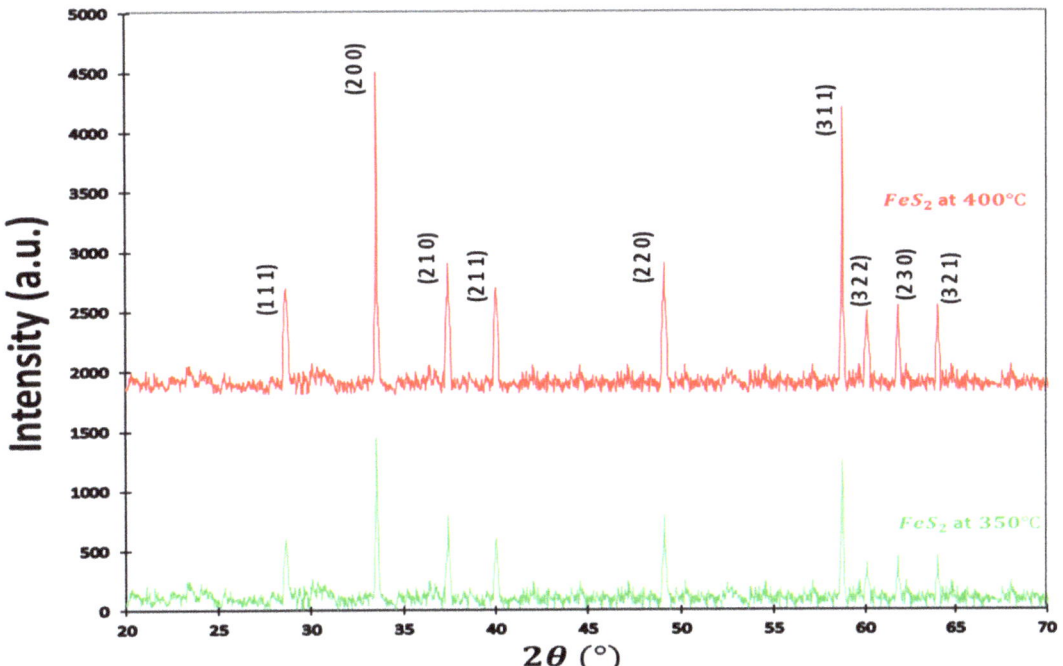

Figure 1. XRD pattern of FeS$_2$ pyrite.

The output parameters are listed in Table 1.

Table 1. Cells parameters.

Tempurature	350 °C	400 °C
Cubic lattice a	5.409 Å	5.417 Å
v (Position paramater)	0.113	0.111
u (Suffer position)	0.387	0.389

2.2.2. Raman Spectra

The Raman spectra of the FeS$_2$ pyrite are presented in Figure 2. As shown, sulfurization was conducted at different times and temperatures (350 °C and 400 °C). We identified that the increase of the sulfurization time and temperature reduced the perfection of formation of other phases, and this resulted in a pure pyrite structure. We observed two strong typical Raman peaks at 342 cm^{-1} and 403 cm^{-1}, corresponding to E^{low} and E^{High} modes and generated by Fe-Fe and S-S vibrations, respectively. We used long sulfurization times for two temperatures to find the growth condition for high throughput and low-cost processing. In the past literature, time sulfurization ranged between 90 and 120 min [63,64] and was still considered short sulfurization between 3 h and 8 h [65,66].

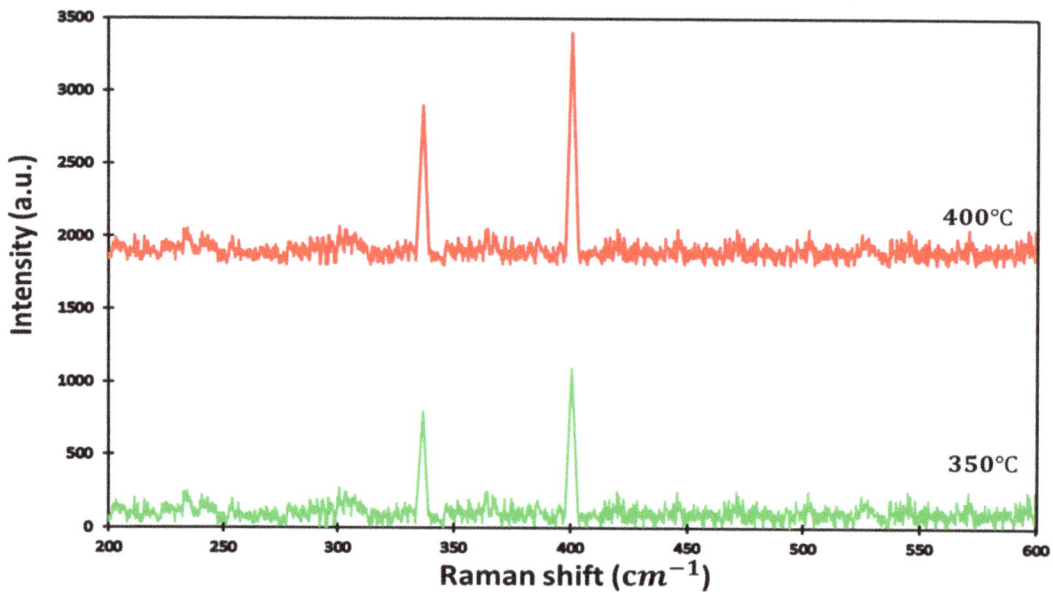

Figure 2. Raman spectra for thin film obtained after sulfurization for time = 14 h and at two sulfurization temperatures.

From our Raman analyses, we calculated the intensity ratio of E^{low} and E^{High}, which are related to sulfur distribution versus the crystallinity of the FeS$_2$ sample. Table 2 presents the difference in the contribution of sulfurization at different temperatures. It can be noticed that sulfurization increased with temperature, and the increase of the band gap highlights the influence of the sulfur position on the formation of intermediate levels in the band gap of the MS$_2$ transition metal.

Table 2. The intensity of the E^{low} mode and the intensity ratio of (E^{low}/E^{High}) modes of the FeS$_2$ sample at different temperatures.

Temperature (°C)	350 °C	400 °C
I(E^{low})	0.66	0.79
I(E^{low})/I(E^{low})	0.73	0.83

2.2.3. Transmission Electron Microscopy

Besides the variation nanoparticle size conforming to the results from transmission electron microscopy (TEM), as in Figure 3, the nature of sulfurization using this method affects the morphology of the resulting FeS$_2$ pyrite. For logical statistics, around 100 particles of a representative sample section were studied. Particle diameters were calculated, assuming an ideal spherical particle shape based on the measured area. In case of FeS$_2$ pyrite at 400 °C Figure 3a–c, the smallest particles in the range of 7–15 nm. Overall, average particle sizes were between 10 nm and 23 nm, and highest size was 39 nm. However, for FeS$_2$ pyrite at 350 °C (Figure 3d,e), we obtained only particles between 10 nm and 23 nm.

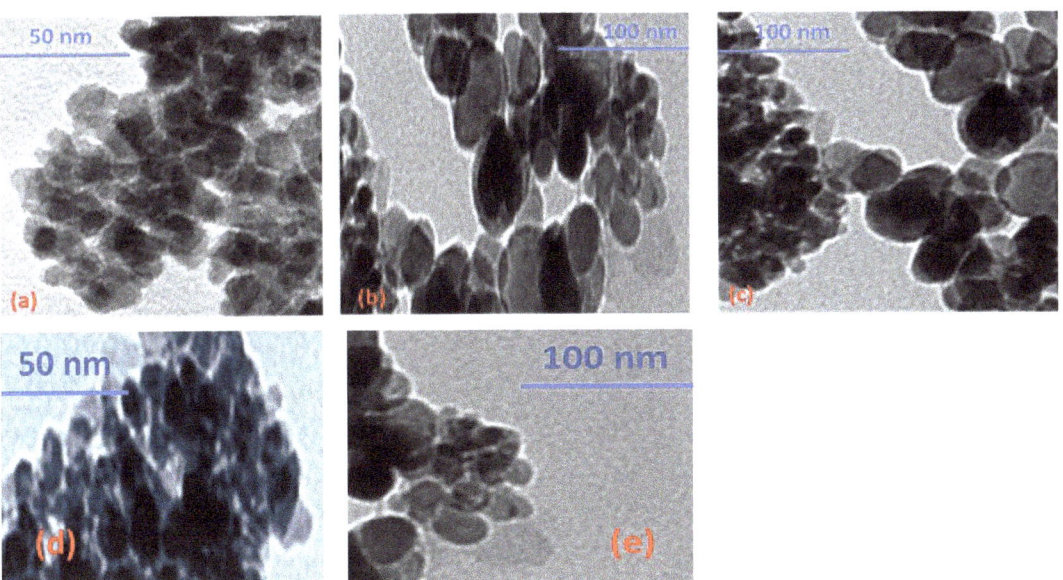

Figure 3. (a–c): TEM image of FeS$_2$ at 400 °C. (d,e): TEM image of FeS$_2$ at 350 °C.

At higher magnification (Figure 3b,c), the surface of layers appeared granular with different grain sizes, and this enabled us to confirm the good crystallinity, showing its photovoltaic performance.

2.2.4. Optical Properties:

The mathematical equation for the optical band gap was given by Tauc et al. [67], and we used it in these studies as well [68,69]:

$$\alpha h\nu = A(h\nu - E_g)^n$$

where $\begin{cases} n = 2 \rightarrow \text{for an indirect transition} \\ n = \frac{1}{2} \rightarrow \text{for an direct transition} \end{cases}$.

where α is the absorbance coefficient, A is a constant, and $h\nu$ is the photon energy. The variation $(\alpha h\nu)^{\frac{1}{2}}$ vs. photon energy $h\nu$ suggest an indirect band gap and for $(\alpha h\nu)^2$ vs. photon energy $h\nu$, depicting a direct experimental transition. Notice that pyrite may have an indirect or direct transition [70,71]. The question is: will the band gap of our sample be indirect or direct? In fact, if iron pyrite layers have a direct band gap, it is very important for them to apply multispectral photovoltaic cells, because a direct band gap means a direct transition.

$(\alpha h\nu)^{\frac{n}{2}}$ were plotted as functions of photon energy $h\nu$, with $n = 4$ and $n = 1$, and it is presented in Figure 4a–c. Only the plot of $(\alpha h\nu)^2$ vs. $h\nu$ has a straight line, indicating that FeS$_2$ pyrite film has a direct band gap energy for different temperatures of fabrication. Table 3 gives the acquired values of the band gap of FeS$_2$ pyrite film according to the two temperatures. FeS$_2$ pyrite had a low band gap when heat treated at 350 °C. Figure 4d shows the absorption coefficients, which were always high, and they were greater than 1.2×10^5 cm^{-1}.

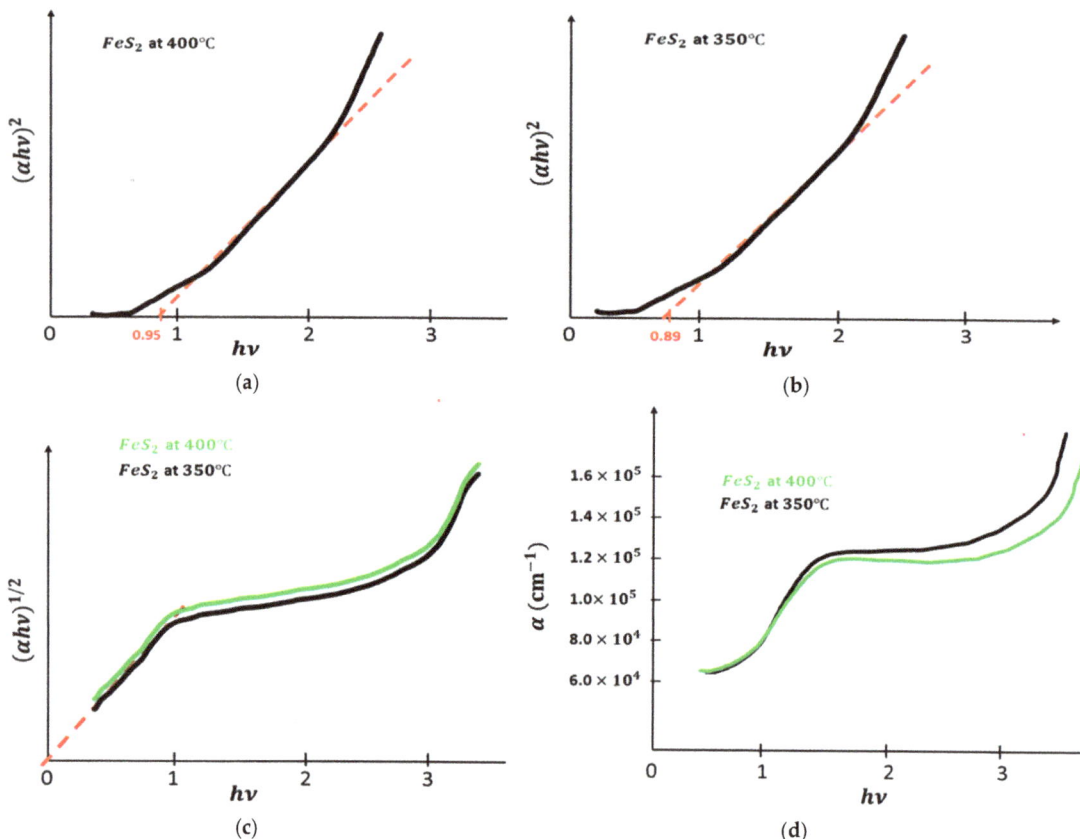

Figure 4. (a): $(\alpha h\nu)^2$ vs. $h\nu$. (b): $(\alpha h\nu)^2$ vs. $h\nu$. (c): $(\alpha h\nu)^{1/2}$ vs. $h\nu$. (d): α vs. $h\nu$.

Table 3. Exprimental band gap.

Temperature	350 °C	400 °C
Band gap (eV)	0.89	0.95

3. Band Structure

3.1. Computational Details

Density–Functional Theory [72,73] was used within the linear muffin-tin orbital method in the atomic-sphere approximation (LMTO-ASA). The LMTO ASA method was explained in detail in several reports [74–76]. In our calculations, we employed the self-consistent band calculations because they are the first principles of calculations utilizing density functional theory (see [72]), utilizing the local density approximation (see [77]), and utilizing numerical techniques based on the treatment of electron ion interaction in the pseudopotential approximation [78]. Moreover, the Hamiltonian Atomic Spheres Approximation is totally specified by the potential parameters. It generates moments from the eigenvectors of the Hamiltonian. Regarding specified potential, there is an individual correspondence between the energy E_v of the wave function φ and the logarithmic derivative D_v at the sphere radius. In essence, it is possible to specify either one. The potential P becomes simple because [79,80]:

$$P(\varepsilon) = \text{const} \frac{D(\varepsilon)+1+1}{D(\varepsilon)-1} \approx \left(\frac{\Delta_1}{\varepsilon - C_1} + \gamma_1\right)^{-1}$$

where γ_1, Δ_1, and C_1 are the "potential parameters" that parameterize P. C_1 defines the band "center of gravity", Δ_1 is the "band width" parameter, which correlates with the bandwidth of l channel if it were uncoupled from the other channels, and γ_1 is the "band distortion parameter", which describes the deformations relative to a universal shape. Generally, small parameterization is a perfect method to study band structure.

First, we obtained the potential parameters for all atomic spheres. The muffin-tin potential constant V_{MTZ} was the crossing point of muffin-tin potential around Fe and S, and it is listed in Table 4. We had 24 symmetry operations. The initial sphere packing was equal to 82.2%, and it was scaled to 92.9%. The role of these empty spheres is to reduce the number of iterations in this system and to reduce the overlap between the spheres centered at Fe and S.

Table 4. The muffin-tin potential constant.

Temperature	350 °C	400 °C
V_{MTZ}	−0.679772	−0.737209

3.2. Pyrite Crystal Structure

FeS$_2$ pyrite has a cubic crystal structure and the space group number 205 (with space-group T_h^6 (Pa3)). In these structure, there are eight S atoms located in eight positions and four Fe atoms in four positions. The lattice parameters for FeS$_2$ pyrite are listed in Table 5.

Table 5. Pyrite cell parameters.

Temperature	350 °C	400 °C
space group	T_h^6	T_h^6
cubic lattice	5.409 Å	5.417 Å
ν position parameter	0.113	0.111
sulfur position	0.387	0.389
Bond length d_{s-s}	2.117 Å	2.083 Å
Bond length d_{Fe-s}	2.264 Å	2.259 Å
Bond angle Fe − S − Fe	115.21°	114.88°
Bond angle Fe − S − S	102.82°	103.28°
Bond angle S − Fe − S	100.09°	100.13°

In FeS$_2$ pyrite, each S atom is coordinated with three Fe atoms, for which the dimer pairs S_S are in tetrahedral sites, and each Fe atom is coordinated with six S atoms in octahedral sites.

Moreover, these structures contain pairs of sulfur S$_2$ molecules, contrary to individual S atoms presented in the Figure 5a image of the overlapped unit cell of FeS$_2$ pyrite. To study the deviation from tetrahedral and octahedral geometries, we describe the correlation of the S-S bond length and cubic lattice. The relationships between our cell parameters are presented in Table 5. The structure ν of the pyrite is between 0.10 and 0.13 [81]. Our work demonstrated that the value of ν ranges between 0.111 and 0.113, showing significant effects of increased temperature conditioning FeS$_2$ pyrite. We noticed that the Fe sites had a small trigonal distortion, for which the S-Fe-S bond angles were 100.09° at 350 °C and

100.13° at 400 °C, and the three Fe-S-Fe bonds were between 114.88° and 115.21°, for which the S sites were distorted from tetrahedral symmetry.

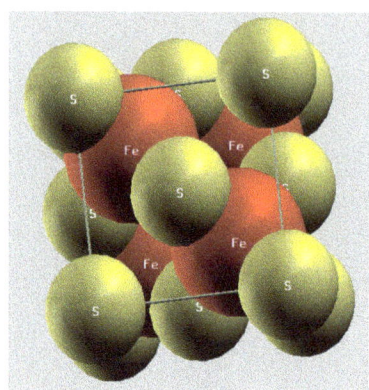

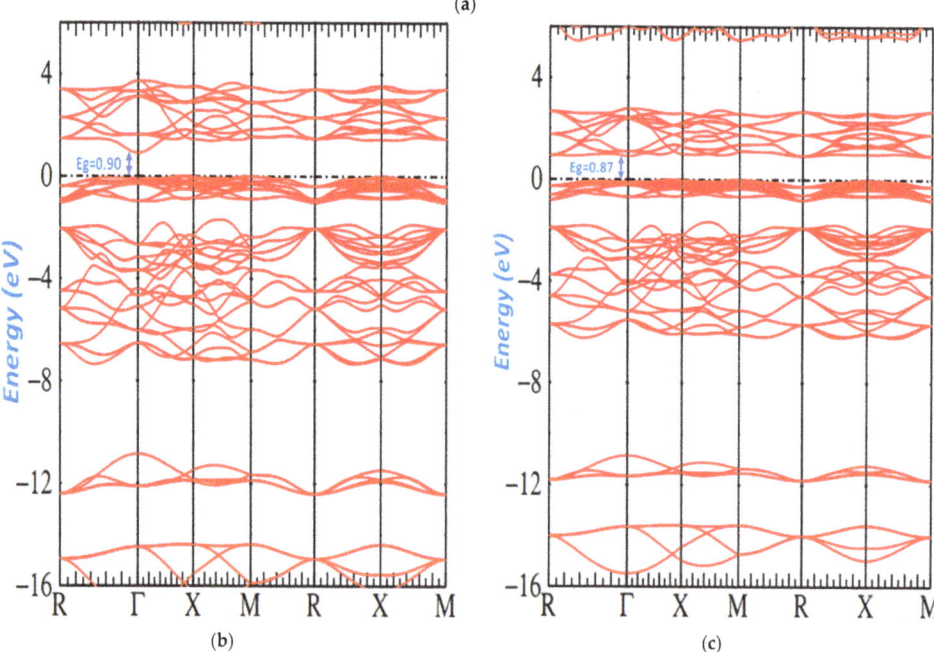

Figure 5. (**a**): Image of overlap of FeS$_2$ pyrite. (**b**): Band structure of FeS$_2$ pyrite. (**c**): Band structure of FeS$_2$ pyrite at the temperature 350 °C and at the temperature 400 °C.

3.3. Energy Bands of FeS$_2$ Pyrite

The DFT energy bands for samples at different temperatures are shown in Figure 5b,c. The two figures provide the band structure on a form of energy that shows a general electronic structure. These figures present the region around the Fermi energy, which clearly depicts the details of a low conduction band and the highest valence band.

However, our band structure indicates that FeS$_2$ has a band gap semiconductor. We had four FeS$_2$ units in each unit cell and accommodated 40 occupied valence bands. The minimum conduction band was in Γ, and the maximum valence band was at Γ. The direct transition was observed at Γ and had the values of 0.87 eV and 0.90 eV.

Our obtained results are different than those of Zhoa et al. [59] and Temmerman [58], who found 0.59 eV and 0.64 eV. Their gap was smaller than the experimental gap of 0.95 eV. Moreover, we were successful because our obtained gaps were significantly consistent with our optical gaps.

The calculated gap, optical gap, minimum band conduction CB_{max}, maximum valence band VB_{max}, and Fermi energy are summarized in Table 6. The bands are relative to bonding and antibonding pairs of S_2 orbitals. In the range between −14 and 8 eV corresponding S3s, the structure of sulfur and these S3 states are predominant. The S 3p state is presented with a small addition of an Fe 3d function, starting at approximately −6 eV. Basically, these bands have Fe t_{2g} hybridized with p orbitals, and they are below the Fermi energy. The lowest conduction band contained a combination of Fe e_g orbitals. They were above Fermi energy. S3p Fe 3d is a preeminent character on the conduction band.

Table 6. Band structure of FeS$_2$ Parameters.

Temperature	350 °C	400 °C
Calculated gap E_g	0.064 Ry = 0.87040 eV	0.06550 Ry = 0.90508 eV
optical gap E_g	0.89 eV	0.95 eV
minimum band conductionCCB_{min}	−0.052792 Ry	0.066017 Ry
maximum band valence VB_{max}	−0.116791 Ry	−0.000533
Fermi enrgy E_f	−0.116791 Ry	−0.000533

Figure 6a,b presents the density of the states calculated, and it shows the states above and below the Fermi level of iron pyrite. It discloses the importance of hybridization between Fe and S states and the effect of temperature on the fabrication of pyrite.

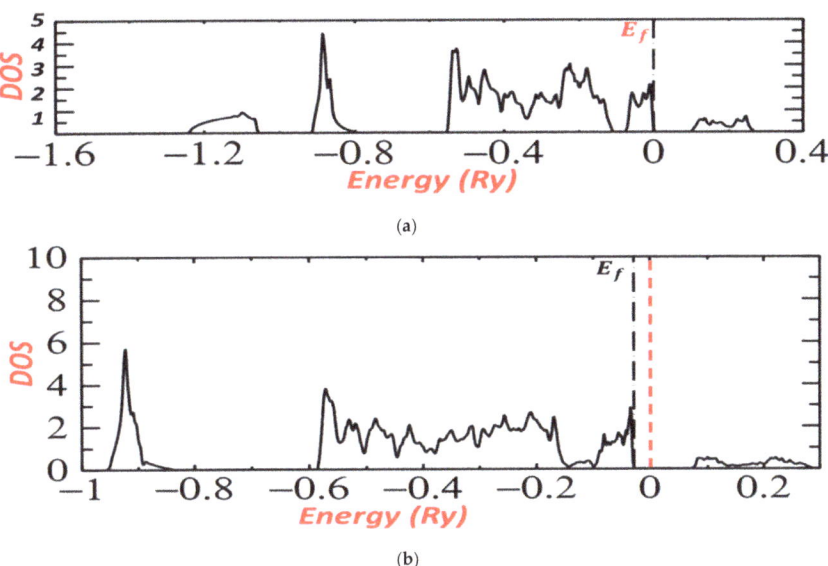

Figure 6. (**a**): DOS for FeS$_2$ pyrite at the temperature 400 °C. (**b**): DOS for FeS$_2$ pyrite at the temperature 350 °C.

For the two graphs, level t2g lies between −0.1 Ry and 0, below the Fermi level, but for FeS$_2$ pyrite prepared at 400 °C, it is near 0 and near Fermi level, which implies the good crystallite and electronic properties of iron pyrite prepared at 400 °C.

For both graphs, we noted that the conduction band was made entirely of Fe eg with Sp, marking that the conduction band was pure Sp while the valence bands were completely derived from the Fe t2g.

4. Pyrite in Photovoltaics: Modeling the ITO/ZnO/FeS$_2$/ MoO$_3$/Au/Ag Device

The synthesized FeS$_2$ pyrite samples were evaluated for the application of photodetector devices. We modeled ITO/ZnO/FeS$_2$/MoO$_3$/Au/Ag to study the improvement in solar cell characteristics realized by the increase of temperature in two cases of preparation of FeS$_2$ pyrite. We chose this application because it holds numerous benefits due its ability to be prepared at mild conditions, its low cost of chemicals, its mechanical flexibility, its better tuning, and due to it being a suitable alternative to silicone-based solar cells [82,83].

The device structure is presented in Figure 7a. ITO film was cut mechanically to obtain a 2.5 cm × 2.5 cm substrate. All substrates were cleaned in isopropanol, water, and soap for 15 min. For layer parameters, a washed indium tin oxide (ITO) glass substrate was managed by ultraviolet-ozone for 15 min. The ZnO layers were spin coated with 60 mg ml^{-1} ZnO/CHC$_3$ solution annealed at 250 °C for 15 min in the air to form a ZnO layer of 100 nm. The MoO$_3$ (20 nm) Au (30 nm) Ag (90 nm) layers were placed successively by thermal evaporation.

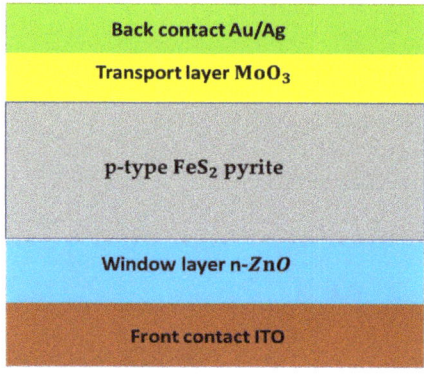
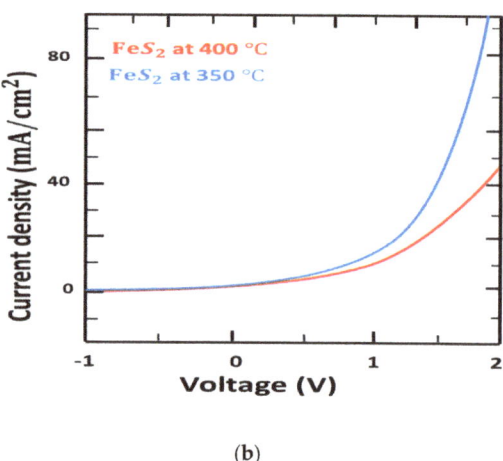

(a) (b)

Figure 7. (a): Device structure of ITO/ZnO/FeS$_2$/ MoO$_3$/Au/Ag. (b): Current density versus voltage (J–V) characteristics of fabricated solid solar cells.

In recent years, ZnO has become the prime candidate for organic photovoltaic cells [84] since its efficient improvement in stability. Here, we used the n-p layer heterojunction p-type FeS$_2$ pyrite solar cell using an n-type window layer. Additionally, MoO$_3$ thin film can react as an effective electron-blocking layer or hole transporting to reduce the recombination of holes and electrons [85].

The schematic illustration characterization by exercising voltage from -1 to 2 V under a dark current using the modeled ITO/ZnO/FeS$_2$/MoO$_3$/Au/Ag structure, as presented in Figure 7b. The reported I-V characteristics and calculations for p-type FeS$_2$ pyrite at 350 °C were similar [86], principally below the onset voltage and the I-V curves above 1.5 V. However, the I-V curves for p-type FeS$_2$ pyrite at 400 °C showed a large difference above 1.4 V, for which the onset voltage was around 1.3 eV. This result corresponds well with the fabrication of FeS$_2$ pyrite, indicating that the nanostructures composed of FeS$_2$ pyrite at 400 °C are excellent in building n-p junction ZnO/FeS$_2$ pyrite.

We concluded that, to obtain high-performing photovoltaic cells on ITO/ZnO/FeS$_2$/ MoO$_3$/Au/Ag, it is necessary to focus on the quality of nanoparticles and nanostructures of FeS$_2$ pyrite, which improved by increasing the temperature of preparation (400 °C). We also mentioned to the effect of sulfur position, distance sulfur–sulfur, and temperature to band gap of FeS$_2$ pyrite.

5. Conclusions

This work was an inclusive study of FeS$_2$ pyrite. Our study reported on the increase of temperature of preparation, characterization, and calculation of band gap of FeS$_2$ pyrite. Our experiment demonstrated the effect of temperature of preparation on the favorable optical and electrical properties of FeS$_2$ pyrite. Our results confirmed that p-type FeS$_2$ pyrite is a good choice for fabricating solar cells. We also found the gap energy and sulfur-sulfur distance for samples with different temperatures of preparation, which were prepared by spray pyrolysis. We proved the correlation between growth parameters and the calculated band structure. The optical gap energy obtained in this work is in good agreement with the gap energy calculated by the LMTO-ASA method.

Our findings proved a significant powerful dependency between gap energy and distance sulfur–sulfur. Moreover, we concluded that the excellent crystallinity, nanoparticles, and nanostructures of FeS$_2$ pyrite confirm a more efficient photovoltaic application.

Finally, it is important to note that the high-performing photovoltaic cells on ITO/ZnO/FeS$_2$/MoO$_3$/Au/Ag positively improved the quality of nanoparticles and nanostructures of FeS$_2$ pyrite.

Author Contributions: Conceptualization, R.S. and R.A.A.; methodology, R.S. and R.A.A.; software, R.S. and R.A.A.; validation, R.S. and R.A.A.; formal analysis, R.S. and R.A.A.; investigation, R.S. and R.A.A.; resources, R.S.; data curation, R.S. and R.A.A.; writing—original draft preparation, R.S. and R.A.A.; writing—review and editing, R.S. and R.A.A.; visualization, R.S. and R.A.A.; supervision, R.S.; project administration, R.S. and R.A.A. All authors have read and agreed to the published version of the manuscript.

Funding: This research received no external funding.

Institutional Review Board Statement: Not applicable.

Informed Consent Statement: Not applicable.

Data Availability Statement: Not applicable.

Acknowledgments: The Author R. Abumousa would like to acknowledge the support of Prince Sultan University for paying the Article Processing Charges (APC) of this publication.

Conflicts of Interest: The authors declare no conflict of interest.

References

1. Rao, C.N.R.; Deepak, F.L.; Gundiah, G.; Govindaraj, A. Effect of Mn Doping on Solvothermal Synthesis of CdS Nanowires. *Prog. Solid State Chem.* **2003**, *31*, 5. [CrossRef]
2. Tang, K.B.; Qian, Y.T.; Zeng, J.H.; Yang, X.G. Solvothermal Route to Semiconductor Nanowires. *Adv. Mater.* **2003**, *15*, 448. [CrossRef]
3. Kumar, S.; Nann, T. Shape control of II–VI semiconductor nanomaterials. *Small* **2006**, *2*, 316. [CrossRef]
4. Altermatt, P.P.; Kiesewetter, T.; Ellmer, K.; Tributsch, H. Specifying targets of future research in photovoltaic devices containing pyrite (FeS$_2$) by numerical modelling. *Sol. Energy Mater. Sol. Cells* **2002**, *71*, 181–195. [CrossRef]
5. Ennaoui, A.; Fiechter, S.; Goslowsky, H.; Tributsch, H. Photoactive synthetic polycrystalline pyrite (FeS$_2$). *J. Electrochem. Soc.* **1985**, *132*, 1579–1582. [CrossRef]
6. Ennaoui, A.; Fiechter, S.; Pettenkofer, C.; Alonso-Vante, N.; Buker, K.; Bronold, M.; Hopfner, C.; Tributsch, H. Iron disulfide for solar-energy conversion. *Sol. Energy Mater. Sol. Cells* **1993**, *29*, 289–370. [CrossRef]
7. Ennaoui, A.; Tributsch, H. Iron sulphide solar cells. *Sol. Cells* **1984**, *13*, 197–200. [CrossRef]
8. Ennaoui, A.; Tributsch, H. Energetic characterization of the photoactive FeS$_2$ (pyrite) interface. *Sol. Energy Mater.* **1986**, *14*, 461–474. [CrossRef]
9. Smestad, G.; Ennaoui, A.; Fiechter, S.; Tributsch, H.; Hofmann, W.; Birkholz, M. Photoactive thin film semiconducting iron pyrite prepared by sulfurization of iron oxides. *Sol. Energy Mater.* **1990**, *20*, 149–165. [CrossRef]
10. Sai, R.; Ezzaouia, H.; Nofal, M.M. Electronic structure of iron pyrite by the LMTO_ASA method. *Results Phys.* **2021**, *22*, 103950. [CrossRef]

11. Huckaba, A.J.; Sanghyun, P.; Grancini, G.; Bastola, E.; Taek, C.K.; Younghui, L.; Bhandari, K.P.; Ballif, C.; Ellingson, R.J.; Nazeeruddin, M.K. Exceedingly Cheap Perovskite Solar Cells Using Iron Pyrite Hole Transport Materials. *ChemistrySelect* **2016**, *1*, 5316–5319. [CrossRef]
12. Bragg, W.L. The analysis of crystals by the X-ray spectrometer. *Proc. R. Soc. London. Ser. A. Contain. Pap. A Math. Phys. Character* **1914**, *89*, 468–489.
13. Abass, A.; Ahmed, Z.; Tahir, R. Absorption edge measurements in chemically deposited pyrite FeS_2 thin layers. *J. Appl. Phys.* **1987**, *61*, 2339–2341. [CrossRef]
14. Bhandari, K.P.; Roland, P.J.; Kinner, T.; Cao, Y.; Choi, H.; Jeong, S.; Ellingson, R.J. Analysis and characterization of iron pyrite nanocrystals and nanocrystalline thin films derived from bromide anion synthesis. *J. Mater. Chem. A* **2015**, *3*, 6853–6861. [CrossRef]
15. Birkholz, M.; Fiechter, S.; Hartmann, A.; Tributsch, H. Sulfur deficiency in iron pyrite (FeS_2-x) and its consequences for band-structure models. *Phys. Rev. B* **1991**, *43*, 11926. [CrossRef]
16. Cabén-Acevedo, M.; Faber, M.S.; Tan, Y.; Hamers, R.J.; Jin, S. Synthesis and Properties of Semiconducting Iron Pyrite (FeS_2) Nanowires. *Nano Lett.* **2012**, *12*, 1977–1982. [CrossRef] [PubMed]
17. Cervantes, P.; Slanic, Z.; Bridges, F.; Knittle, E.; Williams, Q. The band gap and electrical resistivity of FeS_2-pyrite at high pressures. *J. Phys. Chem. Solids* **2002**, *63*, 1927–1933. [CrossRef]
18. Chen, X.; Wang, Z.; Wang, X.; Wan, J.; Liu, J.; Qian, Y. Single-source approach to cubic FeS_2 crystallites and their optical and electrochemical properties. *Inorg. Chem.* **2005**, *44*, 951–954. [CrossRef]
19. Christian Nweze, S.E. Optical Properties of Semiconducting Pyrite Deposited by Aerosol Assisted Chemical Vapour Deposition (AACVD) Method. *Adv. Phys. Theor. Appl.* **2014**, *34*, 21–29.
20. Dong, Y.; Zheng, Y.; Duan, H.; Sun, Y.; Chen, Y. Formation of pyrite (FeS_2) thin nano-films by thermal-sulfurating electrodeposition films at different temperature. *Mater. Lett.* **2005**, *59*, 2398–2402. [CrossRef]
21. Ferrer, I.; Nevskaia, D.; De las Heras, C.; Sanchez, C. About the band gap nature of FeS_2 as determined from optical and photoelectrochemical measurements. *Solid State Commun.* **1990**, *74*, 913–916. [CrossRef]
22. Ferrer, I.; Sanchez, C. Characterization of FeS_2 thin films prepared by thermal sulfidation of flash evaporated iron. *J. Appl. Phys.* **1991**, *70*, 2641–2647. [CrossRef]
23. Ho, C.; Huang, Y.; Tiong, K. Characterization of near band-edge properties of synthetic p-FeS_2 iron pyrite from electrical and photoconductivity measurements. *J. Alloys Compd.* **2006**, *422*, 321–327. [CrossRef]
24. Kim, H.T.; Nguyen, T.P.N.; Kim, C.-d.; Park, C. Formation mechanisms of pyrite (FeS_2) nano-crystals synthesized by colloidal route in sulfur abundant environment. *Mater. Chem. Phys.* **2014**, *148*, 1095–1098. [CrossRef]
25. Kou, W.W.; Seehra, M.S. Optical absorption in iron pyrite (FeS_2). *Phys. Rev. B* **1978**, *18*, 7062. [CrossRef]
26. Liu, Y.; Meng, L.; Zhang, L. Optical and electrical properties of FeS_2 thin films with different thickness prepared by sulfurizing evaporated iron. *Thin Solid Film.* **2005**, *479*, 83–88. [CrossRef]
27. Lucas, J.M.; Tuan, C.-C.; Lounis, S.D.; Britt, D.K.; Qiao, R.; Yang, W.; Lanzara, A.; Alivisatos, A.P. Ligand-Controlled Colloidal Synthesis and Electronic Structure Characterization of Cubic Iron Pyrite (FeS_2) Nanocrystals. *Chem. Mater.* **2013**, *25*, 1615–1620. [CrossRef]
28. Middya, S.; Layek, A.; Dey, A.; Ray, P.P. Synthesis of Nanocrystalline FeS_2 with Increased Band Gap for Solar Energy Harvesting. *J. Mater. Sci. Technol.* **2014**, *30*, 770–775. [CrossRef]
29. Ouertani, B.; Ouerfelli, J.; Saadoun, M.; Bessais, B.; Ezzaouia, H.; Bernede, J.C. Characterization of FeS_2-pyrite thin films synthesized by sulphuration of amorphous iron oxide films pre-deposited by spray pyrolysis. *J. Mater. Charact.* **2005**, *54*, 431–437. [CrossRef]
30. Prince, K.; Matteucci, M.; Kuepper, K.; Chiuzbaian, S.; Bartkowski, S.; Neumann, M. Core-level spectroscopic study of FeO and FeS_2. *Phys. Rev. B* **2005**, *71*, 085102. [CrossRef]
31. Puthussery, J.; Seefeld, S.; Berry, N.; Gibbs, M.; Law, M. Colloidal Iron Pyrite (FeS_2) Nanocrystal Inks for Thin-Film Photovoltaics. *J. Am. Chem. Soc.* **2010**, *133*, 716–719. [CrossRef]
32. Raturi, A.; Ndjeli, L.; Rabah, K. FeS_2 thin films prepared by spray pyrolysis. *Renew. Energy* **1997**, *11*, 191–195. [CrossRef]
33. Santos-Cruz, D.; Mayén-Hernández, S.A.; de Moure-Flores, F.; Arias-Cerón, J.S.; Santos-Cruz, J. Evaporated iron disulfide thin films with sulfurated annealing treatments. *Mater. Sci. Semicond. Process.* **2016**, *42*, 383–389. [CrossRef]
34. Schlegel, A.; Wachter, P. Optical properties, phonons and electronic structure of iron pyrite (FeS_2). *J. Phys. C Solid State Phys.* **1976**, *9*, 3363. [CrossRef]
35. Seefeld, S.; Limpinsel, M.; Liu, Y.; Farhi, N.; Weber, A.; Zhang, Y.; Berry, N.; Kwon, Y.J.; Perkins, C.L.; Hemminger, J.C. Iron Pyrite Thin Films Synthesized from an Fe(acac)3 Ink. *J. Am. Chem. Soc.* **2013**, *135*, 4412–4424. [CrossRef]
36. Subedi, I.; Bhandari, K.P.; Ellingson, R.J.; Podraza, N.J. Near infrared to ultraviolet optical properties of bulk single crystal and nanocrystal thin film iron pyrite. *Nanotechnology* **2016**, *27*, 295702. [CrossRef]
37. Sun, K.; Su, Z.; Yang, J.; Han, Z.; Liu, F.; Lai, Y.; Li, J.; Liu, Y. Fabrication of pyrite FeS_2 thin films by sulfurizing oxide precursor films deposited via successive ionic layer adsorption and reaction method. *Thin Solid Film.* **2013**, *542*, 123–128. [CrossRef]
38. Wu, R.; Zheng, Y.; Zhang, X.; Sun, Y.; Xu, J.; Jian, J. Hydrothermal synthesis and crystal structure of pyrite. *J. Cryst. Growth* **2004**, *266*, 523–527. [CrossRef]
39. Yu, Q.; Cai, S.; Jin, Z.; Yan, Z. Evolutions of composition, microstructure and optical properties of Mn-doped pyrite (FeS_2) films prepared by chemical bath deposition. *Mater. Res. Bull.* **2013**, *48*, 3601–3606. [CrossRef]

40. Karguppikar, A.; Vedeshwar, A. Electrical and optical properties of natural iron pyrite (FeS$_2$). *Phys. Status Solidi A* **1988**, *109*, 549–558. [CrossRef]
41. Antonov, V.; Germash, L.; Shpak, A.; Yaresko, A. Electronic structure, optical and X-ray emission spectra in FeS$_2$. *Phys. Status Solidi B* **2009**, *246*, 411–416. [CrossRef]
42. Bullett, D. Electronic structure of 3d pyrite- and marcasite-type sulphides. *J. Phys. C Solid State Phys.* **1982**, *15*, 6163. [CrossRef]
43. Cai, J.; Philpott, M.R. Electronic structure of bulk and (0 0 1) surface layers of pyrite FeS$_2$. *Comput. Mater. Sci.* **2004**, *30*, 358–363. [CrossRef]
44. Choi, S.; Hu, J.; Abdallah, L.; Limpinsel, M.; Zhang, Y.; Zollner, S.; Wu, R.; Law, M. Pseudodielectric function and critical-point energies of iron pyrite. *Phys. Rev. B* **2012**, *86*, 115207. [CrossRef]
45. Edelbro, R.; Sandstro, Å.; Paul, J. Full potential calculations on the electron bandstructures of Sphalerite, Pyrite and Chalcopyrite. *Appl. Surf. Sci.* **2003**, *206*, 300–313. [CrossRef]
46. Folkerts, W.; Sawatzky, G.; Haas, C.; De Groot, R.; Hillebrecht, F. Electronic structure of some 3d transition-metal pyrites. *J. Phys. C Solid State Phys.* **1987**, *20*, 4135. [CrossRef]
47. Herbert, F.; Krishnamoorthy, A.; Van Vliet, K.; Yildiz, B. Quantification of electronic band gap and surface states on FeS$_2$(100). *Surf. Sci.* **2013**, *618*, 53–61. [CrossRef]
48. Hu, J.; Zhang, Y.; Law, M.; Wu, R. Increasing the band gap of iron pyrite by alloying with oxygen. *J. Am. Chem. Soc.* **2012**, *134*, 13216–13219. [CrossRef]
49. Hu, J.; Zhang, Y.; Law, M.; Wu, R. First-principles studies of the electronic properties of native and substitutional anionic defects in bulk iron pyrite. *Phys. Rev. B* **2012**, *85*, 085203. [CrossRef]
50. Hung, A.; Muscat, J.; Yarovsky, I.; Russo, S.P. Density-functional theory studies of pyrite FeS$_2$ (111) and (210) surfaces. *Surf. Sci.* **2002**, *513*, 511–524.
51. Kolb, B.; Kolpak, A.M. Ultrafast band-gap oscillations in iron pyrite. *Phys. Rev. B* **2013**, *88*, 235208. [CrossRef]
52. Muscat, J.; Hung, A.; Russo, S.; Yarovsky, I. First-principles studies of the structural and electronic properties of pyrite FeS$_2$. *Phys. Rev. B* **2002**, *65*, 054107. [CrossRef]
53. Opahle, I.; Koepernik, K.; Eschrig, H. Full potential band structure calculation of iron pyrite. *Comput. Mater. Sci.* **2000**, *17*, 206–210. [CrossRef]
54. Opahle, I.; Koepernik, K.; Eschrig, H. Full-potential band-structure calculation of iron pyrite. *Phys. Rev. Bs* **1999**, *60*, 14035. [CrossRef]
55. Qiu, G.; Xiao, Q.; Hu, Y.; Qin, W.; Wang, D. Theoretical study of the surface energy and electronic structure of pyrite FeS$_2$ (100) using a total-energy pseudopotential method, CASTEP. *J. Colloid Interface Sci.* **2004**, *270*, 127–132. [CrossRef]
56. Rosso, K.M.; Becker, U.; Hochella, M.F. Atomically resolved electronic structure of pyrite {100} surfaces: An experimental and theoretical investigation with implications for reactivity. *Am. Mineral.* **1999**, *84*, 1535–1548. [CrossRef]
57. Sun, R.; Chan, M.; Ceder, G. First-principles electronic structure and relative stability of pyrite and marcasite: Implications for photovoltaic performance. *Phys. Rev. B* **2011**, *83*, 235311. [CrossRef]
58. Temmerman, W.; Durham, P.; Vaughan, D. The electronic structures of the pyrite-type disulphides (MS2, where M = Mn, Fe, Co, Ni, Cu, Zn) and the bulk properties of pyrite from local density approximation (LDA) band structure calculations. *Phys. Chem. Miner.* **1993**, *20*, 248–254. [CrossRef]
59. Zhao, G.; Callaway, J.; Hayashibara, M. Electronic structures of iron and cobalt pyrites. *Phys. Rev. B* **1993**, *48*, 15781. [CrossRef]
60. Eyert, K.H.H.V.; Fiechter, S.; Tributsch, H. Electronic structure of FeS$_2$: The crucial role of electron-lattice interaction. *Phys. Rev. B* **1998**, *57*, 6350–6359. [CrossRef]
61. Voigt, B.; Moore, W.; Manno, M.; Walter, J.; Jeremiason, J.D.; Aydil, E.S.; Leighton, C. Transport evidence for sulfur vacancies as the origin of unintentional n-type doping in pyrite FeS$_2$. *ACS Appl. Mater. Interfaces* **2019**, *11*, 15552–15563. [CrossRef]
62. Rahman, M.; Boschloo, G.; Hagfeldt, A.; Edvinsson, T. On the mechanistic understanding of photovoltage loss in iron pyrite solar cells. *Adv. Mater.* **2020**, *32*, 1905653. [CrossRef] [PubMed]
63. Jung, H.R.; Shin, S.W.; Gurav, K.; Gang, M.G.; Lee, J.Y.; Moon, J.H.; Kim, J.H. Evolution of detrimental secondary phases in unstable Cu2ZnSnS4 films during annealing. *Electron. Mater. Lett.* **2016**, *12*, 139–146. [CrossRef]
64. Fan, D.; Zhang, R.; Zhu, Y.; Peng, H.; Zhang, J. Structural development and dynamic process in sulfurizing precursors to prepare Cu$_2$ZnSnS$_4$ absorber layer. *J. Alloy. Comp.* **2014**, *583*, 566–573. [CrossRef]
65. Paal, M.; Nkrumah, I.; Ampong, F.K.; Ngbiche, D.U.; Nkum, R.K.; Boakye, F.K. The effect of deposition time and sulfurization temperature on the optical and structural properties of iron sulfide thin films deposited from acidic chemical baths. *Sci. J. Univ. Zakho* **2020**, *8*, 97–104. [CrossRef]
66. Morrish, R.; Silverstein, R.; Wolden, C.A. Synthesis of Stoichiometric FeS$_2$ through Plasma-Assisted Sulfurization of Fe$_2$O$_3$ Nanorods. *J. Am. Chem. Soc.* **2012**, *134*, 17854–17857. [CrossRef]
67. Tauc, J.; Menth, A. States in the gap. *J. Non-Cryst. Solids* **1972**, *8*, 569–585. [CrossRef]
68. Qin, H.; Jia, J.; Lin, L.; Ni, H.; Wang, M.; Meng, L. Pyrite FeS$_2$ nanostructures: Synthesis, properties and applications. *Mater. Sci. Eng. B* **2018**, *236*, 104–124. [CrossRef]
69. Xia, J.; Lu, X.; Gao, W.; Jiao, J.; Feng, H.; Chen, L. Hydrothermal growth of Sn4+-doped FeS$_2$ cubes on FTO substrates and its photoelectrochemical properties. *Electrochim. Acta* **2011**, *56*, 6932–6939. [CrossRef]

70. Ben Achour, Z.; Ktari, T.; Ouertani, B.; Touayar, O.; Bessais, B.; Brahim, J.B. Effect of doping level and spray time on zinc oxide thin films produced by spray pyrolysis for transparent electrodes applications. *Sens. Actuators A* **2007**, *134*, 447–451. [CrossRef]
71. Ouertani, B.; Ouerfelli, J.; Saadoun, M.; Zribi, M.; Ben Rabha, M.; Bessaıs, B.; Ezzaouia, H. Optical and structural properties of FeSe2 thin films obtained by selenization of sprayed amorphous iron oxide films. *Thin Solid Film.* **2006**, *511–512*, 457–462. [CrossRef]
72. Hohenberg, P.; Kohn, W. Inhomogeneous Electron Gas. *Phys. Rev.* **1964**, *136*, B864. [CrossRef]
73. Kohn, W.; Sham, L.J. Self-Consistent Equations Including Exchange and Correlation Effects. *Phys. Rev.* **1965**, *140*, A1133. [CrossRef]
74. Anderson, O.K. Linear methods in band theory. *Phys. Rev. B* **1975**, *12*, 123060. [CrossRef]
75. Jan, J.P.; Skriver, H.L.J. Relativistic bandstructure and Fermi surface of PdTe2 by the LMTO method. *Phys. F Met. Phys.* **1977**, *7*, 1719. [CrossRef]
76. Skriver, H.L. *The LMTO Method*; Springer: Berlin/Heidelberg, Germany, 1984; Volume 6, p. 2066.
77. Hedin, L.; Lundqvist, B.I. Explicit local exchange-correlation potentials. *J. Phys. C* **1971**, *4*, 2064. [CrossRef]
78. Louie, S.G.; Ho, K.-M.; Cohen, M.L. Self-consistent mixed-basis approach to the electronic structure of solids. *Phys. Rev. B* **1979**, *19*, 1774. [CrossRef]
79. Andersen, O.K.; Jepsen, O.; Krier, G. *Lectures on Methods of Electronic Structure Calculations*; Kumar, V., Andersen, O.K., Mookerjee, A., Eds.; World Scientific Publishing Co.: Singapore, 1994; pp. 63–124.
80. Andersen, O.K.; Arcangeli, C.; Tank, R.W.; Saha-Dasgupta, T.; Krier, G.; Jepsen, O.; Dasgupta, I. *Tight-Binding Approach to Computational Materials Science*; Colombo, L., Gonis, A., Turchi, P., Eds.; Materials Research Society: Pittsburgh, PA, USA, 1998; Volume 491, pp. 3–34.
81. Hulliger, F.; Mooser, E. Semiconductivity in pyrite, marcasite and arsenopyrite phases. *J. Phys. Chem. Solids.* **1965**, *26*, 429. [CrossRef]
82. Dyer-Smith, C.; Nelson, J.; Li, Y. Organic Solar Cells. In *McEvoy's Handbook of Photovoltaics*; ELSEVIER: Amsterdam, The Netherlands, 2018; pp. 567–597.
83. Che, X.; Li, Y.; Qu, Y.; Forrest, S.R. High fabrication yield organic tandem photovoltaics combining vacuum-and solution-processed subcells with 15% efficiency. *Nat. Energy* **2018**, *3*, 422–427. [CrossRef]
84. Si, H.; Liao, Q.; Zhang, Z.; Li, Y.; Yang, X.; Zhang, G.; Kang, Z.; Zhang, Y. An Innovative design of perovskite solar cells with Al_2O_3 Inserting at ZnO/perovskite interface for improving the performance and stability. *Nano Energy* **2016**, *22*, 223–231. [CrossRef]
85. Wang, D.Y.; Jiang, Y.T.; Lin, C.C.; Li, S.S.; Wang, Y.T.; Chen, C.C.; Chen, C.W. Solution-Processable Pyrite FeS_2 Nanocrystals for the Fabrication of Heterojunction Photodiodes with Visible to NIR Photodetection. *Adv. Mater.* **2012**, *24*, 3415. [CrossRef] [PubMed]
86. Moon, D.G.; Cho, A.; Park, J.H.; Ahn, S.; Kwon, H.; Cho, Y.S.; Ahn, S. Iron pyrite thin films deposited via non-vacuum direct coating of iron-salt/ethanol-based precursor solutions. *J. Mater. Chem. A* **2014**, *2*, 17779. [CrossRef]

Disclaimer/Publisher's Note: The statements, opinions and data contained in all publications are solely those of the individual author(s) and contributor(s) and not of MDPI and/or the editor(s). MDPI and/or the editor(s) disclaim responsibility for any injury to people or property resulting from any ideas, methods, instructions or products referred to in the content.

Article

Electrostatic Forces in Control of the Foamability of Nonionic Surfactant

Stoyan I. Karakashev [1,*], Nikolay A. Grozev [1], Svetlana Hristova [2], Kristina Mircheva [1] and Orhan Ozdemir [3]

[1] Department of Physical Chemistry, Sofia University, 1 James Bourchier Blvd, 1164 Sofia, Bulgaria
[2] Department of Medical Physics and Biophysics, Medical Faculty, Medical University–Sofia, Zdrave Str. 2, 1431 Sofia, Bulgaria
[3] Department of Mining Engineering, Istanbul University-Cerrahpaşa, Buyukcekmece, Istanbul 34320, Turkey
* Correspondence: fhsk@chem.uni-sofia.bg

Abstract: Can the DLVO theory predict the foamability of flotation frothers as MIBC (methyl isobutyl carbinol)? The flotation froth is a multi-bubble system, in which the bubbles collide, thus either coalescing or rebounding. This scenario is driven by the hydrodynamic push force, pressing the bubbles towards each other, the electrostatic and van der Waals forces between the bubbles, and the occurrence of the precipitation of the dissolved air between the bubbles. We studied the foamability of 20 ppm MIBC at constant ionic strength I = 7.5×10^{-4} mol/L at different pH values in the absence and presence of modified silica particles, which were positively charged, thus covering the negatively charged bubbles. Hence, we observed an increase in the foamability with the increase in the pH value until pH = 8.3, beyond which it decreased. The electrostatic repulsion between the bubbles increased with the increase in the pH value, which caused the electrostatic stabilization of the froth and subsequently an increase in the foamability. The presence of the particles covering the bubbles boosted the foamability also due to the steric repulsion between the bubbles. The decrease in the foamability at pH > 8.3 can be explained by the fact that, under such conditions, the solubility of carbon dioxide vanished, thus making the aqueous solution supersaturated with carbon dioxide. This caused the precipitation of the latter and the emergence of microbubbles, which usually make the bubbles coalesce. Of course, our explanation remains a hypothesis.

Keywords: frothers; foamability; DLVO; interfacial forces; fine particles; zeta potential

1. Introduction

The production of foam/froth is a complex process of the formation of myriads of bubbles in an aqueous medium, thus resulting in the froth observed on the top of the liquid [1–3]. There are many works that study the correlation of the foam stability and/or the foamability of surfactant aqueous solutions (e.g., [4–30]) with, for example, the foam lamellae elastic moduli/Gibbs elasticity [3–5,8,11,25], the state of the surfactant adsorption layer [1,6,11,26,31,32], the presence of particles [1,6,7,23,27–29], the behavior of foam films [1,12,30,32–38], the existence of special stimuli-responsive reagents [39–43], etc., but probably due to the complexity of the object, none of them sought to investigate the correlation between the electrostatic repulsion between the bubbles and the foamability of froths. Yet, the electrostatic stabilization of hydrophobic dispersions is the first principle of stabilization according to the celebrated DLVO theory [44–48]. For this reason, the thin film pressure balance method (TFPB) has been applied to study the dependence of the disjoining pressure between film surfaces on the thickness of the foam films [49–52]. Therefore, foam (emulsion) films with stronger electrostatic repulsion between film surfaces should correspond to more stable foams (or emulsion). Unfortunately, the opposite trend has been established for the case of foams (emulsions) stabilized by ionic surfactants in the presence of different added electrolytes [52]. Our approach is holistic and inductive. For this reason, we prefer to study a simpler system, i.e., the effect of the electrostatic repulsion between the bubbles on the foamability of an aqueous solution of nonionic frother (e.g., MIBC). The intrinsic negative surface potential of the bubbles in water at normal pH value is still not well understood, but it is proved that it depends on the pH value [53]. Their isoelectric point is at pH ≈ 4 [1,54]. Hence, we decided to perform a simplified experiment, in which we controlled the zeta potential of the bubbles by varying the pH values of an aqueous medium at constant ionic strength $I = 7.5 \times 10^{-4}$ mol/L in 20 ppm methyl isobutyl carbinol (MIBC). The experiment on foamability was conducted by means of a dynamic foam analyzer (DFA), producing froth by sparging the air through a porous frit, thus measuring its height at different gas delivery rates. In addition, we expanded our study by introducing particles that were oppositely charged to the bubbles and varied the zeta potential of both the particles and the bubbles. To our knowledge, such a study has never been conducted.

2. Materials and Methods

2.1. Materials

All of the chemicals and silica particles were purchased from Sigma-Aldrich (Darmstadt, Germany). The silica particles were 10 μm radii. Amino-3-methoxy silane (APTMS), ethanol and sodium hydroxide were used for the chemical modification of the silica particles. The buffers with pH = 4, pH = 7, pH = 9, and pH = 10 were used for the preparation of the solutions, by diluting with deionized water (DI) until reaching electroconductivity 92.4 μS/cm, which corresponds to ionic strength $I = 7.5 \times 10^{-4}$ mol/L. DI water was produced by a water purification system (Elga Lab Water Ltd., High Wycombe, UK).

Pre-treatment with amino-3-methoxy silane (APTMS): We used the procedure described in ref. [55] to adjust the isoelectric point (IEP) of the silica particles. The surface of the particles was covered with amino groups by means of this method. Thus, the isoelectric point of the particles changed from pH ≈ 2.5 [56] to pH ≈ 9.2. This is due to a chemical reaction between amino-3-methoxy silane (APTMS) and Si–OH groups. The silica particles were positioned in 1 mol/L NaOH (T= 60 °C) and stirred. After that, they were flushed with DI water and soaked in a mixture of 100 mL ethanol + 2 cm^3 amino-3-methoxy silane (APTMS) for 24 h (T= 60 °C). Finally, they were rinsed with DI water. Figure 1 shows the chemical reaction on the surface of the silica particles. Figure 1 indicates the presence of amino groups on the chains, which were attached to the silica particles. Due to their presence, the isoelectric point was observed at pH = 9.2. We tested the foamability of 20 ppm MIBC at different pH values and constant ionic strength ($I = 7.5 \times 10^{-4}$ mol/L) in the absence and presence of 1 wt.% modified -10 μm silica particles.

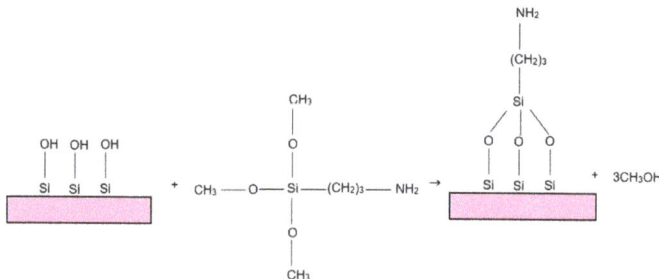

Figure 1. Chemical reaction on the surface of the silica particles.

2.2. Methods

The froth tests were conducted by means of a dynamic foam analyzer (DFA—100, Krüss Optronics GmbH, Hamburg, Germany). The froth was produced by sparging the air with a preliminary adjusted flow rate through a porous bottom in a glass column. The apparatus was controlled using a computer, with which different features of the experiment were initially set, for example, the time and flow rate of the gas delivery. The froth was scanned with a scanline camera, which delivered the image to the computer. Hence, the height of the froth was monitored in time. We conducted the froth test with a gas delivery rate in the range of 0.2–0.5 L/min. The zeta potential of the modified silica particles at different pH values and constant ionic strength were measured by means of Zetasizer (Malvern Panalytical, Worcestershire, UK).

3. Results and Discussion

3.1. Zeta Potential Measurements

The intrinsic isoelectric point (IEP) of the silica particles was at pH ≈ 2.5 [56]. The isoelectric point (IEP) of the modified silica particles was at pH ≈ 9.2. Figure 2 shows the zeta potential of the modified silica particles and microbubbles [57] versus pH at I = 10^{-3} mol/L. One can see that the microbubbles and the modified silica particles were oppositely charged (ξ_b = −22 mV, ξ_p = 60 mV) at pH = 5.8. The procedure did not affect the hydrophobicity of the silica particles (CA ≈ 30°).

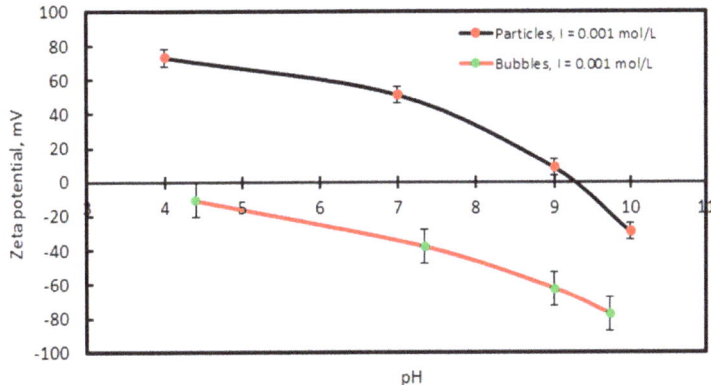

Figure 2. Zeta potential of modified silica particles and microbubbles [53] versus pH at I = 10^{-3} mol/L. The graphic of the zeta potential of bubbles is reproduced with the permission of Elsevier with license number 5372051222249.

The experiment on the foamability of the aqueous solution of 20 ppm MIBC at constant ionic strength in the absence and presence of the modified silica particles was used to analyze two basic effects: (i) the electrostatic repulsion between the bubbles; (ii) the electrostatic attraction between the bubbles and the particles.

The electrostatic disjoining pressure, assuming constant surface potential [58], can be calculated by the following formula:

$$\Pi_{el} = \frac{\varepsilon \varepsilon_0 \kappa^2}{2\pi} \frac{2\Psi_{s1}\Psi_{s2} \cosh(\kappa h) - (\Psi_{s1}^2 + \Psi_{s2}^2)}{\sinh^2(\kappa h)} \tag{1}$$

where ε and ε_0 are the static dielectric permittivities of water and free space, F is the Faraday constant, $\kappa = \sqrt{2F^2 c_0 / \varepsilon \varepsilon_0 RT}$ is the Debye constant (in SI unit), c_0 is the electrolyte concentration), R and T are gas constant and temperature, Ψ_{s1} and Ψ_{s2} are the surface potential values of the first (air/water) and the second (water/solid) surfaces, and h is the thickness of the wetting film. Equation (1) is valid for the different/or opposite surface potential values of the two films' surfaces. Another well-known formula for the electrostatic disjoining pressure, assuming the superposition approximation [59] and surface potential values with the same sign reads:

$$\Pi_{el} = 64cR_g T \tanh\left(\frac{F\Psi_{s1}}{4R_g T}\right) \tanh\left(\frac{F\Psi_{s2}}{4R_g T}\right) \exp(-\kappa h) \quad (2)$$

The van der Waals disjoining pressure, Π_{vdW}, as a function of the film thickness, h, can be calculated by the following formula [60]:

$$\Pi_{vdW} = -\frac{A(h,\kappa)}{6\pi h^3} + \frac{1}{12\pi h^2}\frac{dA(h,\kappa)}{dh} \quad (3)$$

where $A(h,\kappa)$ is the Hamaker–Lifshitz function, which depends on the film thickness and the Debye constant, κ, due to the electromagnetic retardation effect and is described as

$$A(h,\kappa)_{132} = \frac{3k_B T}{4}(1+2\kappa h)e^{-2\kappa h} + \frac{3\hbar\omega}{16\sqrt{2}}\frac{(n_1^2 - n_3^2)(n_2^2 - n_3^2)}{(n_1^2 - n_2^2)}\left\{\frac{I_2(h)}{\sqrt{n_2^2 + n_3^2}} - \frac{I_1(h)}{\sqrt{n_1^2 + n_3^2}}\right\} \quad (4)$$

$$I_j(h) = \left[1 + \left(\frac{h}{\lambda_i}\right)^q\right]^{-\frac{1}{q}} \quad (5)$$

$$\lambda_i = \frac{2\sqrt{2}c}{\omega\pi}\sqrt{\frac{1}{n_3^2(n_i^2 + n_3^2)}} \quad (6)$$

where $\hbar = 1.055 \times 10^{-34}$ Js/rad is the Planck constant (divided by 2π); ω is the absorption frequency in the UV region, typically around 2.068×10^{16} rad/s for water; and n_1, n_2, and n_3 are the characteristic refractive indices of the two dispersion phases (air and mineral) and the medium (water). For example, $n_1^2 = 1$ for air, $n_3^2 = 1.887$ for water, and $n_2^2 = 2.359$ for crystalline quartz. Moreover, $c = 3.10^8$ m/s is the speed of light in vacuum and $q = 1.185$, λ_1 and λ_2 are characteristic wavelengths of the first (air/water) and second (water/solid) surfaces of the wetting film.

3.2. Foamability of 20 ppm MIBC at Different pH Values and Constant Ionic Strength $I = 7.5 \times 10^{-4}$ mol/L

As shown in Figure 2, the absolute value of the zeta potential of the bubbles increased with the increase in the pH value.

Figure 3 shows the height of the froth column of the aqueous solution of 20 ppm MIBC vs. the pH value at constant ionic strength $I = 7.5 \times 10^{-4}$ mol/L and different gas delivery rates. This ionic strength corresponded to the Debye length $1/\kappa = 11.14$ nm. This indicates that the bubbles could repel each other at approx. 33.5 nm distance from each other. Three times the Debye length practically corresponded to the thickness of the diffuse layer. One can see in the figure that the height of the froth increased with the increase in the pH value until reaching the maximum at a certain pH value in the range of pH = 8 to pH = 9. This is easy to explain by increasing the electrostatic repulsion between the bubbles—their absolute value of the zeta potential increased with the increase in the pH value.

Figure 4 presents the DLVO curves of the electrostatic, van der Waals, and the total disjoining pressure versus the distance between two bubbles at ionic strength $I = 7.5 \times 10^{-4}$ mol/L and pH = 7.34. One can see in the figure that the existence of the potential barrier hindered the ability of the bubbles to approach each other and thus coalesce. For example, the bubbles with radii 500 μm would stop thinning at about 43 nm from each other. The increase in the froth height corresponded to an increase in the foamability of the MIBC solution. The decrease in the froth height at pH > 9 was unexpected, but we suggest an explanation below.

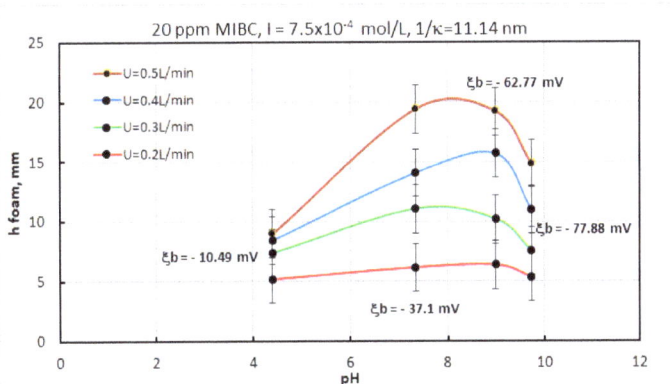

Figure 3. Height of the froth column of aqueous solution of 20 ppm MIBC vs. the pH value at constant ionic strength $I = 7.5 \times 10^{-4}$ mol/L at different gas delivery rates; the zeta potential of the bubbles at each pH value is depicted in the figure. The error bar is ±2 mm. This behavior correlates with the DLVO theory [44,45,61].

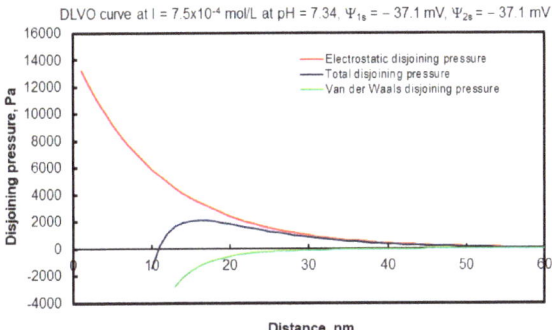

Figure 4. DLVO curves of the electrostatic, van der Waals, and total disjoining pressure of two approaching bubbles at ionic strength $I = 7.5 \times 10^{-4}$ mol/L and pH = 7.34.

3.3. Foamability of 20 ppm MIBC + 1 wt. % Modified Silica Particles at Different pH Values and Constant Ionic Strength $I = 7.5 \times 10^{-4}$ mol/L

Table 1 shows the values of the zeta potential of the bubbles and particles in the 20 ppm solution of MIBC. One can see the negative value of the zeta potential of the bubbles and the positive value of the zeta potential of the modified silica particles. The electrostatic attraction between the bubbles and particles would enable the bubbles to be covered with fine particles, as long as the bubbles were significantly larger than the particles. This resulted in two basic effects: (i) a decrease in the electrostatic repulsion between the bubbles and (ii) steric repulsion between the bubbles. Figure 5 shows the froth height of the 20 ppm aqueous solution of MIBC + 1 wt.% modified silica particles vs. pH and at different gas delivery rates. In Figure 5, one can see the same basic trend of reaching the maximum value of the froth height at a range of pH = 8 to pH = 9 and the decrease in the froth height at pH > 9. The froth height levels in Figure 5 were higher than the ones in Figure 3. This is expectable because of the attachment of the fine particles to the bubbles due to the electrostatic attraction. It generates additional steric repulsive force between the bubbles. Yet, a maximum froth height at a certain pH value in the range of 8 < pH < 9 can be seen. The froth height dropped at pH > 9.

Table 1. Values of pH and the corresponding zeta potentials of the bubbles and the modified silica particles in suspension at 7.5×10^{-4} mol/L ionic strength.

pH	4.40	7.42	8.88	9.40
Zeta potential bubbles, mV	−10.49	−36.93	−57.82	−66.56
Zeta potential particles, mV	72.25	45.55	12.43	−4.84

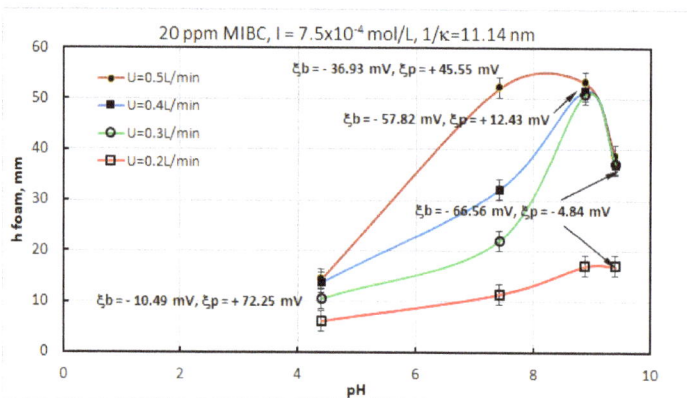

Figure 5. Height of the froth column of aqueous solution of 20 ppm MIBC + 1 wt.% modified fine silica particles vs. the pH value at constant ionic strength $I = 7.5 \times 10^{-4}$ mol/L at different gas delivery rates; the zeta potential of the bubbles at each pH value is depicted in the figure. The error bar is ±2 mm.

Figure 6 shows the DLVO curves of the electrostatic, van der Waals, and the total disjoining pressure vs. distance between the bubbles and the modified silica particles at ionic strength $I = 7.5 \times 10^{-4}$ mol/L and at pH = 4.4. One can see that the total disjoining pressure became negative at a distance below 20 nm.

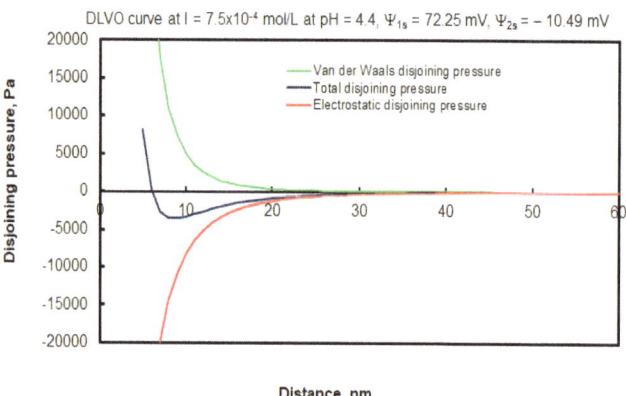

Figure 6. DLVO curves of the electrostatic, van der Waals, and total disjoining pressure of bubble approaching silica particle at ionic strength $I = 7.5 \times 10^{-4}$ mol/L.

The existence of a maximum froth height at 8 < pH < 9 and the drop in the froth height at pH > 9 correlated with the solubility of carbon dioxide, which reached its minimum at pH = 8.36 [62–64]. Water at pH > 8.36 was free of dissolved carbon dioxide. Hence, at pH > 8.36, the water became super-saturated with carbon dioxide. This is an excellent condition for gas precipitation during froth generation. We could not find any studies on the effect of the precipitation of gas on dispersed systems, but such an effect exists. For example, an increase in gas concentration breaks the emulsion films, while the films are durable at decreased gas concentrations [65]. Surfactant-free emulsions can become very durable if they are degassed [66]. These works do not report on the precipitation of gas but only on the effect of gas concentration in an aqueous medium on the stability of thin emulsion films and emulsion. Research that addresses this effect on thin foam films and froths is lacking in the literature.

4. Conclusions

We arrived at the following conclusions with this work:

1. The foamability of 20 ppm aqueous solutions of methyl isobutyl carbinol (MIBC) depended on the pH value and consequently the electrostatic repulsion between the bubbles at ionic strength $I = 7.5 \times 10^{-4}$ mol/L ($1/\kappa = 11.14$ nm) until reaching a maximum between $8 < \mathrm{pH} < 9$. The increase in the electrostatic repulsion between the bubbles prevented their coalescence and hence boosted the foamability.

2. The electrostatic attraction between the bubbles and 1 wt. % modified silica particles increased the froth height, compared with the particle-free sample, until reaching its maximum at $8 < \mathrm{pH} < 9$. This was due to the steric repulsion between the bubbles. The latter additionally prevented the coalescence between the bubbles.

3. The existence of the maximum froth height at $8 < \mathrm{pH} < 9$ can be explained by the lack of the solubility of carbon dioxide at $\mathrm{pH} = 8.36$. Hence, at $\mathrm{pH} > 8.36$, the water was free of dissolved carbon dioxide. The solution under such conditions was pre-saturated with carbon dioxide. This is an excellent condition for gas precipitation during froth generation. The gas precipitation of the dissolved gas contributes to the coalescence of the bubbles. This is likely the reason for the drop in the froth height at $\mathrm{pH} > 8.36$. This explanation remains a hypothesis until it can be either proven or rejected.

The next steps of our studies are to vary the strength of the electrostatic repulsions by performing experiments with different constant ionic strength values.

Author Contributions: Experimental methodology, O.O.; validation, S.I.K. and N.A.G.; investigation, S.I.K., N.A.G., K.M. and S.H.; resources S.I.K., N.A.G. and S.H.; writing—review and editing, S.I.K.; supervision, O.O. and S.I.K.; project administration, S.I.K. All authors have read and agreed to the published version of the manuscript.

Funding: This paper is supported by European Union's Horizon 2020 research and innovation program under Grant Agreement No. 821265, project FineFuture (Innovative technologies and concepts for fine particle flotation: unlocking future fine-grained deposits and Critical Raw Materials resources for the EU).

Institutional Review Board Statement: Not applicable.

Informed Consent Statement: Not applicable.

Data Availability Statement: The authors confirm that the data supporting the findings of this study are available within the article.

Conflicts of Interest: The authors declare no conflict of interest.

References

1. Exerowa, D.; Kruglyakov, P.M. *Foam and Foam Films: Theory, Experiment, Application*; Marcel Dekker: New York, NY, USA, 1997; p. 796.
2. Weaire, D.; Hutzler, S. *Physics of Foams*; Oxford University Press: Oxford, UK, 1999; p. 368.
3. Asadzadeh Shahir, A.; Arabadzhieva, D.; Petkova, H.; Karakashev, S.I.; Nguyen, A.V.; Mileva, E. Effect of Under-Monolayer Adsorption on Foamability, Rheological Characteristics, and Dynamic Behavior of Fluid Interfaces: Experimental Evidence for the Guggenheim Extended Interface Model. *J. Phys. Chem. C* **2017**, *121*, 11472–11487. [CrossRef]
4. Wantke, K.D.; Fruhner, H. The relationship between foam stability and surface rheological properties. In Progress and Trends in Rheology V, Proceedings of the Fifth European Rheology Conference Portorož, Slovenia, 6–11 September 1998; Steinkopff: Steinkopf, South Africa; pp. 315–316.

5. Karakashev, S.I.; Tsekov, R.; Manev, E.D.; Nguyen, A.V. Soap Bubble Elasticity: Analysis and Correlation with Foam Stability. In *Plovdiv University "Paisii Hilendarski" Scientific Papers*; Plovdiv Univeristy Press: Plovdiv, Bulgaria, 2010; p. 37.
6. Karakashev, S.I.; Smoukov, S.K.; Raykundaliya, N.; Grozev, N.A. Duality of foam stabilization. *Colloids Surf. A* **2021**, *619*, 126521. [CrossRef]
7. Karakashev, S.I.; Ozdemir, O.; Hampton, M.A.; Nguyen, A.V. Formation and stability of foams stabilized by fine particles with similar size, contact angle and different shapes. *Colloids Surf. A Physicochem. Eng. Asp.* **2011**, *382*, 132–138. [CrossRef]
8. Karakashev, S.I.; Manev, E.D.; Nguyen, A.V. Effect of Thin Film elasticity on Foam stability. *Annu. Sofia Univ.* **2011**, *102*, 153–163.
9. Karakashev, S.I.; Manev, E.D. Frothing behavior of nonionic surfactant solutions in the presence of organic and inorganic electrolytes. *J. Colloid Interface Sci.* **2001**, *235*, 194–196. [CrossRef]
10. Amani, P.; Karakashev, S.I.; Grozev, N.A.; Simeonova, S.S.; Miller, R.; Rudolph, V.; Firouzi, M. Effect of selected monovalent salts on surfactant stabilized foams. *Adv. Colloid Interface Sci.* **2021**, *295*, 102490. [CrossRef]
11. Sett, S.; Karakashev, S.I.; Smoukov, S.K.; Yarin, A.L. Ion-specific effects in foams. *Adv. Colloid Interface Sci.* **2015**, *225*, 98–113. [CrossRef]
12. Manev, E.; Pugh, R.J. The influence of tetraalkylammonium counterions on the drainage and stability of thin films and foams stabilized by dilute aqueous solutions of sodium dodecyl sulfate. *J. Colloid Interface Sci.* **1997**, *186*, 493–497. [CrossRef]
13. Angarska, J.K.; Manev, E.D. Effect of surface forces and surfactant adsorption on the thinning and critical thickness of foam films. *Colloids Surf. A Physicochem. Eng. Asp.* **2001**, *190*, 117–127. [CrossRef]
14. Manev, E.D.; Sazdanova, S.V.; Rao, A.A.; Wasan, D.T. Foam stability—The effect of a liquid crystalline phase on the drainage and transition behavior of foam films. *J. Dispers. Sci. Techn.* **1982**, *3*, 435–463. [CrossRef]
15. Manev, E.D.; Sazdanova, S.V.; Wasan, D.T. Emulsion and foam stability—The effect of film size on film drainage. *J. Colloid Interface Sci.* **1984**, *97*, 591–594. [CrossRef]
16. Jachimska, B.; Małysa, K. Foam rupture under combined action of gravity and centrifugal force. *Bull. Pol. Acad. Sci. Tech. Sci.* **2000**, *48*, 166–179.
17. Lotun, D.; Pilon, L. Physical modeling of slag foaming for various operating conditions and slag compositions. *ISIJ Int.* **2005**, *45*, 835–840. [CrossRef]
18. Karakashev, S.I.; Georgiev, P.; Balashev, K. Foam production—Ratio between foaminess and rate of foam decay. *J. Colloid Interface Sci.* **2012**, *379*, 144–147. [CrossRef]
19. Karakashev, S.I.; Georgiev, P.; Balashev, K. On the growth of pneumatic foams. *Eur. Phys. J. E* **2013**, *36*, 13. [CrossRef]
20. Karakashev, S.I.; Grozdanova, M.V. Foams and antifoams. *Adv. Colloid Interface Sci.* **2012**, *176–177*, 1–17. [CrossRef] [PubMed]
21. Zhang, H.; Miller, C.A.; Garrett, P.R.; Raney, K.H. Defoaming effect of calcium soap. *J. Colloid Interface Sci.* **2004**, *279*, 539–547. [CrossRef]
22. Samanta, S.; Ghosh, P. Coalescence of air bubbles in aqueous solutions of alcohols and nonionic surfactants. *Chem. Eng. Sci.* **2011**, *66*, 4824–4837. [CrossRef]
23. Rio, E.; Drenckhan, W.; Salonen, A.; Langevin, D. Unusually stable liquid foams. *Adv. Colloid Interface Sci.* **2014**, *205*, 74–86. [CrossRef]
24. Narsimhan, G.; Ruckenstein, E. Effect of Bubble Size Distribution on the Enrichment and Collapse in Foams. *Langmuir* **1986**, *2*, 494–508. [CrossRef]
25. Wantke, K.; Małysa, K.; Lunkenheimer, K. A relation between dynamic foam stability and surface elasticity. *Colloids Surf. A* **1994**, *82*, 183–191. [CrossRef]
26. Karakashev, S.I.; Manev, E.D. Correlation in the properties of aqueous single films and foam containing a nonionic surfactant and organic/inorganic electrolytes. *J. Colloid Interface Sci.* **2003**, *259*, 171–179. [CrossRef] [PubMed]
27. Vilkova, N.G.; Elaneva, S.I.; Karakashev, S.I. Effect of hexylamine concentration on the properties of foams and foam films stabilized by Ludox. *Mendeleev Comm.* **2012**, *22*, 227–228. [CrossRef]
28. Lam, S.; Velikov, K.P.; Velev, O.D. Pickering stabilization of foams and emulsions with particles of biological origin. *Curr. Opin. Colloid Interface Sci.* **2014**, *19*, 490–500. [CrossRef]
29. Fameau, A.L.; Lam, S.; Velev, O.D. Multi-stimuli responsive foams combining particles and self-assembling fatty acids. *Chem. Sci.* **2013**, *4*, 3874–3881. [CrossRef]
30. Exerowa, D.; Churaev, N.V.; Kolarov, T.; Esipova, N.E.; Panchev, N.; Zorin, Z.M. Foam and wetting films: Electrostatic and steric stabilization. *Adv. Colloid Interface Sci.* **2003**, *104*, 1–24. [CrossRef] [PubMed]
31. Manev, E.D.; Karakashev, S.I. Effect of Adsorption of short-chained organic ions on the stability of foam from aqueous solution of a non-ionic surfactant. *Annu. Sofia Univ.* **2001**, *92/94*, 167–173.
32. Manev, E.; Scheludko, A.; Exerowa, D. Critical thicknesses of rupture and of the formation of black spots in microscopic aqueous films. *Izv. Otd. Khim. Nauk. Bulg. Akad. Nauk.* **1972**, *5*, 585–595.
33. Exerowa, D.; Gochev, G.; Platikanov, D.; Liggieri, L.; Miller, R. *Foam Films and Foams: Fundamentals and Applications*; CRC Press: Boca Raton, FL, USA, 2018.
34. Qu, X.; Wang, L.; Karakashev, S.I.; Nguyen, A.V. Anomalous thickness variation of the foam films stabilized by weak non-ionic surfactants. *J. Colloid Interface Sci.* **2009**, *337*, 538–547. [CrossRef]
35. Manev, E.; Pugh, R. Diffuse layer electrostatic potential and stability of thin aqueous films containing a nonionic surfactant. *Langmuir* **1991**, *7*, 2253–2260. [CrossRef]

36. Manev, E.; Sheludko, A.; Ekserova, D. Effect of surfactant concentration on the critical thicknesses of liquid films. *Colloid Polym. Sci.* **1974**, *252*, 586–593. [CrossRef]
37. Manev, E.D. Effect of the disjoining pressure and surfactant diffusion on the thinning rate of aniline foam films. *Annu. Sofia Univ.* **1979**, *70 Pt. 2*, 97–109.
38. Bergeron, V. Forces and structure in thin liquid soap films. *J. Phys. Condes. Matter* **1999**, *11*, R215–R238. [CrossRef]
39. Jia, W.; Xian, C.; Wu, J. Temperature-sensitive foaming agent developed for smart foam drainage technology. *RSC Adv.* **2022**, *12*, 23447–23453. [CrossRef]
40. Lencina, M.M.S.; Fernández Miconi, E.; Fernández Leyes, M.D.; Domínguez, C.; Cuenca, E.; Ritacco, H.A. Effect of surfactant concentration on the responsiveness of a thermoresponsive copolymer/surfactant mixture with potential application on "Smart" foams formulations. *J. Colloid Interface Sci.* **2018**, *512*, 455–465. [CrossRef] [PubMed]
41. Liang, M.Q.; Yin, H.Y.; Feng, Y.J. Smart aqueous foams: State of the art. *Acta Phys. Chim. Sin.* **2016**, *32*, 2652–2662. [CrossRef]
42. Schnurbus, M.; Stricker, L.; Ravoo, B.J.; Braunschweig, B. Smart Air-Water Interfaces with Arylazopyrazole Surfactants and Their Role in Photoresponsive Aqueous Foam. *Langmuir* **2018**, *34*, 6028–6035. [CrossRef] [PubMed]
43. Wang, J.; Liang, M.; Tian, Q.; Feng, Y.; Yin, H.; Lu, G. CO_2-switchable foams stabilized by a long-chain viscoelastic surfactant. *J. Colloid Interface Sci.* **2018**, *523*, 65–74. [CrossRef] [PubMed]
44. Derjaguin, B.; Landau, L. Theory of the stability of strongly charged lyophobic sols an of the adhesion of strongly charged particles in solutions of electrolytes. *Acta Phys. Chim.* **1941**, *14*, 633–662. [CrossRef]
45. Verwey, E.J.W.; Overbeek, J.T.G. Theory of the stability of lyophobic colloids. *J. Colloid Sci.* **1955**, *10*, 224–225. [CrossRef]
46. Verwey, E.J.W. Theory of the Stability of Lyophobic Colloids. *J. Phys. Colloid Chem.* **1947**, *51*, 631–636. [CrossRef] [PubMed]
47. Derjagiun, B.V. On the repulsive forces between charged colloid particles and on the theory of slow coagulation and stability of lyophobic sols. *Trans. Farad. Soc.* **1940**, *36*, 203–215. [CrossRef]
48. Derjaguin, B. A theory of interaction of particles in presence of electric double layers and the stability of lyophobic colloids and disperse systems. *Acta Phys. Chim.* **1939**, *10*, 333–346.
49. Mysels, K.J.; Jones, M.N. Direct Measurement of Variation of Double-Layer Repulsion with Distance. *Disc. Farad. Soc.* **1966**, *42*, 42–50. [CrossRef]
50. Bergeron, V. Measurement of forces and structure between fluid interfaces. *Curr. Opin. Colloid Interface Sci.* **1999**, *4*, 249–255. [CrossRef]
51. Schelero, N.; von Klitzing, R. Ion specific effects in foam films. *Curr. Opin. Colloid Interface Sci.* **2015**, *20*, 124–129. [CrossRef]
52. Ivanov, I.B.; Slavchov, R.I.; Basheva, E.S.; Sidzhakova, D.; Karakashev, S.I. Hofmeister Effect on Micellization, Thin Films and Emulsion Stability Adv. *Colloid Interface Sci.* **2011**, *168*, 93–104. [CrossRef]
53. Karakashev, S.I.; Grozev, N.A. The law of parsimony and the negative charge of the bubbles. *Coatings* **2020**, *10*, 1003. [CrossRef]
54. Exerowa, D. Effect of adsoprtion, ionic strenght and pH on the potential of diffuse electrical layer. *Kolloid Z.* **1969**, *232*, 703–710. [CrossRef]
55. Anirudhan, T.S.; Jalajamony, S.; Sreekumari, S.S. Adsorption of heavy metal ions from aqueous solutions by amine and carboxylate functionalised bentonites. *Appl. Clay Sci.* **2012**, *65–66*, 67–71. [CrossRef]
56. Zurita, L.; Carrique, F.; Delgado, A.V. The primary electroviscous effect in silica suspensions. Ionic strength and pH effects. *Colloids Surf. A* **1994**, *92*, 23–28. [CrossRef]
57. Yang, C.; Dabros, T.; Li, D.Q.; Czarnecki, J.; Masliyah, J.H. Measurement of the zeta potential of gas bubbles in aqueous solutions by microelectrophoresis method. *J. Colloid Interface Sci.* **2001**, *243*, 128–135. [CrossRef]
58. Kralchevsky, P.A.; Danov, K.D.; Denkov, N.D. Chemical Physics of Colloid Systems and Interfaces. In *Handbook of Surface and Colloid Chemistry*, 4th ed.; Birdi, K.S., Ed.; CRC Press Taylor & Francis Group: Boca Raton, FL, USA, 2016; pp. 247–413.
59. Kralchevsky, P.A.; Danov, K.D.; Denkov, N.D. Chemical Physics of Colloid Systems and Interfaces. In *Handbook of Surface and Colloid Chemistry*; Birdi, K.S., Ed.; CRC Press: Boca Raton, FL, USA, 2008; pp. 179–379.
60. Nguyen, A.V.; Schulze, H.J. *Colloidal Science of Flotation*; Marcel Dekker: New York, NY, USA, 2004; p. 840.
61. Israelachvili, J.N. *Intermolecular and Surface Forces*; Academic Press: London, UK, 1992; p. 291.
62. Boyd, C.E. pH, Carbon Dioxide, and Alkalinity. In *Water Quality: An Introduction*; Boyd, C.E., Ed.; Springer US: Boston, MA, USA, 2000; pp. 105–122.
63. König, M.; Vaes, J.; Klemm, E.; Pant, D. Solvents and Supporting Electrolytes in the Electrocatalytic Reduction of $CO2$. *iScience* **2019**, *19*, 135–160. [CrossRef] [PubMed]
64. Zeebe, R.E.; Wolf-Gladrow, D. CO_2 *in Seawater: Equilibrium, Kinetics, Isotopes*; Elsevier Science: Amsterdam, The Netherlands, 2001; p. 360.
65. Karakashev, S.I.; Nguyen, A.V. Do Liquid Films Rupture due to the So-Called Hydrophobic Force or Migration of Dissolved Gases? *Langmuir* **2009**, *25*, 3363–3368. [CrossRef] [PubMed]
66. Pashley, R.M. Effect of Degassing on the Formation and Stability of Surfactant-Free Emulsions and Fine Teflon Dispersions. *J. Phys. Chem. B* **2003**, *107*, 1714–1720. [CrossRef]

Disclaimer/Publisher's Note: The statements, opinions and data contained in all publications are solely those of the individual author(s) and contributor(s) and not of MDPI and/or the editor(s). MDPI and/or the editor(s) disclaim responsibility for any injury to people or property resulting from any ideas, methods, instructions or products referred to in the content.

Article

$Co_{0.6}Ni_{0.4}S_2$/rGO Photocatalyst for One-Pot Synthesis of Imines from Nitroaromatics and Aromatic Alcohols by Transfer Hydrogenation

Hongming Zhang [1,2,3], Jiahe Zhuang [1], Xiangrui Feng [1] and Ben Ma [1,*]

[1] School of Environmental and Chemical Engineering, Jiangsu Ocean University, Lianyungang 222005, China
[2] Jiangsu Institute of Marine Resources Development, Jiangsu Ocean University, Lianyungang 222005, China
[3] School of Safety Science and Emergency Management, Wuhan University of Technology, Wuhan 430070, China
* Correspondence: maben@jou.edu.cn

Abstract: $Co_{0.6}Ni_{0.4}S_2$/rGO catalysts exhibit excellent photocatalytic performance for one-step synthesis of *N*-benzylideneaniline from nitrobenzene and benzyl alcohol by transfer hydrogenation, and the selectivity and yield of *N*-benzylideneaniline can reach as high as 93% and 77.2%, respectively. The reaction process for the synthesis of imines can be divided into two steps: benzyl alcohol is oxidized to benzaldehyde, while nitrobenzene is reduced to aniline; benzaldehyde and aniline are condensed to form imines. Under visible light irradiation, photo-induced electrons in $Co_{0.6}Ni_{0.4}S_2$/rGO photocatalyst play an important role in activating nitrobenzene and benzaldehyde. Photo-induced holes are mainly responsible for the partial dehydrogenation of benzyl alcohol to benzaldehyde. Next, aniline molecules condense with benzaldehyde molecules to synthesize imine. The photocatalytic system provides an environmentally friendly for the synthesis of imines and supplies an alternative approach for hydrogen auto-transfer reactions.

Keywords: imine; photocatalytic; $Co_{0.6}Ni_{0.4}S_2$/rGO; transfer hydrogenation

Citation: Zhang, H.; Zhuang, J.; Feng, X.; Ma, B. $Co_{0.6}Ni_{0.4}S_2$/rGO Photocatalyst for One-Pot Synthesis of Imines from Nitroaromatics and Aromatic Alcohols by Transfer Hydrogenation. *Coatings* 2022, *12*, 1799. https://doi.org/10.3390/coatings12121799

Academic Editor: Emerson Coy

Received: 28 October 2022
Accepted: 21 November 2022
Published: 23 November 2022

Publisher's Note: MDPI stays neutral with regard to jurisdictional claims in published maps and institutional affiliations.

Copyright: © 2022 by the authors. Licensee MDPI, Basel, Switzerland. This article is an open access article distributed under the terms and conditions of the Creative Commons Attribution (CC BY) license (https:// creativecommons.org/licenses/by/ 4.0/).

1. Introduction

Imines are important building blocks for the synthesis of agrochemical, biologically active compounds, pharmaceutical and fine chemicals [1,2]. Conventionally, imines are synthesized by oxidative dehydrogenation of secondary amines in the presence of molecular oxygen or atmosphere oxygen [3]. The use of O_2 instead of other strong oxidants such as o-iodoxybenzoic acid and chromate not only is more environmentally friendly, but also reduces the cost of the reaction. In addition, high selectivity is readily achieved by dehydrogenation reaction since secondary amines cannot be dehydrogenated into nitriles. However, the steric hindrance around the N-H limits the conversion efficiency of substrates [4]. The synthesis of asymmetric imines is limited because it is difficult to selectively oxidize two types of α-CH with similar properties on asymmetric diphenylamine. Condensation of amines and aldehydes over Lewis acid catalysts is another important method for imines synthesis [5,6]. This route usually needs the assistance of over-stoichiometric amounts of base to form target products. In addition, the instability of reactants under normal conditions (both aldehydes and anilines are easily oxidized by oxygen) also increases the risk of the reaction.

Transfer hydrogenation, a fundamental step in a variety of chemistry processes, involves oxidation and reduction processes [7]. The photocatalytic selective oxidation of benzyl alcohol to benzaldehyde and reduction of nitrobenzene to aniline are typical oxidation and reduction processes, respectively. Therefore, photocatalytic synthesis of imines from aromatic alcohols (the hydrogen donor) and nitroaromatics (the hydrogen acceptor) by transfer hydrogenation is an excellent synthetic strategy, which can not only improve the utilization of atoms, but also avoid the separation of intermediates. In addition, the substrates used in this synthesis method have a wide range of sources, which can greatly expand the variety of synthetic imines. Up to now, the catalysts currently used in the reaction process include precious metals [8,9], non-noble metal oxides [10] and transition metal compounds [11,12], which exhibit high catalytic activity for one-pot synthesis of imine directly from nitrobenzene and benzyl alcohol by transfer hydrogenation strategy.

Ternary transition metal sulfide semiconductors have received tremendous attention due to their appropriate physical and chemical properties [13], and show excellent photocatalytic performance for photocatalytic pollutants degradation, photocatalytic hydrogen evolution, and photocatalytic carbon dioxide reduction [14–17]. Herein, we report the synthesis of $Co_xNi_{1-x}S_2/rGO$ through a facile solvothermal synthesis. The as-prepared noble-metal-free $Co_{0.6}Ni_{0.4}S_2/rGO$ material is proved to be a highly active photocatalyst for one-pot system of imines from nitroaromatics and aromatic alcohols.

2. Materials and Methods

2.1. Catalyst Preparation

$Co_{0.6}Ni_{0.4}S_2/rGO$ composites were prepared in a hydrothermal autoclave with a 100 mL Teflon lining. Firstly, 3 mmol cobaltous acetate ($Co(CH_3COO)_2 \cdot 4H_2O$), 2 mmol nickel acetate ($Ni(CH_3COO)_2 \cdot 4H_2O$) and 50 mg reduced graphene oxide were dispersed in a mixture of deionized water (20 mL) and ethylene glycol (20 mL). After vigorous stirring for 0.5 h at room temperature, the soliquoid was labeled solution A. Secondly, 5 mmol sodium hyposulfite ($Na_2S_2O_3 \cdot 5H_2O$) and 300 mg oxalic acid ($H_2C_2O_4$) were dispersed in 20 mL of deionized water, and the solution was marked as solution B after stirring for 0.5 h. Next, solution B was slowly dropped into solution A while keeping stirring for another 0.5 h. After that, the soliquoid was transferred into the Teflon-lined hydrothermal autoclave, and maintained at 180 °C. After 18 h, the reaction system was naturally cooled to room temperature. Deionized water and anhydrous ethanol were used to clean the precipitate, respectively. Finally, $Co_{0.6}Ni_{0.4}S_2/rGO$ composites were obtained via drying in a vacuum environment at 40 °C for 12 h. By changing the molar ratio of cobalt acetate and nickel acetate in the preparation process, CoS_2/rGO, $Co_{0.8}Ni_{0.2}S_2/rGO$, $Co_{0.4}Ni_{0.6}S_2/rGO$, $Co_{0.2}Ni_{0.8}S_2/rGO$, and NiS_2/rGO catalysts could be obtained under the same condition.

2.2. Catalyst Characterization

The physical and chemical properties of the samples were measured by X-ray diffractometer (XRD, Rigaku D-max/RB, Rigaku, Tokyo, Japan), high resolution transmission electron microscopy (HRTEM, JEM-2100F, Bruker, Billerica, MA, USA), field-emission scanning electron microscopy (FESEM, JSM-5600, Jeol, Tokyo, Japan), X-ray photoelectron spectroscopy (XPS, Kratos XSAM800, Kratos, Manchester, Britain), UV-visible absorption spectrometer (UV-vis, UV-3600, Shimadzu, Tokyo, Japan), fluorescence spectrophotometer (PL, F-700 FL 220–240 eV, Tokyo, Japan) and so on.

2.3. Photocatalytic Synthesis of Imines

The photocatalytic reactions were conducted in a 20 mL Pyrex glass tube under nitrogen atmosphere. In a typical photocatalytic experiment process, 0.5 mmol of nitrobenzene, 1.5 mmol of benzyl alcohol, 40 mg of $Co_{0.6}Ni_{0.4}S_2/rGO$ catalysts and 10 mL of toluene were introduced into the glass vessel with at a stirring speed of 300 rpm. The suspension was purged with nitrogen flow for 20 min until an adsorption-desorption equilibrium was established between photocatalytic and reactants. The reaction system was irradiated using a 350 W Xe lamp in a nitrogen atmosphere of 1 atmosphere, and the light intensity of the liquid surface was measured to be 280 mW/cm^2. The reaction temperature was carefully controlled at 90 °C for 6 h by circulating cool water. After reaction, the reaction mixtures were analyzed by a gas chromatograph (Agilent 7820A, Agilent, Santa Clara, CA, USA).

The conversion of reactants, selectivity and yield of target products were calculated following Equations (1)–(3):

$$Conv.(\%) = \frac{C_0 - C_t}{C_0} \times 100\% \tag{1}$$

$$Select.(\%) = \frac{C_p}{C_0 - C_t} \times 100\% \tag{2}$$

$$Yield(\%) = \frac{C_p}{C_0} \times 100\% \tag{3}$$

where C_0 and C_t represent the concentrations of reactant at initial and reaction time t, respectively, and C_p is the concentration of target product. It should be worth noting that the selectivity and yield of imine are calculated based on reduction product of nitrobenzene.

3. Results

3.1. Catalyst Characterization

The surface chemical composition of $Co_{0.6}Ni_{0.4}S_2$/rGO samples were analyzed by the XPS measurements. The main peaks at 854.0, 778.9, 285.0 and 163.1 eV in the survey spectrum are, respectively, attributed to Ni 2p, Co 2p, C 1s and S 2p, proving the coexistence of Co, Ni, S and C in the $Co_{0.6}Ni_{0.4}S_2$/rGO sample (Figure 1A) [15,18]. In the Co 2p XPS spectrum (Figure 1B), two main peaks at 778.6 and 793.6 eV are corresponding to Co $2p_{3/2}$ and Co $2p_{1/2}$ [19]. The binding energies at 780.9 and 797.0 eV can be attributed to Co^{2+} photoelectron peak in cobalt oxide, indicating that a small amount of cobalt is oxidized on the surface [20]. The binding energies of Ni $2p_{3/2}$ (853.3 eV) and Ni $2p_{1/2}$ (870.6 eV) in Ni 2p XPS spectrum (Figure 1C) agree well with the values reported in the previous work [21]. In addition, peaks at 857.0 and 874.5 eV are in accordance with the Ni^{2+} in nickel oxide, indicating that a small amount of nickel is also oxidized on the surface [22]. Figure 1D shows the XPS spectrum of S_2^{2-}, the binding energies at 162.7 and 163.7 eV are the characteristic of S_2^{2-} species in transition metal disulfides, and the binding energies at 161.6 and 169.5 eV, and 165.3 and 171.1 eV are assigned to S^{2-} and SO_4^{2-}, respectively [23–25]. In summary, element Co, Ni and S are mainly in the form of disulphide in sample, and there is a small amount of oxidation state on the surface.

XRD analysis was applied to investigate the crystalline structure of as-prepared samples, and the results are demonstrated in Figure 2A. No obvious diffraction peaks of reduced graphene oxide are observed in all samples. The strong diffraction peaks of CoS_2 are consistent with a cubic phase (JCPDS Card No. 89-1492), implying the purity of as-synthesized CoS_2 [26]. The as-prepared NiS_2/rGO samples are the mixture of NiS/rGO and NiS_2/rGO, which are assigned to a hexagonal phase (JCPDS Card No. 89-1955) and a cubic phase (JCPDS Card No. 88-1709), respectively [27,28]. It is maybe because that NiS_2 is reduced by oxalic acid to NiS. Interestingly, the characteristic diffraction peaks of NiS (30.0, 34.4, 45.5 and 53.2°) disappear in $Co_xNi_{1-x}S_2$/rGO samples. It means that the stability of the sample is improved in the presence of Co. Both physical and chemical similarities of Co and Ni enable the structural compatibility [29]. There are new peaks (38.1 and 50.3°) observed in $Co_xNi_{1-x}S_2$/rGO samples, indicating the formation of solid solutions.

The light-absorbing ability of $Co_{0.6}Ni_{0.4}S_2$ and $Co_{0.6}Ni_{0.4}S_2$/rGO samples were determined by UV-vis diffuse reflectance spectra. $Co_{0.6}Ni_{0.4}S_2$ particles and their composites show strong light absorption in the range of 400–600 nm (Figure 2B). According to the literature, the absorption band in the range of 400–600 nm can be ascribed to the d-d transitions of Co (III) and Ni (III) ions in the catalysts [30]. After the introduction of graphene carrier, the light absorption capacity of $Co_{0.6}Ni_{0.4}S_2$/rGO composites is stronger than that of pure $Co_{0.6}Ni_{0.4}S_2$, indicating that visible light can be exploited more efficient by composites in the presence of rGO [31].

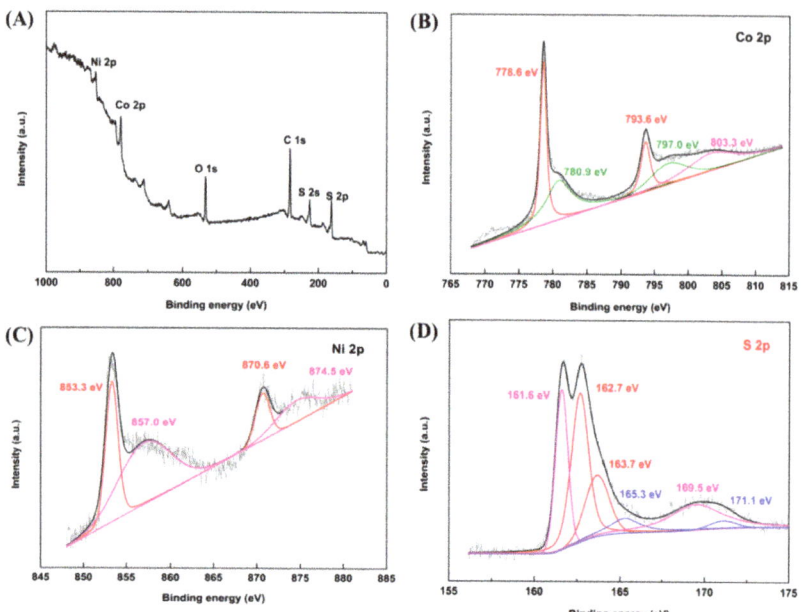

Figure 1. XPS spectra of $Co_{0.6}Ni_{0.4}S_2/rGO$ samples: (**A**) the wide-scan XPS spectrum, (**B**) Co 2p, (**C**) Ni 2p, (**D**) S 2p.

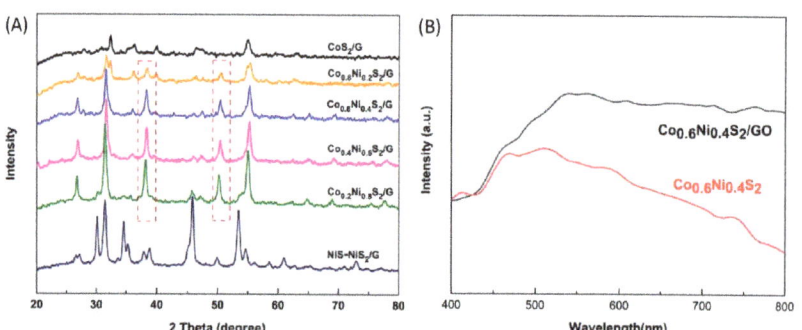

Figure 2. XRD pattern of $Co_xNi_{1-x}S_2/rGO$ (x = 0, 0.2, 0.4, 0.6, 0.8 and 1) (**A**), and UV-vis diffuse reflectance spectra of $Co_{0.6}Ni_{0.4}S_2$ and $Co_{0.6}Ni_{0.4}S_2/rGO$ samples (**B**).

TEM and SEM are common tools for researchers to study the crystal structure and microstructure of solid catalysts. Figure 3 shows representative field-emission scanning electron microscopy (FESEM) images of $Co_{0.6}Ni_{0.4}S_2/rGO$ sample. From Figure 3A, $Co_{0.6}Ni_{0.4}S_2/rGO$ subspheroidal particles with an average particle size of about 400 nm are well distributed on the surface of graphene, and the interplanar crystal spacing is 0.24 nm. Figure 3B is the FESEM of $Co_{0.6}Ni_{0.4}S_2$ subspheroidal particles, and the surface of particles is particularly rough. It can be explained that samples are etched by oxalic acid during the preparation. The rough surface not only increases the surface area, but also provides more adsorption and active positions for the reactants, which can help improve the catalytic activity. It can be observed that Co, Ni and S elements distribute almost uniformly in the surface of $Co_{0.6}Ni_{0.4}S_2$ from the elemental distribution maps of Co, Ni and S (Figure A1A–D), consistent with the EDX results (Figure A2).

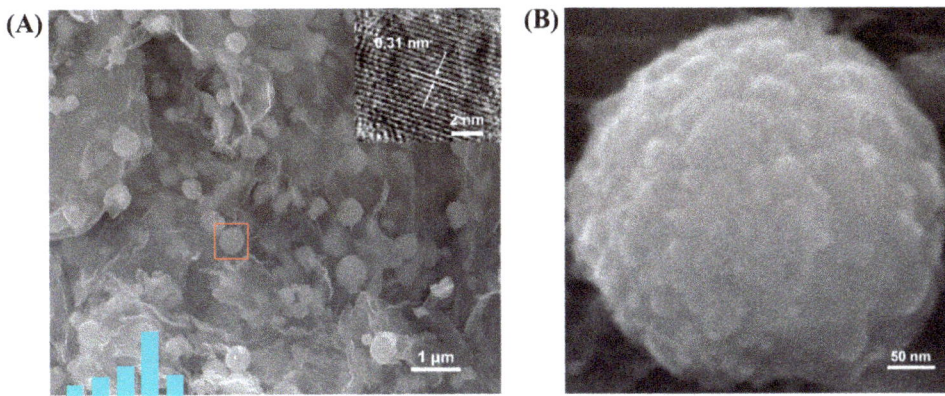

Figure 3. SEM images (**A**,**B**) of Co$_{0.6}$Ni$_{0.4}$S$_2$/rGO, and the inset pictures in picture (**A**) are the size distributions of Co$_{0.6}$Ni$_{0.4}$S$_2$ particles and HRTEM images.

3.2. The Performance of Photocatalytic Synthesis of Imines

The photocatalytic performances of various catalysts for the synthesis of N-benzylideneaniline from nitrobenzene and benzyl alcohol were tested (Figure 4). It is clear that catalysts have greatly affected the yield of N-benzylideneaniline from Table 1. No imines are detected using pure graphene as photocatalytic (entry1 Table 1). CoS$_2$/rGO and NiS$_2$/rGO show poor photocatalytic activity under the giving conditions and the yields of N-benzylideneaniline are only 32.2 and 27.9%, respectively (entries 2 and 7, Table 1). Expectedly, all Co$_x$Ni$_{1-x}$S$_2$/rGO catalysts show better photocatalytic activity for the synthesis of N-benzylideneaniline than that of pure CoS$_2$/rGO and NiS$_2$/rGO catalysts. In addition, the highest yield of N-benzylideneaniline (77.2%) is achieved over Co$_{0.6}$Ni$_{0.4}$S$_2$/rGO sample (entry 4, Table 1). Therefore, Co$_{0.6}$Ni$_{0.4}$S$_2$/rGO is used as the catalyst for the following experiments. For comparison, the reaction was conducted without light irradiation (entry 8, Table 1), and the yield of N-benzylideneaniline is only 29.1%. This phenomenon shows that the light plays a significant role in this reaction under the giving conditions.

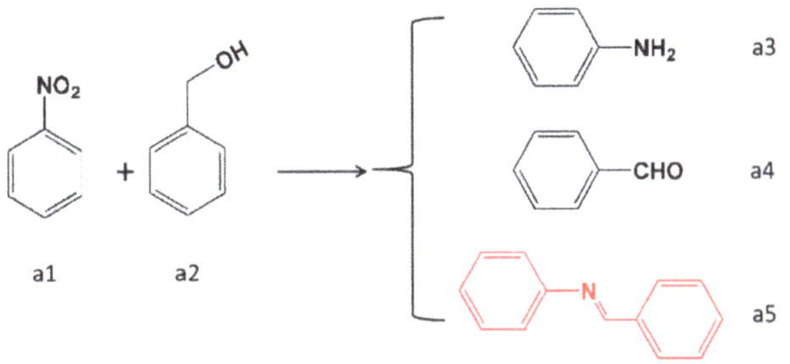

Figure 4. Photocatalytic One-Pot Synthesis of Imine from Nitrobenzene and Benzyl alcohol.

The conversion of nitrobenzene, the selectivity of aniline, N-benzylideneaniline and N-phenylbenzylamine dependence on the irradiation intensity are shown in Figure 5. Nitrobenzene is first converted to aniline, and the higher of light intensity results in the higher conversion of nitrobenzene under the same reaction conditions. When the light intensity exceeds 280 mW/cm^2, most of the aniline has been converted into imine. It is worth noting that no imine is detected when the light intensity is less than 320 mW/cm^2. However, increasing the intensity of light from 320 to 360 mW/cm^2, the selectivity of N-phenylbenzylamine rises obviously, suggesting that a part of N-benzylideneaniline has been reduced to N-phenylbenzylamine in reaction process. When the intensity of light is enhanced, more photo-generated electrons can be transferred to higher energy level, and have enough energy to overcome the energy barrier of N-phenylbenzylamine synthesis, resulting in the increase of N-phenylbenzylamine selectivity. Obviously, the energy barrier of N-benzylideneaniline synthesis is less than that of N-phenylbenzylamine synthesis from N-benzylideneaniline.

Table 1. Experimental results of photocatalytic synthesis of N-benzylideneaniline over $Co_xNi_{1-x}S_2/rGO$ (x = 1, 0.8, 0.6, 0.4, 0.2 and 0) catalyst.

Entry	Catalyst	Conv.(%)	Select. (%)			Yield (%)
		a1	a2	a3	a5	a5
1	rGO	1	14	0	0	0
2	CoS_2/rGO	62	76	44	52	32.2
3	$Co_{0.8}Ni_{0.2}S_2/rGO$	77	83	9	88	67.8
4	$Co_{0.6}Ni_{0.4}S_2/rGO$	83	91	6	93	77.2
5	$Co_{0.4}Ni_{0.6}S_2/rGO$	74	79	11	86	63.6
6	$Co_{0.2}Ni_{0.8}S_2/rGO$	57	62	20	77	43.9
7	NiS_2/rGO	41	55	27	68	27.9
8 [a]	$Co_{0.6}Ni_{0.4}S_2/rGO$	35	44	15	83	29.1

[a] The reaction was conducted without light irradiation.

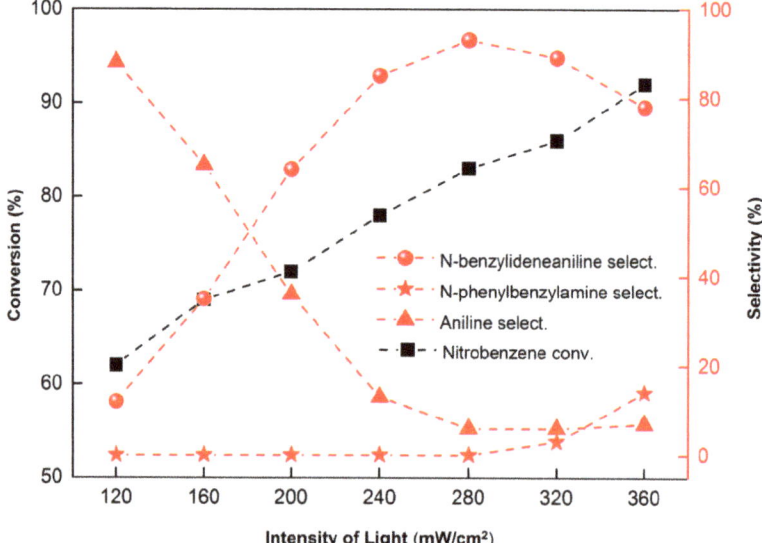

Figure 5. The dependence of photocatalytic activity for N-benzylideneaniline synthesis on the irradiation intensity over $Co_{0.6}Ni_{0.4}S_2/rGO$.

3.3. Impact of Solvent on Photocatalytic Activity

Generally speaking, there is an important influence on photocatalytic activity for solvents [32,33]. The reaction of imines synthesis was conducted in various solvent. Nonpolar solvents promote the synthesis of imines, and yield of imines are 77.2% and 76.5% in toluene and mesitylene (entries1 and 2, Table 2), respectively. Interestingly, both the conversion of reactants and selectivity of imine are suppressed in polar solvents (entries 3–5, Table 2). It can be explained that the solvent with high polarity index has a large interaction with the reaction substrates, which weakens the interaction between the substrates and the catalysts, thus reducing the catalytic activity [34]. For condensation process, -NH_2 in aniline and -CHO in benzaldehyde are less nucleophilic and electrophilic in polar solvents than that in non-polar solvents [35]. Therefore, imines are more easily synthesized in nonpolar solvents. In addition, the by-product water also has a great influence on the reaction. When 30 mg of anhydrous magnesium sulfate (MgSO$_4$) was added to THF and ethanol solvent, respectively, both the conversion of reactants and yield of N-benzylideneaniline enhances gradually (entries 7 and 8, Table 2). However, the adding of MgSO$_4$ has little effect on the reaction in toluene (entry 6, Table 2). This may be because MgSO$_4$ can remove water dissolved in the solvent, resulting in the reaction proceeds in the positive direction. In other words, synthesis of imine from condensation of amines and aldehydes is reversible reaction.

Table 2. The performances of photocatalytic synthesis of N-benzylideneaniline in various solvents.

Entry	Solvent	Conv.(%)		Select. (%)	Yield (%)
		Nitrobenzene	Benzyl Alcohol	Imine	Imine
1	Toluene	83	91	93	77.2
2	Mesitylene	85	94	90	76.5
3	THF	69	76	58	40.0
4	Ethanol	35	42	19	6.7
5	Water	11	25	-	Trace
6 [a]	Toluene	85	95	94	79.9
7 [a]	THF	82	88	71	58.2
8 [a]	Ethanol	49	55	32	15.7

[a] 30 mg of anhydrous magnesium sulfate (MgSO$_4$) was added to the reaction.

3.4. Reductive Coupling of Several Other Substrates to Imine

To demonstrate the general applicability of the Co$_{0.6}$Ni$_{0.4}$S$_2$/rGO photocatalyst, the synthesis of imines from a wide range of aromatic alcohols and nitroarenes were investigated (Figure 6), and results are summarized in Table 3. It can be seen that substituent groups on reactant have significantly affected both selectivity and yield of imines. For instance, when nitrobenzene reacts with p-chlorophenyl alcohol or 3-bromobenzyl alcohol, the conversations of nitrobenzene and aromatic alcohols, and the yield of target products are higher than that in the reaction system of benzyl alcohol and nitrobenzene (entries 1 and 2, Table 3). The functional groups with strong electronegativity contribute to the synthesis of imines [11]. However, substitutions with electron-donating (-CH$_3$, -CH$_2$CH$_3$) on aromatic alcohols exhibit poor both conversion of reactant and yield of imines, and the yields are only 59.3 and 22.4%, respectively (entries 3 and 4, Table 3). When nitroaromatics reacts with benzyl alcohol, electron-withdrawing groups (-Cl and -Br) have slight effect on both conversion of reactants and yield of imines (entries 5 and 6, Table 3), and electron-donating group (-CH$_3$) shows poor selectivity of imine (entry 7, Table 3). For the nitroaromatics and p-substituted benzyl alcohols coupled system, substitutions with electron-donating and electron-withdrawing inhibit the occurrence of coupling reaction, and the selectivity of imines are significantly reduced (entries 8 and 9, Table 3). In a word, Co$_{0.6}$Ni$_{0.4}$S$_2$/rGO catalysts show the high yield of other imines from nitroaromatics and aromatic alcohols.

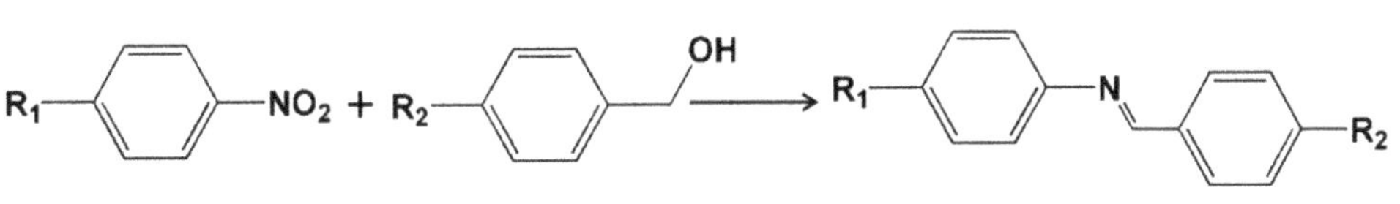

Figure 6. Photocatalytic One-Pot Synthesis of Imine from Nitroaromatics and Aromatic Alcohols.

Table 3. The photocatalytic synthesis of Imine from several other substrates over the $Co_{0.6}Ni_{0.4}S_2/rGO$ catalyst.

Entry	Reactant		Conv. (%)	Select. (%)	Yield (%)
	R_1	R_2	a6	a7	a8	a8
1	H	Cl	90	99	94	84.6
2	H	Br	85	94	91	77.4
3	H	CH_3	69	77	86	59.3
4	H	CH_3CH_2	51	60	44	22.4
5	Cl	H	82	96	92	75.4
6	Br	H	80	93	94	75.2
7	CH_3	H	93	97	80	74.4
8	Cl	Cl	88	94	63	55.4
9	CH_3	CH_3	66	70	54	35.6

3.5. Stability of Catalyst

Stability is an important criterion to evaluate the properties of heterogeneous catalysts. According to our previous research, the activity of transition metal sulfides is greatly reduced after several cycles, such as FeS_2 and CoS_2. In our work, the $Co_{0.6}Ni_{0.4}S_2/rGO$ catalyst shows better stability for synthesis of N-benzylideneaniline after five successive reactions under the protection of graphene (Figure 7). FESEM analysis was performed on the catalyst for recycled catalysts. As can be seen from Figure A3, there are no significant changes observed for $Co_{0.6}Ni_{0.4}S_2/rGO$ catalysts in synthesis of N-benzylideneaniline.

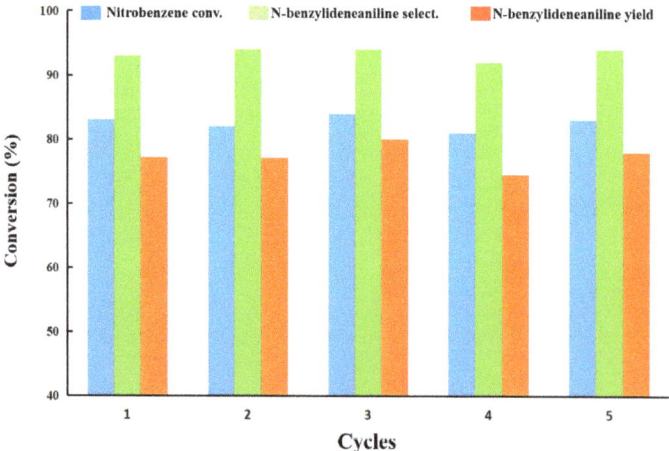

Figure 7. Photocatalytic stability of $Co_{0.6}Ni_{0.4}S_2/rGO$ in five cycles for synthesis of N-benzylideneaniline.

3.6. Proposed Mechanism

As a semiconductor material, $Co_{0.6}Ni_{0.4}S_2$ can produce photo-generated electrons and holes under visible light illumination. This phenomenon has been proved by photoluminescence (PL) spectra of $Co_{0.6}Ni_{0.4}S_2$ and $Co_{0.6}Ni_{0.4}S_2/rGO$ (Figure 8). When photo-generated electrons are excited from its valence band (VB) to conduction band (CB), photo-generated holes are left at the VB. In addition, graphene can effectively suppress the recombination of photo-generated electrons and holes [25]. In order to investigate the photocatalytic mechanism for imines synthesis by transfer hydrogenation, the scavenger of electrons (5,5-dimethyl-1-pyrroline N-oxide, DMPO) and holes (triethanolamine, TEA) were added to the experiment. When 1.0 mL of DMPO was employed to trap the photogenerated electrons of $Co_{0.6}Ni_{0.4}S_2$, the conversion of nitrobenzene decreased sharply from 83% to 26%, and the conversion of benzyl alcohol increased from 91% to 98%, which can be explained that more photo-generated holes can participate in the dehydrogenation of aromatic alcohols to aromatic aldehydes. Adding 1.0 mL of TEA to the reaction system, the conversion of benzyl alcohol and nitrobenzene decreased to 38% and 21%, respectively, which is almost the same as that in the dark reaction (Figure 9). On the basis of the results described above, it can be speculated that the dehydrogenation of benzyl alcohol is the prerequisite reaction, and protons (H$^+$) liberated in the dehydrogenation reaction are used as hydrogen sources for reducing nitrobenzene effectively to aniline.

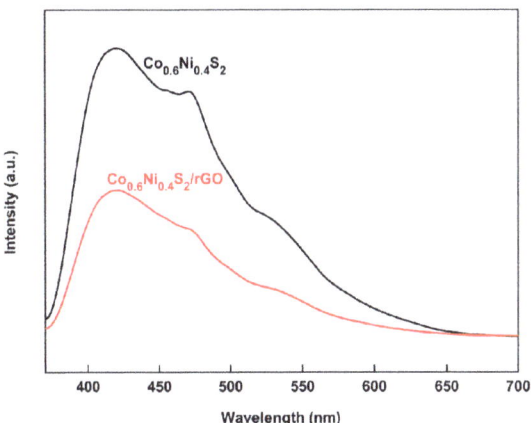

Figure 8. Photoluminescence spectra of $Co_{0.6}Ni_{0.4}S_2$ and $Co_{0.6}Ni_{0.4}S_2$/rGO (excitation wavelength, 280 nm).

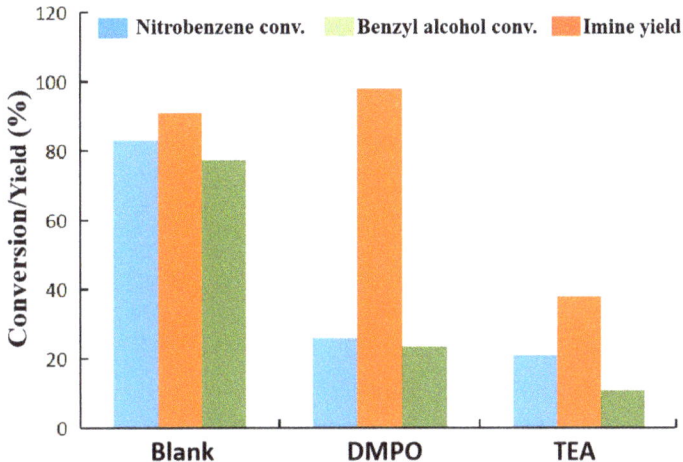

Figure 9. The conversions of nitrobenzene and benzyl alcohol, and the yield of imine by adding different additives.

Using aniline (1 mmol) and benzaldehyde (1 mmol) as reactants, the conversion of aniline was 74% in the same reaction condition for 3 h. When 0.8 mL of TEA was added to the reaction, the conversion of aniline was reduced to 46%. By changing the amount of benzaldehyde, the initial concentration ratio of benzaldehyde and aniline can be adjusted to 0.2, 0.5, 1.5 and 2. In addition, the addition of TEA had little effect on reaction of the initial concentration ratio of 0.2 and 0.5 (Figure 10). However, TEA could significantly reduce the conversion of the reaction, when the initial concentration ratio was 1.5 and 2. On the contrary, adding 0.9 mL of DMPO to the reaction which the initial concentration ratio was 0.2, 0.5 and 1, the conversion of aniline was significantly lower than the blank experiment. However, DMPO had little effect on the reaction which the initial concentration ratio was 1.5 and 2.

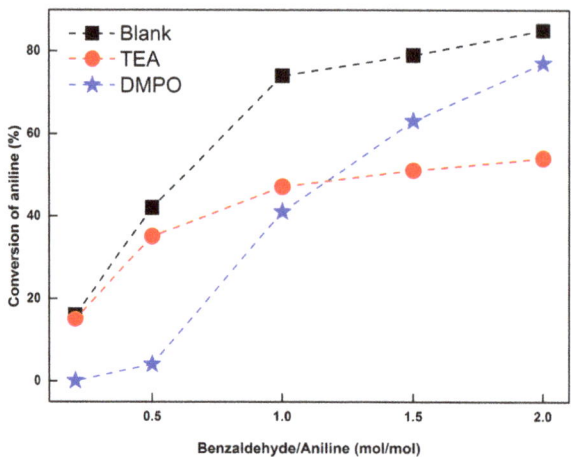

Figure 10. The conversions of nitrobenzene and benzyl alcohol, and the yield of imine by adding different additives.

According to the reaction formula, the amount of aniline and benzaldehyde needed for the formation of imine is the same. When an amount of TEA is added to reaction, a part of photo-generated hole is trapped, and the activation of aniline is inhibited, resulting in reducing the conversion of aniline. This phenomenon is especially obvious in the reaction of benzaldehyde excess. On the contrary, DMPO can inhibit the activation of benzaldehyde by capturing photo-generated electrons, which has a great negative impact on the reaction of aniline excess. Therefore, the reaction process for the synthesis of imines from nitrobenzene and benzyl alcohol by hydrogen transfer can be divided into two steps: in the first step, under the action of photo-generated electron and hole, benzyl alcohol is dehydrated to benzaldehyde, while nitrobenzene is reduced to aniline; in the second step, benzaldehyde and aniline are condensed to synthesize imines.

4. Conclusions

The present work demonstrates a novel photocatalytic route for one-step synthesis of imines from nitroaromatics and aromatic alcohol by transfer hydrogenation using $Co_{0.6}Ni_{0.4}S_2$/rGO catalyst. Under visible light irradiation, the reaction for the synthesis of imines can be divided into two steps. Benzyl alcohol is oxidized to benzaldehyde, while nitrobenzene is reduced to aniline; meanwhile, benzaldehyde and aniline are condensed to synthesize imines. The photo-induced electrons in $Co_{0.6}Ni_{0.4}S_2$/rGO photocatalyst play an important role in activating nitrobenzene and benzaldehyde. Photo-induced holes are mainly responsible for the partial oxidation of benzyl alcohol to benzaldehyde, and the H^+ liberated are used as hydrogen sources to reduce photo-induced electrons activated nitrobenzene to aniline. Next, aniline molecules condense with benzaldehyde molecules to synthesize imine. Under the synergistic action of electrons and holes, $Co_{0.6}Ni_{0.4}S_2$/rGO photocatalyst shows good stability, commendable activity and excellent selectivity, and the yield N-benzylideneaniline can reach as high as 77.2%. As a dynamic equilibrium process, the removal of by-product water is helpful to improve the yield of imine. The photocatalytic system provides an environmentally friendly and effective pathway for the synthesis of imines, and also supplies an alternative approach for hydrogen auto-transfer reactions.

Author Contributions: Methodology, H.Z. and B.M.; software, J.Z. and X.F.; investigation, J.Z. and X.F.; writing—original draft preparation, H.Z.; writing—review and editing, H.Z. and B.M.; funding acquisition, H.Z. and B.M. All authors have read and agreed to the published version of the manuscript.

Funding: This work was financially supported by Natural Science Foundation of Jiangsu Province (Grant No. BK20201030), Open Research Fund of Jiangsu Institute of Marine Resources Development (Grant No. JSIMR202118), Open-end Funds of Jiangsu Key Laboratory of Function Control Technology for Advanced Materials, Jiangsu Ocean University (Grant No. jsklfctam201807).

Institutional Review Board Statement: Not applicable.

Informed Consent Statement: Not applicable.

Data Availability Statement: All data that support the findings of this study are included within the article.

Conflicts of Interest: The authors declare no conflict of interest.

Appendix A.

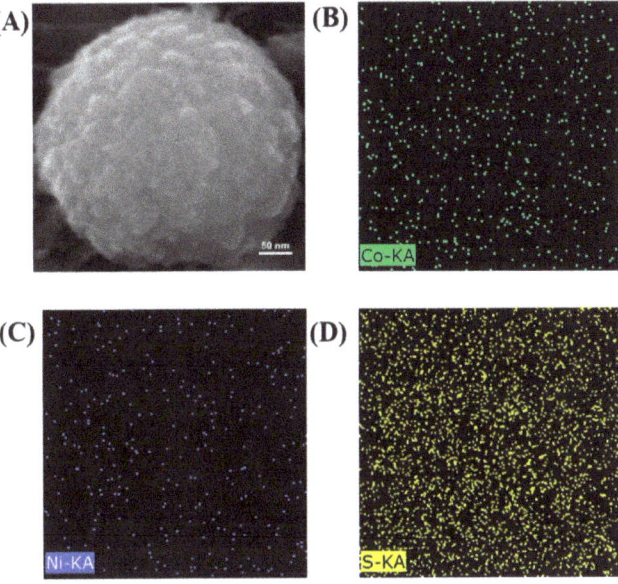

Figure A1. SEM image (**A**) and the elemental distribution maps of Co (**B**), Ni (**C**) and S (**D**) of $Co_{0.6}Ni_{0.4}S_2$/rGO.

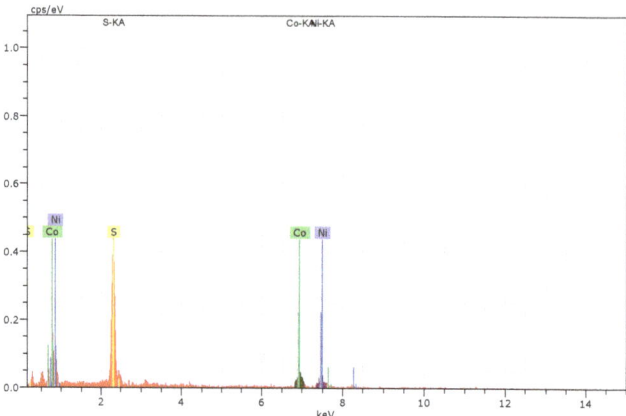

Figure A2. EDX spectra of Co$_{0.6}$Ni$_{0.4}$S$_2$/rGO sample.

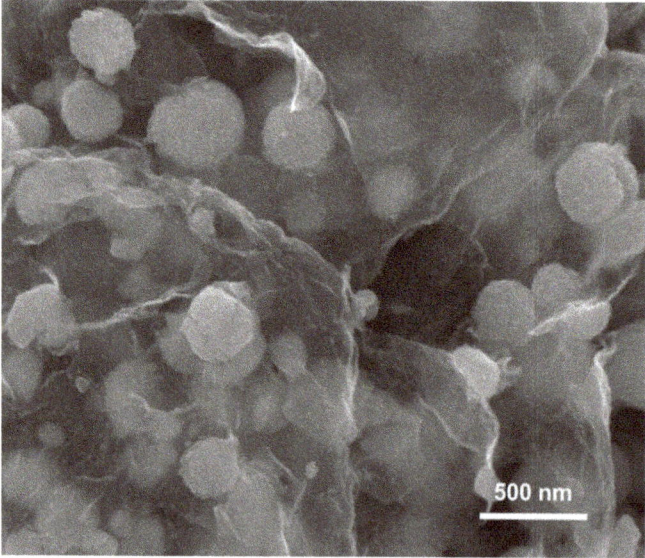

Figure A3. FESEM images of recycled Co$_{0.6}$Ni$_{0.4}$S$_2$/rGO catalysts in synthesis of imine.

References

1. Liu, D.; Yang, P.; Zhang, H.; Liu, M.J.; Zhang, W.F.; Xu, D.M.; Gao, J. Direct reductive coupling of nitroarenes and alcohols catalysed by Co-N-C/CNT@AC. *Green Chem.* **2019**, *21*, 2129–2137. [CrossRef]
2. Zhou, P.; Zhang, Z.H. One-pot Reductive Amination of carbonyl Compounds with Nitro Compounds by Transfer Hydrogenation over Co-N$_x$ as catalyst. *ChemSusChem* **2017**, *10*, 1892–1897. [CrossRef] [PubMed]
3. Chen, P.Q.; Guo, Z.F.; Liu, X.; Lv, H.; Che, Y.; Bai, R.; Cui, Y.H.; Xing, H.Z. A visible-light-responsive metal-organic framework for highly efficient and selective photocatalytic oxidation of amines and reduction of nitroaromatics. *J. Mater. Chem. A* **2019**, *7*, 27074–27080. [CrossRef]
4. Chen, B.; Wang, L.Y.; Gao, S. Recent Advances in Aerobic Oxidation of Alcohols and Amines to Imines. *ACS Catal.* **2015**, *5*, 5851–5876. [CrossRef]
5. Zhang, B.; Guo, X.W.; Liang, H.J.; Ge, H.B.; Gu, X.M.; Chen, S.; Yang, H.M.; Qin, Y. Tailoring Pt-Fe$_2$O$_3$ interfaces for selective reductive coupling reaction to synthesize imine. *ACS Catal.* **2016**, *6*, 6560–6566. [CrossRef]
6. Ch, V.; Sandupatla, R.; Lingareddy, E.; Raju, D. One-pot synthesis of imines by direct coupling of alcohols and amines over magnetically recoverable CdFe$_2$O$_4$ nanocatalyst. *Mater. Lett.* **2021**, *302*, 130417. [CrossRef]

7. Yang, H.M.; Cui, X.J.; Dai, X.C.; Deng, Y.Q.; Shi, F. Carbon-catalysed reductive hydrogen atom transfer reactions. *Nat. Commun.* **2015**, *6*, 6478. [CrossRef] [PubMed]
8. Sahoo, D.P.; Patnaik, S.; Rath, D.; Mohapatra, P.; Mohanty, A.; Parida, K. Influence of Au/Pd alloy on an amine functionalised ZnCr LDH-MCM-41 nanocomposite: A visible light sensitive photocatalyst towards one-pot imine synthesis. *Catal. Sci. Technol.* **2019**, *9*, 2493–2513. [CrossRef]
9. Sankar, M.; He, Q.; Dawson, S.; Nowicka, E.; Lu, L.; Bruijnincx, P.C.A.; Beale, A.M.; Kiely, C.J.; Weckhuysen, B.M. Supported bimetallic nano-alloys as highly active catalysts for the one-pot tandem synthesis of imines and secondary amines from nitrobenzene and alcohols. *Catal. Sci. Technol.* **2016**, *6*, 5473–5482. [CrossRef]
10. Hirakawa, H.; Katayama, M.; Shiraishi, Y.; Sakamoto, H.; Wang, K.; Ohtani, B.; Ichikawa, S.; Tanaka, S.; Hirai, T. One-Pot Synthesis of Imines from Nitroaromatics and Alcohols by Tandem Photocatalytic and Catalytic Reactions on Degussa (Evonik) P25 Titanium Dioxide. *ACS Appl. Mater. Interfaces* **2015**, *7*, 3797–3806. [CrossRef]
11. Wu, Y.H.; Ye, X.J.; Zhang, S.J.; Meng, S.G.; Fu, X.L.; Wang, X.C.; Zhang, X.M.; Chen, S.F. hotocatalytic synthesis of Schiff base compounds in the coupled system of aromatic alcohols and nitrobenzene using $Cd_xZn_{1-x}S$ photocatalysts. *J. Catal.* **2018**, *359*, 151–160. [CrossRef]
12. Ye, X.; Chen, Y.; Lin, C.; Ding, R.; Wang, X.; Zhang, X.; Chen, S. One-pot synthesis of Schiff base compounds via photocatalytic reaction in the coupled system of aromatic alcohols and nitrobenzene using $CdIn_2S_4$ photocatalyst. *Dalton Trans.* **2018**, *47*, 10915–10924. [CrossRef] [PubMed]
13. Qin, J.F.; Yang, M.; Chen, T.S.; Dong, B.; Hou, S.; Ma, X.; Zhou, Y.Z.; Yang, X.L.; Nan, J.; Chai, Y.M. Ternary metal sulfides MoCoNiS derived from metal organic frameworks for efficient oxygen evolution. *Int. J. Hydrog. Energ.* **2020**, *45*, 2745–2753. [CrossRef]
14. Huang, D.L.; Wen, M.; Zhou, C.Y.; Li, Z.H.; Cheng, M.; Chen, S.; Xue, W.J.; Lei, L.; Yang, Y.; Xiong, W.P.; et al. $Zn_xCd_{1-x}S$ based materials for photocatalytic hydrogen evolution, pollutants degradation and carbon dioxide reduction. *Appl. Catal. B-Environm.* **2020**, *267*, 118651. [CrossRef]
15. Dai, D.S.; Xu, H.; Ge, L.; Han, C.C.; Gao, Y.Q.; Li, S.S.; Lu, Y. In-situ synthesis of CoP co-catalyst decorated $Zn_{0.5}Cd_{0.5}S$ photocatalysts with enhanced photocatalytic hydrogen production activity under visible light irradiation. *Appl. Catal. B-Environm.* **2017**, *217*, 429–436. [CrossRef]
16. Liu, X.L.; Liang, X.Z.; Wang, P.; Huang, B.B.; Qin, X.Y.; Zhang, X.Y.; Dai, Y. Highly efficient and noble metal-free NiS modified $Mn_xCd_{1-x}S$ solid solutions with enhanced photocatalytic activity for hydrogen evolution under visible light irradiation. *Appl. Catal. B-Environm.* **2017**, *203*, 282. [CrossRef]
17. Yang, S.P.; Zhang, C.H.; Cai, Y.Q.; He, X.W.; Niu, H.Y. Synthesis of $Cd_xZn_{1-x}S@Fe_3S_4$ magnetic photocatalyst nanoparticles for the photodegradation of methylene blue. *J. Alloys Compd.* **2018**, *735*, 1955–1961. [CrossRef]
18. Zhang, H.H.; Guan, B.; Gu, J.N.; Li, Y.; Ma, C.; Zhao, J.; Wang, T.Y.; Cheng, C.J. One-step synthesis of nickel cobalt sulphides particles: Tuning the composition for high performance supercapacitors. *RSC Adv.* **2016**, *6*, 58916–58924. [CrossRef]
19. Liu, Y.R.; Shang, X.; Gao, W.K.; Dong, B.; Chi, J.Q.; Li, X.; Yan, K.L.; Chai, Y.M.; Liu, Y.Q.; Liu, C.G. Ternary $CoS_2/MoS_2/RGO$ electrocatalyst with CoMoS phase for efficient hydrogen evolution. *Appl. Surf. Sci.* **2017**, *412*, 138–145. [CrossRef]
20. Li, Y.H.; Cao, L.J.; Qiao, L.; Zhou, M.; Yang, Y.; Xiao, P.; Zhang, Y.H. Ni-Co sulfide nanowires on nickel foam with ultrahigh capacitance for asymmetric Supercapacitors. *J. Mater. Chem. A* **2014**, *2*, 6540–6548. [CrossRef]
21. Shang, X.; Li, X.; Hu, W.H.; Dong, B.; Liu, Y.R.; Han, G.Q.; Chai, Y.M.; Liu, Y.Q.; Liu, C.G. In situ growth of Ni_xS_y controlled by surface treatment of nickel foam as efficient electrocatalyst for oxygen evolution reaction. *Appl. Surf. Sci.* **2016**, *378*, 15–21. [CrossRef]
22. Yang, F.; Yao, J.Y.; He, H.C.; Zhou, M.; Xiao, P.; Zhang, Y.H. Ni-Co oxides nanowire arrays grown on ordered TiO_2 nanotubes with high performance in supercapacitors. *J. Mater. Chem. A* **2013**, *1*, 594–601. [CrossRef]
23. Ma, B.; Tong, X.L.; Guo, C.X.; Guo, X.N.; Guo, X.Y.; Keil, F.J. Pyrite Nanoparticles: An Earth-Abundant Mineral Catalyst for Activation of Molecular Hydrogen and Hydrogenation of Nitroaromatics. *RSC Adv.* **2016**, *6*, 55220–55224. [CrossRef]
24. Ji, N.; Wang, X.Y.; Werdenthaler, C.; Spliethoff, B.; Rinaldi, R. Iron(II) Disulfides as Precursors of Highly Selective Catalysts for Hydrodeoxygenation of Dibenzyl Ether into Toluene. *ChemCatChem* **2015**, *7*, 960–966. [CrossRef]
25. Cai, Y.F.; Pan, Y.G.; Xue, J.Y.; Sun, Q.F.; Su, G.Z.; Li, X. Comparative XPS study between experimentally and naturally weathered pyrites. *Appl. Surf. Sci.* **2009**, *255*, 8750–8760. [CrossRef]
26. Ma, B.; Wang, Y.Y.; Tong, X.L.; Guo, X.N.; Zheng, Z.F.; Guo, X.Y. Graphene-supported CoS_2 Particles: An Efficient Photocatalyst for Selective Hydrogenation of Nitroaromatics in Visible Light. *Catal. Sci. Technol.* **2017**, *7*, 2805–2812. [CrossRef]
27. Surendran, S.; Sankar, K.V.; Berchmans, L.J.; Selvan, R.K. Polyol synthesis of α-NiS particles and its physico-chemical Properties. *Mat. Sci. Semicon. Proc.* **2015**, *33*, 16–23. [CrossRef]
28. Luo, P.; Zhang, H.J.; Liu, L.; Zhang, Y.; Deng, J.; Xu, C.H.; Hu, N.; Wang, Y. Targeted synthesis of unique nickel sulfides (NiS, NiS_2) microarchitectures and the applications for the enhanced water splitting system. *ACS Appl. Mater. Interfaces* **2017**, *9*, 2500–2508. [CrossRef]
29. Chen, G.; Liaw, S.S.; Li, B.S.; Xu, Y.; Dunwell, M.; Deng, S.G.; Fan, H.Y.; Luo, H.M. Microwave-assisted synthesis of hybrid $Co_xNi_{1-x}(OH)_2$ nanosheets: Tuning the composition for high performance supercapacitor. *J. Power Sources* **2014**, *251*, 338–343. [CrossRef]
30. Huirache-Acuña, R.; Pawelec, B.; Rivera-Muñoz, E.M.; GuilLópez, R.; Fierro, J.L.G. Characterization and HDS activity of sulfided Co-Mo-W/SBA-16 catalysts: Effects of P addition and Mo/(Mo + W) ratio. *Fuel* **2017**, *198*, 145–158. [CrossRef]

31. Du, J.M.; Wang, H.M.; Yang, M.K.; Li, K.D.; Zhao, L.X.; Zhao, G.Y.; Li, S.J.; Gu, X.L.; Zhou, Y.L.; Wang, L.; et al. Pyramid-like CdS nanoparticles grown on porous TiO2 monolith: An advanced photocatalyst for H_2 production. *Electrochim. Acta* **2017**, *250*, 99–107. [CrossRef]
32. Ferguson, C.T.J.; Huber, N.; Kuckhoff, T.; Zhang, K.; Landfester, K. Dispersible porous classical polymer photocatalysts for visible light-mediated production of pharmaceutically relevant compounds in multiple solvents. *J. Mater. Chem. A* **2020**, *8*, 1072–1076. [CrossRef]
33. Ma, B.; Wang, Y.Y.; Guo, X.N.; Tong, X.L.; Liu, C.; Wang, Y.W.; Guo, X.Y. Photocatalytic synthesis of 2,5-diformylfuran from 5-hydroxymethylfurfural or fructose over bimetallic Au-Ru nanoparticles supported on reduced graphene oxides. *Appl. Catal. A-Gen.* **2018**, *552*, 70–76. [CrossRef]
34. Ma, J.J.; Yu, X.J.; Liu, X.L.; Li, H.Y.; Hao, X.L.; Li, J.Y. The preparation and photocatalytic activity of Ag-Pd/g-C_3N_4 for the coupling reaction between benzyl alcohol and aniline. *Mol. Catal.* **2019**, *476*, 110533. [CrossRef]
35. Nakai, Y.; Azuma, M.; Muraoka, M.; Kobayashi, H.; Higashimoto, S. One-pot imine synthesis from benzylic alcohols and nitrobenzene on CdS-sensitized TiO_2 photocatalysts: Effects of the electric nature of the substituent and solvents on the photocatalytic activity. *Mol. Catal.* **2017**, *443*, 203–208. [CrossRef]

Review

Recent Advances in CoSe$_x$ and CoTe$_x$ Anodes for Alkali-ion Batteries

Yuqi Zhang, Zhonghui Sun *, Dongyang Qu *, Dongxue Han and Li Niu

College of Chemistry and Chemical Engineering, Guangzhou University, Guangzhou 510006, China; 2112105042@e.gzhu.edu.cn (Y.Z.); dxhan@gzhu.edu.cn (D.H.); lniu@gzhu.edu.cn (L.N.)
* Correspondence: cczhsun@gzhu.edu.cn (Z.S.); dyqu@gzhu.edu.cn (D.Q.)

Abstract: Transition metal selenides have narrow or zero band-gap characteristics and high theoretical specific capacity. Among them, cobalt selenide and cobalt telluride have some typical problems such as large volume changes, low conductivity, and poor structural stability, but they have become a research hotspot in the field of energy storage and conversion because of their high capacity and high designability. Some of the innovative synthesis, doping, and nanostructure design strategies for CoSe$_x$ and CoTe$_x$, such as CoSe-InCo-InSe bimetallic bi-heterogeneous interfaces, CoTe anchoring MXenes, etc., show great promise. In this paper, the research progress on the multistep transformation mechanisms of CoSe$_x$ and CoTe$_x$ is summarized, along with advanced structural design and modification methods such as defect engineering and compositing with MXenes. It is hoped that this review will provide a glimpse into the development of CoSe$_x$ and CoTe$_x$ anodes for alkali-ion batteries.

Keywords: CoTe$_x$; CoSe$_x$; anodes; alkali-metal-ion batteries

Citation: Zhang, Y.; Sun, Z.; Qu, D.; Han, D.; Niu, L. Recent Advances in CoSe$_x$ and CoTe$_x$ Anodes for Alkali-ion Batteries. *Coatings* 2023, 13, 1588. https://doi.org/10.3390/coatings13091588

Academic Editor: Je Moon Yun

Received: 1 August 2023
Revised: 6 September 2023
Accepted: 8 September 2023
Published: 12 September 2023

Copyright: © 2023 by the authors. Licensee MDPI, Basel, Switzerland. This article is an open access article distributed under the terms and conditions of the Creative Commons Attribution (CC BY) license (https://creativecommons.org/licenses/by/4.0/).

1. Introduction

The demand for clean renewable energy sources such as wind and solar power has increased due to the energy problems and environmental pollution from traditional fossil energy. Lithium-ion batteries (LIBs) were born as high-performance energy storage devices to compensate for the inconvenience of storing clean energy [1,2]. Today, lithium resources are increasingly scarce; thus, sodium-ion batteries (SIBs) and potassium-ion batteries (PIBs) have become two of the hopes to solve the problem of resource shortage [3,4]. The prerequisite for further application of sodium/potassium-ion batteries is to prepare negative electrode materials with appropriate working voltage, excellent cycling performance, and high energy density. Anode materials can be divided into three types: intercalation type (e.g., carbon [5] and TiO$_2$ [6]), alloy type (Si, Ge, Sb [7], etc.), and conversion type (Co$_3$O$_4$ [8], Fe$_x$S [9], etc.). Although an intercalated anode has a stable structure and good electrochemical cycling stability, it still has the disadvantages of low energy density and limited specific capacity. In addition, the alloy-type anode has a high capacity, but the large volume change caused by the alloy reaction brings the disadvantage of poor stability. Compared with the former two, the conversion negative electrode has a reasonably high specific capacity and a small volume expansion, making it a potential choice for ion batteries' anodes [10].

In recent years, transition metal chalcogenides (TMCs) have been widely studied as representative conversion-type anode materials due to their stable performance, high theoretical specific capacity, abundant natural resources, diverse types, and low cost. They hold great potential in meeting the demand for high-energy-density anodes in alkali-ion batteries [11]. Compared to transition metal sulfides, transition metal selenides exhibit higher volumetric capacity as anionic battery anodes. Additionally, the weaker M-Se bond compared to M-O and M-S bonds improves the reaction kinetics of transition metal selenides, resulting in higher energy density and electronic conductivity. Moreover, transition metal selenides have larger interlayer spacing, which helps alleviate volume expansion issues during cycling, especially for larger alkali metal ions such as sodium and potassium.

Therefore, they possess advantages in electrochemical energy storage applications [12–17]. Compared to transition metal oxides and sulfides, transition metal tellurides possess lower electronegativity, the highest conductivity, and the highest volumetric capacity. They are typical layered materials with a large interlayer space, which is beneficial for rapid ion transport in electrodes. This results in excellent electrode wetting and ion diffusion kinetics. So far, many selenides (such as $CoSe_x$ [18], $ZnSe_x$ [19], $MoSe_x$ [20], and $FeSe_x$ [21]) and tellurides (e.g., $NiTe_x$ [22], $FeTe_x$ [23]) have been studied as ion battery anode materials.

Cobalt-based materials have become important materials in various fields (such as hydrodesulfurization and hydrodefluorination) due to their unique catalytic activity, electrochemical performance, and ferromagnetism [24–26]. Among them, $LiCoO_2$ is one of the most successful commercial cathode materials for lithium-ion batteries (LIBs) [27]. Based on the success of $LiCoO_2$, research on Na_xCoO_2 as a cathode material for sodium-ion batteries has also been widely conducted [28]. In terms of anode materials, cobalt has various oxidation states, and cobalt-based chalcogenides have diverse types and high theoretical sodium storage capacity, making them highly favored by researchers. Cobalt-based composite anodes, such as cobalt oxide [29], cobalt phosphide, and cobalt–aluminum alloys, have theoretical capacities of 700~1000 mAh g^{-1}, which is 2~3 times that of graphite anodes. They have diverse crystal structures and low cost, receiving increasing attention in recent years. Among them, $CoSe_x$ is widely used as an anode material for alkali-ion batteries due to its large capacity, strong conductivity, and weak electronegativity [18,30–34]. $CoTe_x$ is considered to be a potential choice with excellent electrochemical performance in terms of metallicity, magnetism, electronics, and electrocatalysis [35]. In the past few years, extensive research has shown that cobalt selenides and tellurides can be used as potential anodes with high capacity or improved rate performance. However, their inherent volume changes as conversion anodes, relatively low electronic conductivity, and side reactions can lead to significant reversible capacity loss and low active material utilization. Recently, various energy storage batteries have been working on mitigating these issues and have achieved successes, such as forming hybrid/composite samples to enhance conductivity, optimizing electrode design and electrolyte content, and adjusting voltage windows [31].

At present, some progress has been made in the research of energy storage and conversion of transition metal selenides and tellurides—typically cobalt-based compounds [36]. However, it is necessary to study the application and reaction mechanisms of cobalt selenide and cobalt telluride in the three types of metal-ion battery in a timely and systematic manner. This paper reviews the recent progress in the preparation and application of $CoSe_x/CoTe_x$-based electrodes (Figure 1). The electrode design and reaction mechanism are introduced in detail. Finally, the challenges and opportunities in this field are presented (Figure 2). This review will significantly advance the research on innovative $CoSe_x/CoTe_x$ design to improve its application in energy storage devices.

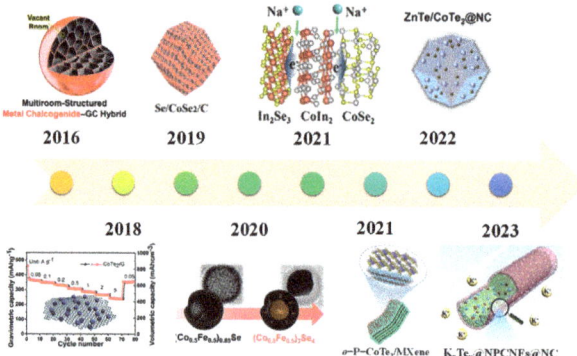

Figure 1. Typical recent advances in the synthesis of $CoSe_x/CoTe_x$–based electrodes, Refs. [37–45].

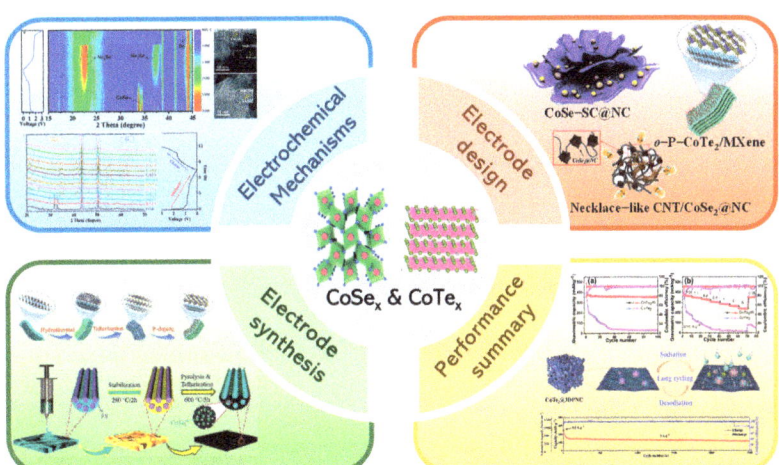

Figure 2. This paper summarizes from four perspectives, including electrochemical mechanisms, electrode design, synthesis, and performance; Refs. [38,42,46–48].

2. Electrochemical Reaction Mechanisms

As anode materials during charging and discharging processes, different metal-based compounds usually have different kinds of reactions, such as intercalation, conversion, and alloying. For example, intercalation–conversion–alloying reactions take place during the charging and discharging of Sn- and Sb-based compounds; Mo-, W-, Fe-, and Ni-based compounds mainly have intercalation and conversion reactions, while Nb-based compounds only have intercalation reactions [49]. The metal-based anode affects its reaction type. TMCs with conversion reactions can accommodate more alkali ions through the multi-electron transfer process and have a higher capacity. In addition, the combination with chalcogenide elements is conducive to the conversion reaction during the charge–discharge process. Due to the weak TM–B (B = Se and Te) bonds, TMBs have a high theoretical capacity, as well as excellent electrical conductivity and convenient ion transport channels. Co_xB_y has been used as a potential anode in alkali-ion batteries. The storage mechanism between Co_xB_y in LIBs/SIBs/PIBs is the conversion reaction that eventually generates Co and A_xB (A = Li, Na, K). In particular, a few intermediate reactions may occur during the Co_xB_y discharge processes, as shown in the following equations:

$$Co_xB_y + A^+ + e^- \rightarrow ACo_xB_y \tag{1}$$

$$Co_xB_y + A^+ + e^- \rightarrow Co_xB_{y-1} + AB \tag{2}$$

Deciphering the electrochemical reaction mechanism of $CoSe_x$ and $CoTe_x$ is significant to enhance their application in batteries. The latest progress on the reaction mechanisms of cobalt selenide and cobalt telluride for the anode material of LIBs, SIBs, and PIBs is summarized below.

2.1. $CoSe_x$

Many types of research on the storage mechanisms of $CoSe_x$ anodes have been conducted. Among them, the storage path of $CoSe_x$ in PIBs is different from that in SIBs and LIBs. Yu et al. [34] studied the potassium-ion storage mechanism of octahedral $CoSe_2$ nanotubes threaded with N-doped carbon through first-principles calculation (Figure 3a–f). The calculation showed that the formation energies of CoK_xSe_x were all positive, indicating that the insertion of K^+ into $CoSe_2$ is difficult to achieve, and it is more inclined to replace Co to generate K_2Se. The reason for this is that potassium atoms are too large to penetrate the lattice of $CoSe_2$, whereas the non-stoichiometric selenide $CoSe_{1-x}$ (Co_3Se_4) has a wider lattice than $CoSe_2$, achieving K^+ embedding. The actual reaction process can be explained as follows:

$$CoSe_2 + \frac{4}{3}K^+ + \frac{4}{3}e^- \rightarrow \frac{1}{3}Co_3KSe_4 \quad (3)$$

$$Co_3KSe_4 + K^+ + e^- \rightarrow 3CoSe + K_2Se \quad (4)$$

$$CoSe + 2K^+ + 2e^- \rightarrow Co + K_2Se \quad (5)$$

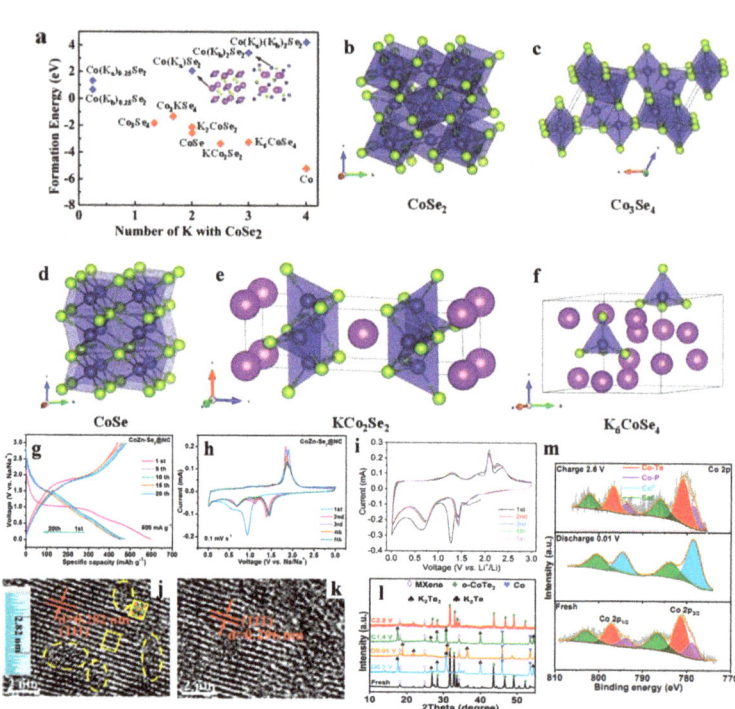

Figure 3. (a) The formation energy of conversion products. (b–f) Schematic molecular structures using in this work; Ref. [34]. (g) The cycling performance and (h) CV curve; Ref. [50]. (i) CoSe/G CV curve; Ref. [51]. (j) HRTEM for o–P–CoTe$_2$/MXene and (k) after discharging to 1.0 V. (l) Ex situ XRD and (m) ex situ XPS Co 2p; Ref. [38].

In contrast to PIBs, Jiang et al. [50] found the appearance of three new peaks in cyclic voltammetry (CV) curves (Figure 3h), indicating that $CoSe_2$ turns into Na_xCoSe_2, CoSe, and Co. The oxidation peaks at 1.53 V and 1.89 V represent the intermediate product Na_xCoSe_2 and $CoSe_2$, which is the fully charged product during the charge process, respectively. The locations of these peaks are highly consistent throughout the five cycles of reduction and oxidation, together with the corresponding position of the charging and discharging platform (Figure 3g), testifying to the reversibility of the charge–discharge process. The actual reaction mechanism can be summarized as follows:

$$CoSe_2 + xNa^+ + xe^- \leftrightarrow Na_xCoSe_2 \quad (6)$$

$$Na_xCoSe_2 + (2-x)Na^+ + (2-x)e^- \leftrightarrow CoSe + Na_2Se \quad (7)$$

$$CoSe + 2Na^+ + 2e^- \leftrightarrow Na_2Se + Co \quad (8)$$

$$2Na_2Se + Co \leftrightarrow CoSe_2 + 4Na^+ + 4e^- \quad (9)$$

In LIBs, Jiang et al. [51] found three prominent peaks of CoSe/G at 1.42, 1.27, and 0.63 V in the first cathodic scan through the first five CV curves in LIBs (Figure 3i). Li$^+$ inserted into the CoSe showed a prominent peak at 1.42 V, corresponding to the formation of Li$_x$CoSe. The solid–electrolyte interphase (SEI) formed at 1.27 V. The wide peak at 0.63 V can be ascribed to the conversion reaction from Li$_x$CoSe to Li$_2$Se and Co. The anode peaks at 1.28 and 2.1 V signify the oxidation reactions that convert Li$_2$Se and Co into CoSe. The reaction process of CoSe anode in LIBs can be expressed as follows:

$$CoSe_2 + xLi^+ + xe^- \leftrightarrow Li_xCoSe \quad (10)$$

$$Li_2CoSe + (2-x)Li^+ + 2e^- \leftrightarrow Co + Li_2Se \quad (11)$$

2.2. CoTe$_x$

As an emerging potential anode, the storage mechanism of CoTe$_x$ is being extensively studied. Xu et al. [38] performed high-resolution transmission electron microscopy (HRTEM), X-ray diffraction (XRD), and X-ray photoelectron spectroscopy (XPS) analyses on CoTe$_2$ nanowires with a Te vacancy (o-P-CoTe$_2$/MXene) in different discharge and charge states to study the potassium storage behavior. When the half-cell discharged to 1.0 V, the HRTEM (Figure 3j) showed that the lattice spacing of the (111) plane expanded significantly to 0.296 nm compared with the initial state (Figure 3k, 0.282 nm), which proves the existence of K$^+$ intercalation above 1.0 V. At 0.5 V, peaks of K$_2$Te$_3$ and Co were observed in the ex situ XRD spectra (Figure 3l). This observation confirmed the conversion reaction from CoTe$_2$ to Co and K$_2$Te$_3$. After further discharging to 0.01 V, a peak of K$_2$Te was observed, indicating that K$_2$Te$_3$ was further converted into the final reduction product (K$_2$Te). When the half-cell was charged to 1.4 V, K$_2$Te converted to K$_2$Te$_3$ and Co. When the anode was further charged to 2.6 V, the peak of o-P-CoTe$_2$/MXene reappeared. Ex situ XPS (Figure 3m) spectra further confirmed that the peak of Co 2p$^{3/2}$ continued to shift from 781.4 eV to 778.4 eV during complete discharge, corresponding to the formation of Co. Therefore, the o-P-CoTe$_2$/MXene anode conversion mechanism in PIBs can be described by the following reactions:

$$CoTe_2 + xK^+ + xe^- \leftrightarrow K_xCoTe_2 \quad (12)$$

$$3K_xCoTe_2 + (4-3x)K^+ + (4-3x)e^- \leftrightarrow 3Co + 2K_2Te_3 \quad (13)$$

$$K_2Te_3 + 4K^+ + 4e^- \leftrightarrow 3K_2Te \quad (14)$$

Recently, Li et al. [41] published a more detailed supplement on the storage mechanism of CoTe$_x$ in PIBs. As shown in the in situ XRD pattern (Figure 4a), after discharging to 1.21 V, the peak of CoTe$_2$ shifted to a lower angle, indicating that K$^+$ was embedded into the CoTe$_2$ lattice and K$_x$CoTe$_2$ formed. Upon further discharging to 0.01 V, the peak of CoTe$_2$ became weaker, corresponding to the conversion of CoTe$_2$ to Co and K$_x$Te$_y$. When charged to 3 V, the peak of CoTe$_2$ reappeared. In situ XPS analysis (Figure 4b) showed that the metallic Co phase appeared in the high-resolution Co 2p spectrum when the cell was discharged to 0.8 V. In the Te 3d spectrum, K$_5$Te$_3$ appeared, indicating that K$_x$CoTe$_2$ was converted to K$_5$Te$_3$ at 0.8 V. When discharging to 0.4 V, the cobalt peak strengthened and a new peak of K$_2$Te appeared in the Te 3d spectrum. Upon further discharging to 0.1 V, K$_5$Te$_3$ was almost transformed into K$_2$Te. The electrochemical reaction mechanism was further studied by transmission electron microscope (TEM), energy-dispersive X-ray spectroscopy (EDS), and HRTEM (Figure 4c–h). The TEM and EDS patterns proved that the structure remains stable and the elements are evenly distributed after the cycle. The discharge HRTEM image shows the (100) crystal plane of Co, the (251) crystal face of K_5Te_3, and the (311) and (200) crystal faces of K_2Te (Figure 4d). In addition, upon full charge, the (011) plane of $CoTe_2$ (Figure 4g) appears, which is consistent with the in situ XRD results. In summary, the above results show that K_xCoTe_2 converted totally to K_5Te_3, and then K_5Te_3 converted to K_2Te during the discharge process, and the authors clearly express the evolution of polytellurides in Figure 4i. Therefore, the relevant reversible reactions are summarized below:

$$CoTe_2 + xK^+ + xe^- \rightarrow K_xCoTe_2 \tag{15}$$

$$3K_xCoTe_2 + (10-3x)K^+ + (10-3x)e^- \rightarrow 3Co + 2K_5Te_3 \tag{16}$$

$$K_5Te_3 + K^+ + e^- \rightarrow 3K_2Te \tag{17}$$

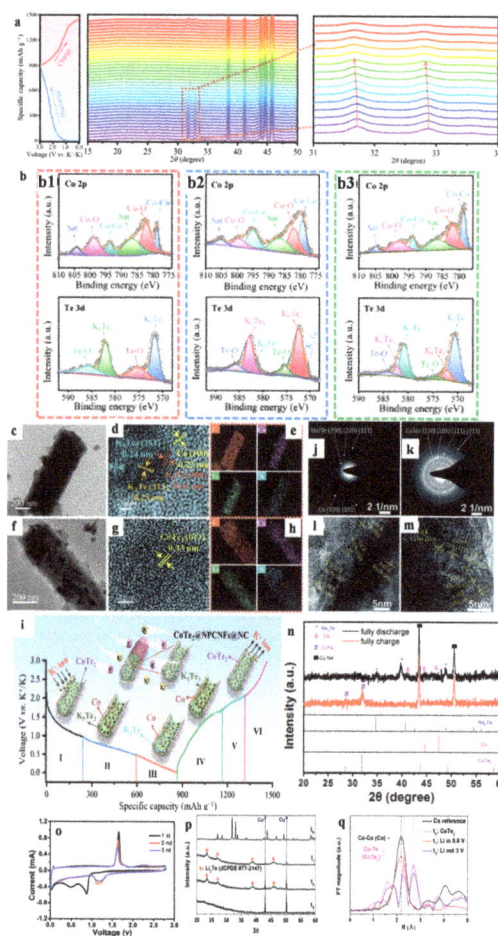

Figure 4. (**a**) In situ XRD patterns. (**b**) XPS of Co 2p and Te 3d at discharged state of (**b1**) 0.8 V, (**b2**) 0.4 V, and (**b3**) 0.1 V. (**c,f**) TEM, (**d,g**) HRTEM, and (**e,h**) EDS images at 0.1 V and 3 V, respectively. (**i**) Schematic diagram of the PIB storage mechanism of the $CoTe_2$@NPCNFs@NC; Ref. [41]. $CoTe_2$/G (**j,k**) SAED patterns, (**l,m**) HRTEM, and (**n**) ex situ XRD at 0.01 V and 2.8 V in the first cycle. (**o**) CV curve; Ref. [42]. (**p**) Ex situ XRD. (**q**) EXAFS after Li–embedded/Li–detached; Ref. [52].

Zhang et al. [42] clarified the sodium-ion storage mechanism of CoTe$_2$ by selected-area electron diffraction (SAED, Figure 4j–k), HRTEM (Figure 4l–m), and ex situ XRD (Figure 4n). At the end of discharge, the peaks of Co and Na$_2$Te could be clearly observed from the ex situ XRD pattern. Moreover, the signals of Co and Na$_2$Te could be detected in the SAED pattern. The crystal lattice of Na$_2$Te and Co also appeared in HRTEM. After charging to 2.8 V, the CoTe$_2$ phase peak could be collected again. A clear CoTe$_2$ ring was observed in the SAED diagram. At the same time, Zhang analyzed the storage mechanism of Na ions from the CV curve (Figure 4o). A reduction peak formed at about 0.83 V, because CoTe$_2$ formed Co nanocrystals and Na$_2$Te with Na ions. The oxidation peak at about 1.65 V can be ascribed to the reformation of CoTe$_2$. It can be summarized as follows:

$$CoTe_2 + 4Na^+ + 4e^- \leftrightarrow Co + 2Na_2Te \tag{18}$$

$$Co + 2Na_2Te \leftrightarrow CoTe_2 + 4Na^+ + 4e^- \tag{19}$$

Ganesan et al. [52] studied the deep material changes of CoTe$_2$ cycling in LIBs by ex situ XRD (Figure 4p) and extended X-ray absorption fine structure (EXAFS, Co-K-Edge, Figure 4q). In the XRD pattern, when discharging to 0.9 V, it can be seen that the peak of CoTe$_2$ changed to the peak of Li$_2$Te, and CoTe$_2$ also changed to Co in EXAFS, which means that the conversion reaction in the discharge process is that CoTe$_2$ changes to Co and Li$_2$Te. Finally, although no recovery of CoTe$_2$ was observed in the XRD pattern at a full charge of 3.0 V, the main CoTe$_2$ EXAFS peak reappeared, confirming the reversibility of the material. Based on the experimental results, the LIB conversion mechanism for CoTe$_2$ is proposed as follows:

$$CoTe_2 + xLi^+ + xe^- \leftrightarrow Co + Li_2Te \tag{20}$$

It is known that many works have fully studied the reaction mechanisms of cobalt selenide and cobalt telluride in LIBs, SIBs, and PIBs. Many in situ and ex situ analytical characterization methods have emerged. In situ XRD, XPS, and ex situ TEM, HRTEM, etc., can well demonstrate the mechanisms of CoSe$_x$ and CoTe$_x$. However, due to emerging anode design methods such as doping and heterojunction, heteroatoms and conductive carbon materials will bring various synergistic effects to cobalt selenide and cobalt telluride. At present, the mechanistic analysis of cobalt selenide and cobalt telluride anodes mainly focuses on the reaction between alkali ions and CoSe$_x$ or CoTe$_x$. In fact, the analysis of the synergistic mechanism between substances in the anode requires more in-depth research, and at the same time, innovative research methods need to be developed.

3. Electrode Design

As conversion-type anodes, CoSe$_x$ and CoTe$_x$ can bring enthralling high weight/volume capacity due to multiple electron transfer processes. However, due to the low conductivity of the converted anode materials and the large volume change in the process of intercalation/deintercalation, their practical application has been hindered. Therefore, effective strategies to solve these problems have been explored through carbon recombination [53], core–shell structures [54], defect engineering [55], etc.

Two approaches used to modify cobalt selenides and cobalt tellurides are the introduction of defects (e.g., point defects and heterogeneous interfaces) and structural design. Point defects like vacancies and dopings can interfere with surrounding atoms, resulting in lattice distortion, forming a good adjusted electronic structure with improved conductivity and diffusion coefficients [56]. The inorganic heterostructure of the anode builds an interfacial electric field as a facilitative transport channel, resulting in rapid Li/Na/K ion reaction kinetics. Considering the excellent prospects of interface engineering, Dong et al. [57] realized a "structural function motifs" design in a ZnO/ZnS anode. The ZnO/ZnS heterostructure regulates the state around the Fermi level, narrowing the band gap and, thus, improving the electronic conductivity. The introduction of defect engineering can create synergistic effects that alter the inherent properties of materials, unlike structural design and composite strategies [58,59]. In addition, Co 3d electrons with the low spintronic structure t2g6eg1 are known to produce a Jahn–Teller effect in $CoSe_2$ [60]. Proper doping of metal ions can alleviate Jahn–Teller distortion to a certain extent [61,62] and promote a more stable electrode structure and better cycling stability. Defect engineering is widely used in $CoSe_2$ and has also been reported in recently emerging $CoTe_2$ materials.

In addition to defects, structural designs such as core–shell structures or $CoSe_x$ and $CoTe_x$ composite carbon materials [42] can take advantage of carbon's high conductivity and deformation resistance, not only reducing the risk of electrode crushing but also improving the rate performance. The latest reports of introducing defects or innovative structural designs to $CoSe_x$ and $CoTe_x$ are summarized below.

3.1. Introducing Defects

The modification methods of defect engineering include entry-point defects and surface defects. In recent studies, the introduction of point defects has been divided into metallic elements and non-metallic elements. Secondly, surface defects generally introduce heterogeneous structures. Recent advances in the combination of the above engineering with cobalt selenide and cobalt telluride are summarized below.

3.1.1. Doping with Metal Elements

Due to the large volume variation during circulation, the structural solidity and circulation function of metal selenides are weak, which limits the feasibility of their application. To overcome the above problems, some studies have prepared polyselenide metal nanostructures by adding inert elements [63]. The activity of the transition-metal-based materials can further be improved through coupling with other metals and non-metals [64]. The reason for the excellent doping performance of metal atoms is that the introduction of two or more metal atoms into a mixed-metal selenide results in the formation of more multicomponent selenides. Some studies have shown that this result can significantly improve the electrical conductivity of discharge products and further provide a buffer matrix to adapt to the expansion stress [65]. They benefit from strain relief and spectacular pseudocapacitance effects, thus reducing the energy barrier for ionic diffusion and showing considerable improvement in electrochemical performance.

For the voltage window of $CoSe_2$, previous studies have usually shortened it to 0.5~3.0 V to stabilize the reversible capacity of long cycles [66,67]. Ji et al. [50] doped a variety of metal ions such as Ni^{2+} and Cu^{2+} into nitrogen-doped carbon-shell-coated $CoSe_2$ particles to solve the problem of capacity loss caused by shortening the window. The doped product $CoM–Se_2@NC$, obtained by carbonizing a bimetallic zeolite imidazole skeleton (ZIFs) precursor, had more outstanding capacity than CoSe2@NC at 0.01~3.0 V. It was shown that the long cycle stability of $CoSe_2$ can be optimized effectively and the inefficient cycle caused by window adjustment can be eliminated under the synergistic action of the participation of another metal ion and the N-doped carbon shell.

Lu et al. [68] introduced Mo into a $CoSe_2$ self-supported nanosheet array (Mo-$CoSe_2$@NC) and investigated the performance of Mo in SIBs. In the XPS diagram (Figure 5a–b), the 3d spectrum of Mo consisted of peaks at 232.3 eV and 235.4 eV, associated with the characteristic spin-orbital bistates of Mo $3d^{5/2}$ and $3d^{3/2}$, respectively, which implies that Mo^{6+} ions enter the $CoSe_2$ lattice [69]. The embedded Mo defect in the $CoSe_2$ lattice not only produced more redox sites, but also induced the generation of mixed phases of o-$CoSe_2$ and c-$CoSe_2$, as evidenced by the coexistence of two hybrid phases of $CoSe_2$ in the SAED diagram (Figure 5d).

Wang [70] formed FeCo-Se in situ on a carbon cloth, and this multilayer array was able to provide sufficient active surfaces and shuttle channels for sodium ion exchange, thereby achieving excellent electrochemical kinetics. XRD (Figure 5g) proved that after the Co-Se was doped with Fe, impurities such as cobalt oxide, iron oxide, iron selenide, and FeCo alloy were not produced, implying that Fe ions only enter the Co-Se lattice. Notably, when more than 50% iron salt was added to the Co-Se material, the morphology of Fe-Se changed from a layered nanosheet array into pyramidal particles (Figure 5e–f), with a decline in electrochemical performance.

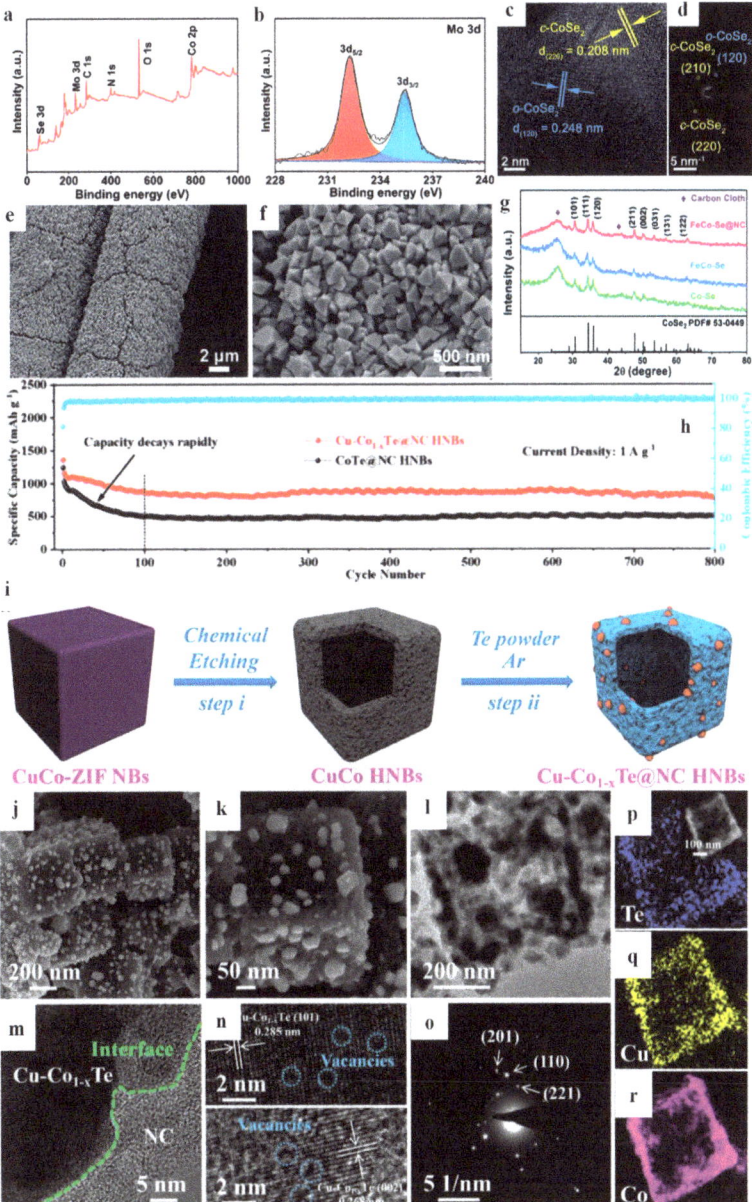

Figure 5. XPS of Mo–CoSe$_2$@NC with (**a**) survey and (**b**) Mo 3d. (**c**) HRTEM and (**d**) SAED; Ref. [68]. (**e**,**f**) Typical low– and high–magnification SEM (scanning electron microscope) images and (**g**) FeCo–Se XRD; Ref. [70]. (**h**) Li–ion batteries' performance. (**i**) Schematic representation of the synthesis of Cu–Co$_{1-x}$Te@NC HNBs with (**j**–**l**) SEM and TEM images, (**m**,**n**) HRTEM images, and (**o**) SAED pattern and (**p**–**r**) elemental mappings, Ref. [71].

3.1.2. Doping with Non-Metallic Elements

For cobalt telluride, Hu et al. [71] prepared Cu-doped cobalt telluride hollow carbon nanocells (Cu-Co$_{1-x}$Te@NC HNBs, Figure 5j–l) by chemically etching CuCo-ZIF nanocells and tellurizing them (Figure 5i), The copper–cobalt tellurium was homogeneously embedded in the nanobox (Figure 5p–r), and Co vacancies were constructed by copper induction (Figure 5n). At the same time, the interaction in the heterogeneous interface between nitrogen-doped carbon and cobalt telluride (Figure 5m) triggered the transfer of the p-band to interface charge transfer, so that the composite exhibited faster ion and electron diffusion kinetics than HNB electrodes, contributing to improved storage performance of lithium-ion batteries (Figure 5h).

Doping with non-metallic elements, especially chalcogenide elements, can not only produce polysulfides with alkali ions to improve the storage capacity, but also induce vacancy formation like metal elements to form a built-in electric field. Yu [34] reported an octahedral CoSe$_2$ sawtooth chain array threaded with N-doped carbon nanotubes (NCNTs) as a high-performance PIB anode. Highly conductive NCNTs form a flexible conductive network that greatly accelerates electron transfer and improves rate performance. Nitrogen-doped NCNTs also provide additional capacity to CS-NCNTs, constituting a versatile composite carbon material.

Wang [33] prepared Co$_{0.85}$Se$_{1-x}$S$_x$ nanoparticles by using S, which is of the same family as Se, as a point doping tool. The sample was formalized in a metal–organic framework, carbonized in graphene and, finally, cured in situ. Density functional theory (DFT) proved that S doping enhanced the adsorption energy of Co to alkali metal ions (Figure 6a–h). The hollow polyhedral skeleton of the double-carbon shell (Co$_{0.85}$Se$_{1-x}$S$_x$ @C/G) with a stable structure and S vacancies reduced the volume change, increased the number of electrochemical active sites, and improved the alkali-ion storage performance.

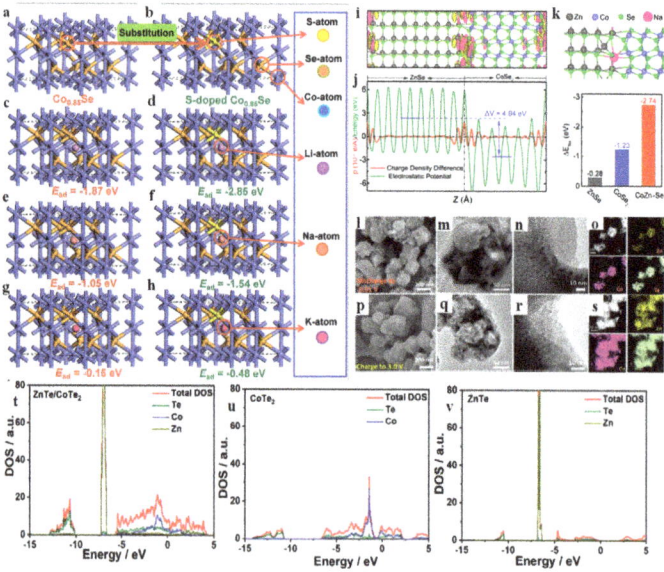

Figure 6. (**a–h**) DFT in crystal structures and alkali–ion adsorption sites for (**a,c,e,g**) Co$_{0.85}$Se and (**b,d,f,h**) S–doped Co$_{0.85}$Se; Ref. [33]. (**i**) The differential charge density between CoSe$_2$ and ZnSe in the phase boundary is calculated. (**j**) Averaged electrostatic potential. (**k**) The Na atom adsorption energy of ZnSe, CoSe$_2$, and CoZn–Se; Ref. [72]. (**l–o**) SEM, TEM, HRTEM, and EDS images, respectively, for a ZnTe/CoTe$_2$@NC anode at 0.01 V and (**p–s**) 3.0 V. (**t–v**) Total density of states (TDOS) for CoTe$_2$, ZnTe, and ZnTe/CoTe$_2$, respectively; Ref. [40].

Moreover, given the high conversion capacity of phosphorus, Ye et al. [73] reported in another work that P was successfully doped into $CoSe_2$. It was found through P 2p spectroscopy that additional Co-P and P-Se bonds could be generated by P doping, which could enhance the structural stability of $CoSe_2$. Xu [38] also used P to induce Te vacancies in o-$CoTe_2$ nanowires. Compared with h-$CoTe_2$, the Co 2p and Te 3d peaks of o-$CoTe_2$ showed significantly higher binding energy on the XPS images after P doping. EPR showed that the characteristic Te vacancy signal of o-P-$CoTe_2$/MXene was stronger and wider, indicating that the vacancy content increased significantly. Brunauer–Emmett–Teller (BET) theory showed that P doping increased the specific surface area. These results show that Te vacancy defects and P doping not only provide the active sites but also jointly enhance the structural stability of $CoTe_2$.

3.1.3. Heterojunctions

Heterostructure refers to the geometric structure and connecting interface formed by the combination of two or more materials through physical or chemical methods [74,75]. In the heterojunction, electrons will transition from a high Fermi level to a low Fermi level, resulting in potential difference due to interfacial polarization at the heterojunction [76] and the internal electric field. The existence of the internal electric field effect is helpful to accelerate the diffusion of metal ions and increase the adsorption energy of ions [77,78]. In addition, lattice distortion and charge redistribution taking place at heterogeneous boundaries can not only improve charge-transfer efficiency and conductivity, but also provide additional active sites for reversible redox reactions [79,80]. Therefore, due to the active role of the heterogeneous interface, heterogeneous anodes can obtain a stable nanostructure and excellent storage performance [81]. Therefore, in view of problems such as the low conductivity and large volume expansion of chalcogen-based metal materials, it is considered to be an effective strategy to construct high-quality heterogeneous structural interfaces to improve their storage performance [82].

Previously, Fang [72] reported a bimetallic selenide heterogeneous structure ($CoSe_2$/ZnSe@C) for SIBs. The phase-boundary charge redistribution of Co/Zn was studied by theoretical calculation. The Na^+ adsorption energy showed that the ZnSe side had a phase boundary with high electron density, which was more conducive to the adsorption of Na^+ ions, accelerating the ion movement and charge conduction, and achieving high reaction kinetics and high electrical conductivity (Figure 6i–k). In addition, the multistep conversion reaction in the Co/Zn heterostructure can effectively relieve the stress of Na^+ insertion. $CoSe_2$/ZnSe@C also showed good OER (oxygen evolution reaction) activity, providing additional evidence that the Co/Zn heterogeneous interface accelerated the redox reaction.

Xiao [83] recently also reported a similar application of heteroselenides in SIBs. He developed ZnSe/$CoSe_2$-CN heterostructures with Se vacancies via controllable in situ selenization. HETEM images showed a clear heterogeneous interface between ZnSe and $CoSe_2$, which can create electric fields to promote the movement of ions and electrons. Electron paramagnetic resonance (EPR) spectra showed an obvious signal caused by the selenium vacancy trapping electrons at g = 2.003, confirming the presence of the Se vacancy in the sample. The more induced active sites and the enhanced electronic conductivity introduced by the Se vacancy can improve the electrochemical properties of the materials.

Analogously, Zhang et al. [40] reported a template-assisted strategy to achieve in situ formation of a bimetallic tellurium heterostructure (ZnTe/$CoTe_2$@NC) on N-doped carbon shells. The results showed that both the electron-rich Te sites and the internal electric fields generated by electron transfer from ZnTe to $CoTe_2$ in Te heterojunctions provided rich cation-adsorption sites and promoted interfacial electron transport during the potassic/de-potassic process. Notably, as Zn led to ZIF-8 crystal formation, with the introduction of Zn^{2+} species into Co-ZIF-67, Zn^{2+} ions acted like an obstacle to Co-ZIF-67 crystals' nucleation and growth. Concretely, Zn delays the Co-ZIF-67 nucleation dynamics, thus resulting in relatively small doping particle sizes. The growth of ZIF-67 dominated by Co leads to the production of larger crystals in the reaction–diffusion process [84,85]. The finer ZnTe/CoTe$_2$ nanoparticles prepared by the heterojunction method showed higher structural stability. Ex situ SEM, HRTEM, and EDS were performed on the cycled ZnTe/CoTe$_2$@NC (Figure 6l–s). The ZnTe and CoTe$_2$ nanoparticles kept their original appearance, with no obvious particle aggregation. In addition, the TDOS results presented in Figure 6t–v highlight the discontinuous band gap of Fermi levels and confirm the semiconductor properties. In contrast, ZnTe/CoTe$_2$ has an obvious hybridization zone in the conduction band, resulting in significantly higher TDOS of ZnTe/CoTe$_2$ than that of both ZnTe and CoTe$_2$. This indicates that the electronic conductivity of ZnTe/CoTe$_2$ is significantly improved.

In addition to zinc–cobalt heterojunctions, He et al. [86] chose Ni to produce coral-like Ni$_x$Co$_{1-x}$Se$_2$ for the first time using a layered structure. The introduction of Ni formed a hierarchical structure to prevent structural fragmentation, accelerated the electron transmission, and shortened the Na$^+$ diffusion path. Zhu et al. [87] used the carbonization and subsequent selenization process design to encapsulate Ni$_3$Se$_4$@CoSe$_2$@C/CNTs in a core–shell three-dimensional interpenetrating two-carbon frame. Ni$_3$Se$_4$@CoSe$_2$ was evenly dispersed into a three-dimensional carbon skeleton structure/carbon nanotube network, greatly enhancing the conductivity and further achieving fast Na+ diffusion.

3.2. Structural Design

Nanomaterials, especially spherical and small nanocrystals, have their isotopic physical properties, large active surface, and high packing density [88–90]. More active sites can be exposed by delicate structures (e.g., sea urchin CoSe$_2$ [88], CoSe microspheres [91]). By using carbon coatings, CoSe$_x$ and CoTe$_x$ can be well maintained in nanoform and improved in terms of cycling stability by using constraint effects (such as CoSe-C@C [92] and o/h-CoTe$_2$ [93]). Combination with structural carbon can produce interfacial interactions, which can effectively improve diffusion kinetics, electrical conductivity, and structural compatibility, thus achieving excellent electrochemical performance [10,94]. For example, three-dimensional conductive carbon network structures with large channels can provide diffusion paths, improve the migration rate of ions in the battery material, and reduce the limitation of poor conductivity of CoSe$_x$ and CoTe$_x$ [66]. Metal–organic frameworks (MOFs) possess exceptionally high surface area and porosity, allowing for the adsorption and storage of gases, liquids, and even small molecules. This property makes them promising for applications such as energy storage, separation, and catalysis. Moreover, by optimizing the composite structure, the electrochemical properties of the composites can be further improved, and the volume strain of the active nanomaterials in the carbon matrix can be reduced. Therefore, carbon materials such as reduced graphene oxide (rGO) and MXenes are not only electrically conductive but also have good ductility, which can greatly enhance the structural compatibility of CoSe$_x$ and CoTe$_x$ anode materials.

However, MOFs, MXenes, and other carbon-based materials have their limitations. For example, MOFs are prone to acid–base interference and structural collapse. MXene etching processes can be complex and hazardous. Additionally, low energy density is a common drawback of these materials. Therefore, recently, researchers have been consolidating the carbon-based structure of transition-metal-based MOFs through annealing and synthesizing high-performance ion battery anodes using vapor-phase methods [95]. High-capacity ion battery anode materials can be obtained through green etching of MXenes using only grinding and annealing via molten salt methods [96]. These structural design approaches are frequently employed in the design of cobalt selenide and cobalt telluride anodes. The latest progress in combining the abovementioned methods with cobalt selenide and cobalt telluride for alkali-metal-ion battery cathodes is summarized below.

3.2.1. Freestanding Electrodes

Common electrodes are made of an active material composited with a binder and conductive carbon on the current collector, and the presence of the binder and collector fluid especially affects the energy density. Therefore, the development of freestanding electrode structures to increase the proportion of active materials and eliminate the negative effects of the binder is a promising direction [97].

Electrospinning is used to produce individual electrode materials, allowing the active material to be embedded in a network of carbon fibers. The carbonized material can be cut so as to be used as an electrode, avoiding mechanical losses caused by mixing adhesives with conductive materials and coatings [98,99]. Self-supported electrodes can alleviate electrode pulverization and construct interwoven electron/ion transport networks, thus improving electrochemical energy storage performance.

Zhan et al. [47] innovatively prepared a self-supported cobalt telluride electrode for SIBs. They synthesized CoTe$_2$@NMCNFs via electrospinning and subsequent in situ tellurization (Figure 7a–f). Material characterizations revealed NMCNFs with N-doped active sites and porous carbon networks with high specific surface area. These positive effects promote electrolyte penetration and mitigate the aggregation effect in the CoTe$_2$ nanoparticles by reducing the volume changes during the cycle (Figure 7g–j). The long cycle of CoTe$_2$@NMCNFs shows a high reversible Na$^+$ storage capacity, long-term cycle stability, and high rate capacity. In general, the overall conductivity of the electrode is greatly improved, and Na$^+$ diffusion is promoted, thanks to the strong three-dimensional crosslinked composite fiber and the absence of voluminous inactive components.

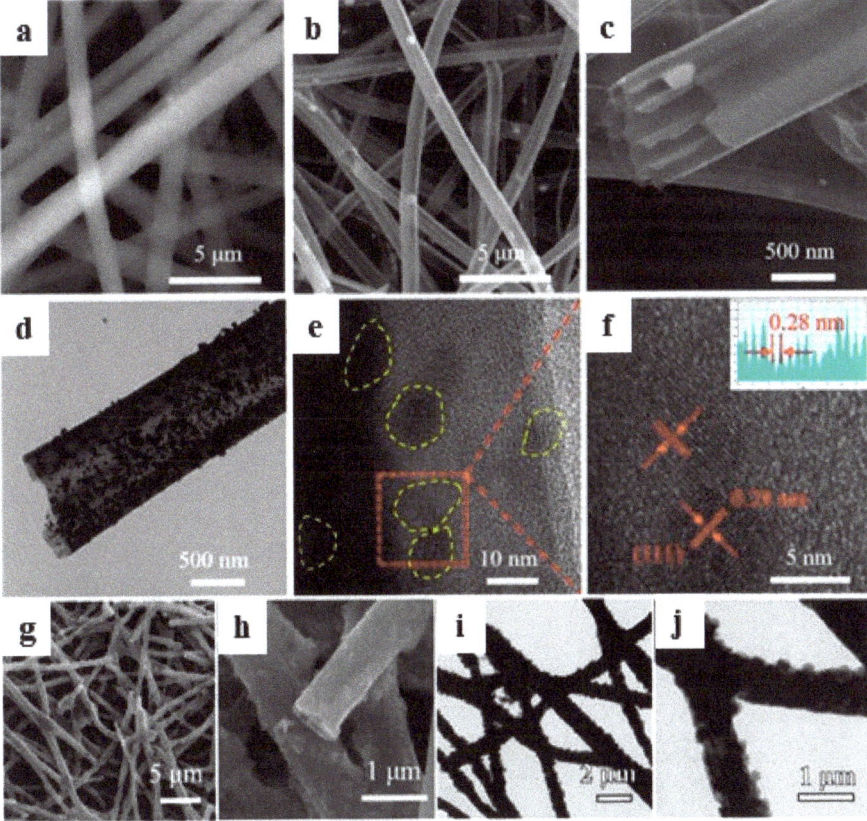

Figure 7. (**a**–**c**) SEM and (**d**–**f**) TEM images of PAN/PS/Co(ac)$_2$·4H$_2$O. (**g**,**h**) SEM and (**i**,**j**) TEM images of CoTe$_2$@NMCNFs after 300 cycles at 200 mA g^{-1}. Ref. [47].

Meanwhile, Park et al. [100] prepared cobalt selenite (p-CoSeO$_3$-CNF) embedded in carbon composite nanofibers by electrospinning. Among them, the heat-treated nanofibers containing cobalt and selenium formed CoSe$_2$ nanocluster intermediates. Then, through the final oxidation step, CoSe$_2$ was converted to amorphous CoSeO$_3$. Part of the carbon fiber was changed into CO$_2$ gas during the oxidation process, leaving pores and forming mesopores in the nanofibers. The carbon and pores limited the volume change of CoSeO$_3$ during repeated sodification/desodification processes, allowing the electrolyte to permeate easily and reducing the diffusion path of sodium ions. The large network carbon structure accelerated electron transport, and the high performance of p-CoSeO$_3$-CNF was demonstrated. Electrospinning has also been reported for SIB and LIB anode materials other than CoSe$_2$ [32,101].

3.2.2. Carbon-Cloth-Based Electrodes

Commercially available carbon cloth (CC) material, consisting of hundreds of fibers woven tightly together, ensures that the fibers are interconnected and can be rolled into different shapes without compromising their integrity [102]. The stable cycling performance of CC-based electrodes makes them a competitive option for industrial applications of metal-ion batteries [103].

According to the different raw materials, carbon fiber is mainly divided into three substrates: asphalt-based carbon fiber, PAN-based carbon fiber, and viscose-based carbon fiber. PAN-based carbon fiber is rich in raw materials and superior to the other two types. Therefore, PAN-based carbon fiber is the most widely used, accounting for more than 90% [104]. The difference between the growth of active substances in PAN-based carbon cloth and the embedding of active substances in electrospinning is that carbon cloth generally penetrates active substances into existing carbon substrates via infiltration methods, hydrothermal methods, and other methods. However, electrospinning uses PAN, solvent, and metal salt to dissolve directly into the active material. Both have their advantages, such as the toughness of carbon cloth and the advantage of electrospinning for in situ doping treatments, which can enhance the synergistic effects of carbon nanowires and active substances [105].

To date, carbon cloth combined with active substances has been used in fuel cells [106], catalysis [107], lithium–sulfur batteries [108,109], supercapacitors [102], etc. There have also been some studies on carbon cloth combined with metallic chalcogenide compounds as negative electrodes for LIBs [110]. For cobalt selenide, Wang et al. [70] grew FeCoSe on carbon cloth, and Lu et al. [48] synthesized Mo-doped CoSe$_2$ nanosheets, proving that the carbon cloth and cobalt selenide have a good synergistic effect and are more conducive to the formation of bimetallic synergies (Figure 8). However, cobalt telluride combined with carbon cloth as a negative electrode for LIBs, SIBs, and PIBs has rarely been reported.

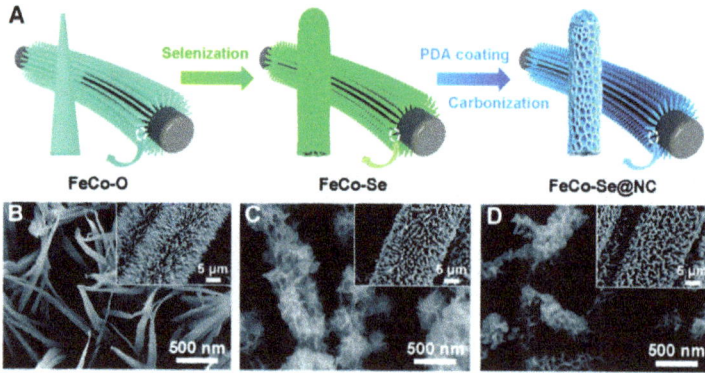

Figure 8. (**A**) Synthesis of FeCo–Se@NC and (**B**–**D**) SEM images of the material in each step; Ref. [68].

3.2.3. MOF-Based Electrodes

MOFs with a high specific surface area and adjustable pore structure are widely used in the fields of calcification and energy conversion due to their stable structure, large specific surface area, and diverse functions [95,111,112]. Under appropriate conditions, MOFs can be used as precursors and/or templates to derive metal chalcogenides with regular topography and excellent electrochemical properties (Figure 9) [113–115]. Notably, MOFs have proven to be ideal templates for the preparation of transition metal selenides and tellurides. In recent reports, it has been shown that transition metal selenide and telluride nanocrystals can be embedded 1–3D porous carbon substrates to form efficient conductive networks [116]. In addition, the relatively high surface area and large pores of the high-dimensional conductive substrate enhance the wettability of the electrolyte to the material and promote ion diffusion, thereby improving the storage performance [117]. Therefore, MOFs as a template for the preparation of nanoscale electroactive materials and porous carbon matrix hybrid structures have broad application prospects.

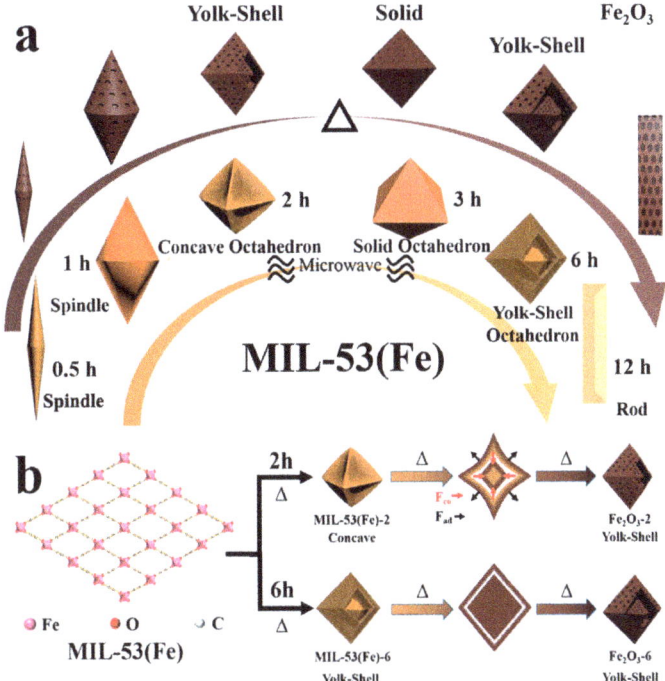

Figure 9. Transition metals have various valence states and can form various complex structures, such as Fe–based MIL–53 with yolk–shell octahedron morphologies: (**a**) process of MIL–53(Fe) growth and (**b**) the formation process of Fe_2O_3–2 and Fe_2O_3–6; Ref. [115].

More recently, Xiao [39] prepared hollow In_2Se_3/$CoSe_2$ nanorods by growing Co-ZIF-67 on the surface of In-MIL-68 and via in situ selenization. During selenization, selenium atoms penetrate from the surface of the nanorod to the inside, while indium atoms diffuse outward in opposite directions and at different speeds, creating the Hilken–Dahl effect and forming hollow structures [118]. At the same time, the mass ratio of ZIF-67 to MIL-68 grown on the surface was studied by changing the concentration of cobalt nitrate. The results showed that the ZIF-67 nanoparticles became smaller and could be uniformly deposited on the surface of MIL-68 with the decrease in the concentration of cobalt nitrate in solution (Figure 10).

Figure 10. SEM images of In-MOF: Co $(NO_3)_2$ = (**a**) 1:8, (**b**) 1:6, (**c**) 1:4, and (**d**) 1:3; Ref. [39].

Wang et al. [119] prepared cobalt selenide ($CoSe_2$-CNS) anchored on carbon nanosheets by selenizing cobalt embedded in 1D-MOF carbon nanosheets (Co-CNS). $CoSe_2$ grown in situ on carbon nanosheets showed close contact between $CoSe_2$ and carbon. At the same time, $CoSe_2$-CNS also has typical two-dimensional structural characteristics, which are conducive to electrolyte penetration and ion/electron transport. In addition, the carbon matrix can alleviate the volume expansion of $CoSe_2$ during the desodification process. As the result of these advantages, $CoSe_2$-CNS provides a stable cycle capacity of 468 mA h g^{-1}, together with a high rate capability of 352 mA h g^{-1} at 10,000 mA g^{-1}.

3.2.4. MXene-Based Electrodes

MXenes, which include transition metal carbides or carbon nitride, have proven to be novel materials with great potential for energy storage [120,121]. MXenes have high electrical conductivity, large interlayer spacing, and excellent mechanical properties, enabling them to act as a fluid collector and buffer deformation at the same time [122]. Their electrical and electronic properties can be customized through modifications to the MXenes' functional groups, solid solution structures, or stoichiometry to meet the needs of different scenarios [123]. As a result, titanium carbide has demonstrated outstanding catalytic activity and selectivity as a co-catalyst in various catalytic reactions, finding wide applications in energy conversion, environmental protection, and organic synthesis [124].

Xu [38] used P-induced o-$CoTe_2$ nanowires to anchor MXene nanosheets (o-P-$CoTe_2$/MXene). The elastic MXene formed a tight interfacial interaction with P-$CoTe_2$ (Figure 11). DFT calculations further showed that the new superstructure had higher electronic conductivity, enhanced the adsorption capacity of K^+ ions, and reduced the energy barrier to ion diffusion. Meanwhile, the entire battery was cycled through with negligible capacity loss even after bending. This proves that the combination with MXenes can greatly improve the stability of the material. Hong [91] found a strong Co-O-Ti covalent interaction in

CoSe$_2$@MXene. It assembled the inner hollow CoSe$_2$ microsphere and the outer MXene via electrostatic self-assembly. This covalent bond facilitates electron/ion transport and enhances the structural durability of the CoSe$_2$@MXene hybrid, thus improving the rate performance and cyclic stability. It has been reported that the polar surface of MXenes can inhibit the polysulfides' shuttle effect [125,126]. Wang [127] was therefore inspired to grow CoSe$_2$ nanorods in situ on the surface of an MXene via a simple one-step hydrothermal method. Importantly, the stability of the CoSe$_2$/MXene cycle in this study was better than that of the pure CoSe$_2$ cycle, demonstrating that Ti$_3$C$_2$T$_x$ can inhibit the undervoltage failure caused by Na$_x$Se instability.

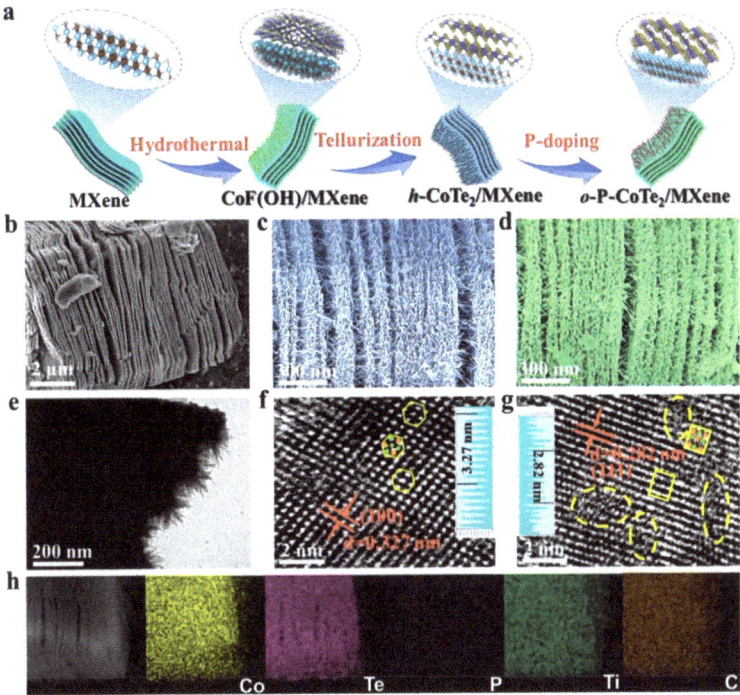

Figure 11. (a) Preparation process. SEM images of (b) MXene, (c) h–CoTe$_2$/MXene, and (d) o–P–CoTe$_2$/MXene. (e,g) TEM and HRTEM images of o–P–CoTe$_2$/MXene; (f) HRTEM of h–CoTe$_2$/MXene. (h) EDS spectrum of o–P–CoTe$_2$/MXene; Ref. [38].

3.3. Summary of Electrode Design

Cobalt selenide, as a high-capacity conversion anode, has been widely studied in alkali-metal-ion batteries. The new electrode design methods, such as doping with composite conductive materials (e.g., MXenes), can show excellent high capacity and high stability. However, the wide application and innovation of cobalt selenide in potassium-ion batteries still need to be improved. Cobalt telluride, as a compound of the same series, is still in its infancy in the design of advanced anode materials. Some novel designs of cobalt telluride show excellent electrochemical performance, which indicates the great potential of cobalt telluride. However, innovative designs of cobalt selenide are still to be developed; for example, co-doping with multiple non-metallic elements, carbon cloth growth, and electrospinning in situ growth methods have not been tried. At the same time, the electrochemical reaction mechanism of various synergistic effects is still unclear, and products with ultra-long periods of stability are still to be developed.

4. Electrode Synthesis

In addition to the elemental composition of the material, the morphology, crystal phase, and structure of the material also have a great influence on its properties. The microstructure of materials, such as their size and morphology, will affect their electron energy density, as well as influencing their optical and electrical properties [90,107,108,128]. Therefore, choosing appropriate preparation methods to synthesize materials with special structures is the key to improving the properties of the materials. Conventional preparation methods of cobalt-based anode materials include, but are not limited to, hydrothermal/solvothermal methods [32,101,113,116], heat treatment methods [52,93,129], sol–gel methods [130], vapor deposition methods [34], and electrospinning [100]. At the same time, the usual method for preparing selenides and tellurides is the gas-phase method [34].

Hydrothermal reaction is a typical synthesis route. Under certain temperature and high-pressure conditions, cobalt-based materials with high phase purity and crystallinity can be synthesized. For example, Lei et al. [131] synthesized uniform $CoTe_2$ nanorods by the hydrothermal method, without any surfactant or mineralizer for 20 h, using ascorbic acid as a reducing agent. The study found that simply adjusting the concentration of sodium hydroxide can produce different forms, including short rod-shaped bodies with many spines and flower-like hierarchies (Figure 12a–f). Hydrothermal methods are widely used in the synthesis of various materials, due to their advantages of fewer experimental steps, changeable morphology, and simple operation. However, hydrothermal methods also have some shortcomings, such as too many uncontrollable factors, easily leading to an uneven appearance, and a generally long hydrothermal reaction time. As a high-temperature treatment, the vapor-phase method can be used alone or in combination with a template method for intermediate processes such as selenides and tellurides. Cobalt-based electrode materials with an abundant pore structure and high electrochemical performance can be obtained by heat treatment. Heat treatment is almost a necessary experimental method for MOF-series materials, but this method has high energy consumption, and for MOF materials the sintering temperature is too high, which can easily lead to the collapse of the structure. The above methods are conducive to the formation of high-functional structures for $CoSe_x/CoTe_x$, but high-temperature pyrolysis is always necessary in the process of in situ carbonization. To achieve energy savings and high yields, it is urgent to explore more effective modification or combination methods such as magnetron sputtering [132], microwave [133], and molten salt etching [96,134] methods. The next section summarizes the current $CoSe_x/CoTe_x$ anode synthesis methods for LIBs, SIBs, and PIBs.

Figure 12. (a–f) SEM for $CoTe_2$ synthesized with different NaOH concentrations: 0.0 M, 0.2 M, 0.5 M, 1.2 M, 1.5 M, and 2.0 M. Ref. [131].

5. Performance Summary

This section focuses on the latest advances in the synthesis of various cobalt-based selenides and tellurides for LIBs, SIBs, and PIBs, and it summarizes the key properties and synthesis methods of various heterostructured anodes, as shown in the following Tables 1 and 2.

Table 1. Review of the properties of $CoSe_x$ electrode materials.

Sample Name	Synthesis Method	Structure	Cycle Performance (mAh g^{-1}, A g^{-1})	Rate Performance (mAh g^{-1}, A g^{-1})	Initial Coulombic Efficiency (%, A g^{-1})	Battery Type	Ref./Year
CoSe$_2$/carbon nanoboxes	Annealing	Nanoboxes	860/0.2/100 th 660/1/100 th	686/2	78.3%/0.2	LIB	[135]/2016
CoSe@NCNTs	Annealing	Nanowires	326/0.5/100 th	278/3	95%/0.1	PIB	[18]/2021
CoSe$_2$-CNS	Annealing	Nanosheets	~250/10/2000 th	352/10		SIB	[103]/2019
CoSe$_2$@NC/CNTs	Annealing	Nanowires	480/1/200 th 369/10/800 th	360/10		SIB	[136]/2022
Mo–CoSe$_2$@NC	Electrodeposition + selenization	Self-supporting nanosheet	672/0.5/200 th	313/5	47.9%/0.5	SIB	[68]/2022
In$_2$Se$_3$/CoSe$_2$–450	Annealing	MIL-68@ZIF-67 core–shell	445.0/0.5/200 th 297.5/5/2000 th 205.5/10/2000 th	371.6/20	61.2%/0.1	SIB	[39]/2021
TNC–CoSe$_2$	Precipitation + sol–gel + annealing	Microcubes	511.2/0.2/200 th	464/6.4	88.26%/0.2	SIB	[130]/2021
ZnSe/CoSe$_2$–CN	Annealing	Spherical particles	547.1/0.5/300 th 422.6/3/300 th	362.1/20	97.7%/0.1	SIB	[83]/2022
CoSe$_2$@MXene	Annealing	Porous hollow shell encapsulated by MXene nanosheets	910/0.2/100 th 1279/1/1000 th	465/5	71.9%/0.2	LIB	[91]/2020
Ni$_{0.47}$Co$_{0.53}$Se$_2$	Solvothermal	Coral-like	321/2/2000 th	277/15	98.5%/0.1	SIB	[86]/2019
p-CoSeO$_3$–CNF	Electrospun + annealing	Composite porous nanofibers	288/0.2/200 th	207/5	56.7%/0.5	SIB	[100]/2021
CoSe$_2$/CoSe	Annealing	Hierarchical hollow raspberry-like superstructure	1361/1/1000 th 579/2/2000 th 315/5/1000 th	406/5		LIB	[54]/2019
Ni$_{0.33}$Co$_{0.67}$Se$_2$	Solvothermal selenization	Hierarchical mesoporous nanospheres	301.9/1/1300 th	314.5/8	85.4%/0.5	SIB	[137]/2023
CoSe$_2$/ZnSe	Annealing	Bimetallic heterostructure	~250/8/4000 th ~200/10/4000 th	263/10	72.3%/0.1	SIB	[72]/2019
CoSSe@C/G	Solvothermal + annealing	Double-carbon shells	636.2/2/1400 th 353/1/200 th 208.1/0.2/100 th	254.5/10 266.6/10 195.7/2	68.09%/0.1 71.2%/0.1 48.3%/0.1	LIB SIB PIB	[33]/2020
NCNF@CS	Chemical vapor deposition + solvothermal	Octahedral threaded N-doped carbon nanotubes	253/0.2/100 th 173/2/600 th	196/2	69.3%/0.2	PIB	[34]/2018
3DG/CoSe$_2$@CNWs	Hydrogel	Multidimensional porous nanoarchitecture	543/0.1/100 302/2/500 th	~320/2	63.5%/0.1	SIB	[30]/2022

Table 1. Cont.

Sample Name	Synthesis Method	Structure	Cycle Performance (mAh g^{-1}, A g^{-1})	Rate Performance (mAh g^{-1}, A g^{-1})	Initial Coulombic Efficiency (%, A g^{-1})	Battery Type	Ref./Year
CoSe$_2$@NC/rGO-5	Two-step co-precipitation+ pyrolysis	Sandwich-like	527.5/0.1/100 th 226/0.5/400 th	206/5 157/10	72.6%/0.1	PIB	[138]/2021

Table 2. Review of the properties of CoTe$_x$ electrode materials.

Sample	Synthesis Method	Structure	Cycle Performance ^{x}	Rate Performance (mAh g^{-1}, A g^{-1})	Initial Coulombic Efficiency (%, A g^{-1})	Battery Type	Ref./Year
CoTe$_2$–C	Annealing	Polyhedron	500/0.1/200 th 480/1/200 th	386/3	70.52%/0.1 56.07%/0.1	LIB SIB	[52]/2020
CoTe$_2$@3DG	Annealing	Polyhedron + rGO	191/0.05/350 th 103/1/4500 th	169/2	66.7%/0.05	SIB	[139]/2023
CoTe$_2$@NMCNFs	Electrospinning + annealing	Carbon fiber network	261.2/0.2/300 th 143.5/2/4000 th	152.4/10	57.1%/0.2	SIB	[47]/2021
CoTe$_2$@3DPNC	Annealing	Dual-type carbon	216.5/0.2/200 th	164.2/5	87.6%/0.2	SIB	[140]/2022
o–P–CoTe$_2$/MXene	Solvothermal annealing	MXene	373.7/0.2/200 th 232.3/2/2000 th	168.2/20	72.9%/0.05	PIB	[38]/2021
CTNRs/rGO	One-pot solvothermal	Nanorods/rGO	306/0.05/100 th	176/2	56%/0.05	SIB	[141]/2019
CoTe$_2$–C	Spray pyrolysis	Microsphere	295.8/0.2/100 th	163.7/2	69.2%/0.2	PIB	[142]/2020
Cu–Co$_{1-x}$Se$_2$@NC	Chemical etching tellurization	Hollow nanoboxes	796/1/800 th	~400/5	~80%/1	LIB	[71]/2022
CoTe$_2$/G	One-pot solvothermal	Nanosheets	356/0.05/100 th	246/5	78.1%/0.05	SIB	[42]/2018
ZnTe/CoTe$_2$@NC	Annealing	Bimetallic heterostructure	317.5/0.5/1000 th 254.5/2/2000 th 165.2/5/5000 th	190.2/5	58.3%/0.1	PIB	[40]/2022
o/h–CoTe$_2$	One-step hydrothermal	Submicron-sized rods	807/0.12/200 th 438/0.6/400 th	305.8/3	75.2%/0.12	LIB	[93]/2023
CoTe@NCD	Simultaneous pyrolysis–tellurium melt impregnation	Carbon dodecahedra	300/0.1/100 th 200/0.1/100 th	207/2 58/2	57%/0.1 50%/0.05	SIB PIB	[129]/2022
CoTe$_2$@NPC-NFs@NC	Electrospinning	N-doped porous carbon nanofibers	409.1/0.05/50 th 198/0.5/600 th 120/2/1000 th	148.9/2	57.2%/0.05	PIB	[41]/2023

In general, cobalt selenide has been shown to be a high-energy anode for LIBs, SIBs, and PIBs, which greatly improves their electrochemical performance through independent electrodes without binders and with innovative heterojunction design. The reaction mechanism has also been studied by theoretical calculation and in situ techniques. As a potential electrode with high area capacity, cobalt telluride has many surprising applications in PIBs, and some electrode design methods that have proven excellent for cobalt selenide, such as heterojunctions and composite MXenes, have also been efficiently practiced for CoTe. However, compared with the metal oxides, sulfides, or selenides, the existing modification strategies are still insufficient. In addition, electrochemical performance can be enhanced by electrode synthesis and preparation. For example, studies on the compatibility of cobalt telluride with electrolytes and binders are still lacking. For practical applications, in future, more investigations and full battery evaluation should be emphasized.

6. Summary and Expectations

Cobalt selenide and cobalt telluride show promising applications in LIBs, SIBs, and PIBs due to their unique crystal structures, diverse structural designs, and high theoretical capacities. In this review, the progress of CoSe and CoTe was comprehensively summarized, not only focusing on the running mechanism and electrochemical performance, but also putting forward challenges to the synthesis strategies. The mechanism of Co-based chalcogenides in energy storage reactions is complex because it may have mechanisms for multilevel transformation. Often, conversion reactions are indispensable, which can lead to high weight capacities. However, disadvantages remain, such as large volume expansion and sluggish kinetics. In response to this series of problems, effective improvement strategies have been reported, including the design of nanocomposite carbon materials and the introduction of defects. Firstly, 0–3D conductive carbon nanocomposites can improve conductivity, create abundant active sites, and enhance electrolyte penetration. In addition, the introduced vacancies and heterogeneous interfaces can accelerate the conduction of ions and electrons. Tables 1 and 2 summarize the latest modified properties and synthesis methods, which can provide a reference point for innovative strategies for cobalt selenide and cobalt telluride materials. Nevertheless, the application of cobalt selenide and cobalt telluride anode materials in alkali-metal-ion batteries still requires extensive efforts. Therefore, we propose some prospective directions to promote further research to enable cobalt selenide and cobalt telluride to achieve higher energy storage efficiency.

6.1. Replenishing Existing Strategies

Heteroatom doping is an effective strategy to increase the numbers of ion storage locations and ion mobility channels, thereby facilitating faster electron transport. However, it is often reported that only one metallic element or non-metallic element (such as the common elements S and P) is selected to be mixed into cobalt selenide or cobalt telluride. At the same time, to improve the conductivity and active sites of $CoSe_x$ and $CoTe_x$, the selection of composite carbon materials has been limited to traditional carbon materials, such as rGO, dopamine, super-C, or metal–organic frameworks. In addition, many reports testify that MXenes have the advantages of large interlayer spacing, good conductivity, and heteroatom doping. In summary, the new strategy of multi-element doping and synthesis and the novel MXene composite method can improve the performance of cobalt telluride and cobalt selenide as anodes for alkali-ion batteries.

6.2. Expanded CBs' Anode Material Synthesis Method

Compared to CoO_x and CoS_x, the available modification strategies are still limited. Currently, constructed multilayer, porous, or three-dimensional mesh structures can provide short transmission channels and large specific surface areas with active sites. This helps to improve material toughness and electrolyte contact, thus improving long-cycle performance. Relevant synthesis methods have been reported for $CoSe_x$, but less often for $CoTe_x$. Based on this, we should further explore the latter's electrochemical performance in LIBs/SIBs/PIBs. For example, three-dimensional nanonetworks of $CoTe_x$ based on hydrogels may be a good synthesis method. Furthermore, binder-free electrodes for $CoTe_x$ are rarely reported. Independent electrodes can be prepared via carbon cloth chemical deposition and electrospinning to avoid the obstruction of poor-conducting binders and low-capacity conductive agents, thus saving resources, reducing impurities, and improving the energy density.

Author Contributions: Writing—review and editing, Y.Z.; supervision, Z.S. and D.Q.; resources, D.H. and L.N. All authors have read and agreed to the published version of the manuscript.

Funding: This work was financially supported by the National Natural Science Foundation of China (22204028, 22104021, 22204159), Young Talent Support Project of Guangzhou Association for Science and Technology (QT-2023-003), Guangdong Basic and Applied Basic Research Fund Project (2022A1515110451), Guangzhou University Graduate Student Innovation Ability Cultivation Funding Program (2022GDJC-M06), and Science and Technology Projects in Guangzhou (202201010245, 2023A03J0029).

Institutional Review Board Statement: Not applicable.

Informed Consent Statement: Not applicable.

Data Availability Statement: Not applicable.

Conflicts of Interest: The authors declare no conflict of interest.

References

1. Grey, C.P.; Hall, D.S. Prospects for Lithium-Ion Batteries and beyond—A 2030 Vision. *Nat. Commun.* **2020**, *11*, 6279. [CrossRef] [PubMed]
2. Huy, V.P.H.; So, S.; Kim, I.T.; Hur, J. Self-Healing Gallium Phosphide Embedded in a Hybrid Matrix for High-Performance Li-Ion Batteries. *Energy Storage Mater.* **2021**, *34*, 669–681. [CrossRef]
3. Zhang, X.; He, Q.; Xu, X.; Xiong, T.; Xiao, Z.; Meng, J.; Wang, X.; Wu, L.; Chen, J.; Mai, L. Insights into the Storage Mechanism of Layered VS_2 Cathode in Alkali Metal-Ion Batteries. *Adv. Energy Mater.* **2020**, *10*, 1904118. [CrossRef]
4. Sun, Y.; Wang, H.; Wei, W.; Zheng, Y.; Tao, L.; Wang, Y.; Huang, M.; Shi, J.; Shi, Z.C.; Mitlin, D. Sulfur-Rich Graphene Nanoboxes with Ultra-High Potassiation Capacity at Fast Charge: Storage Mechanisms and Device Performance. *ACS Nano* **2021**, *15*, 1652–1665. [CrossRef]
5. Hao, Q.; Jia, G.; Wei, W.; Vinu, A.; Wang, Y.; Arandiyan, H.; Ni, B.J. Graphitic Carbon Nitride with Different Dimensionalities for Energy and Environmental Applications. *Nano Res.* **2020**, *13*, 18–37. [CrossRef]
6. Kim, H.; Kim, M.J.; Yoon, Y.H.; Nguyen, Q.H.; Kim, I.T.; Hur, J.; Lee, S.G. Sb_2Te_3–TiC–C Nanocomposites for the High-Performance Anode in Lithium-Ion Batteries. *Electrochim. Acta* **2019**, *293*, 8–18. [CrossRef]
7. Xie, H.; Kalisvaart, W.P.; Olsen, B.C.; Luber, E.J.; Mitlin, D.; Buriak, J.M. Sn–Bi–Sb Alloys as Anode Materials for Sodium Ion Batteries. *J. Mater. Chem. A Mater.* **2017**, *5*, 9661–9670. [CrossRef]
8. Zhou, Y.; Sun, W.; Rui, X.; Zhou, Y.; Ng, W.J.; Yan, Q.; Fong, E. Biochemistry-Derived Porous Carbon-Encapsulated Metal Oxide Nanocrystals for Enhanced Sodium Storage. *Nano Energy* **2016**, *21*, 71–79. [CrossRef]
9. Xiao, Y.; Hwang, J.Y.; Belharouak, I.; Sun, Y.K. Na Storage Capability Investigation of a Carbon Nanotube-Encapsulated $Fe_{1-x}S$ Composite. *ACS Energy Lett.* **2017**, *2*, 364–372. [CrossRef]
10. Zhao, Y.; Wang, L.P.; Sougrati, M.T.; Feng, Z.; Leconte, Y.; Fisher, A.; Srinivasan, M.; Xu, Z. A Review on Design Strategies for Carbon Based Metal Oxides and Sulfides Nanocomposites for High Performance Li and Na Ion Battery Anodes. *Adv. Energy Mater.* **2017**, *7*, 1601424. [CrossRef]
11. Hartmann, F.; Etter, M.; Cibin, G.; Liers, L.; Terraschke, H.; Bensch, W. Superior Sodium Storage Properties in the Anode Material $NiCr_2S_4$ for Sodium-Ion Batteries: An X-ray Diffraction, Pair Distribution Function, and X-ray Absorption Study Reveals a Conversion Mechanism via Nickel Extrusion. *Adv. Mater.* **2021**, *33*, 2101576. [CrossRef] [PubMed]
12. Park, C.; Samuel, E.; Joshi, B.; Kim, T.; Aldalbahi, A.; El-Newehy, M.; Yoon, W.Y.; Yoon, S.S. Supersonically Sprayed Fe_2O_3/C/CNT Composites for Highly Stable Li-Ion Battery Anodes. *Chem. Eng. J.* **2020**, *395*, 125018. [CrossRef]
13. Ni, J.; Sun, M.; Li, L. Highly Efficient Sodium Storage in Iron Oxide Nanotube Arrays Enabled by Built-In Electric Field. *Adv. Mater.* **2019**, *31*, e1902603. [CrossRef] [PubMed]
14. Wang, X.; Liu, Y.; Han, H.; Zhao, Y.; Ma, W.; Sun, H. Polyaniline Coated Fe_3O_4 Hollow Nanospheres as Anode Materials for Lithium Ion Batteries. *Sustain. Energy Fuels* **2017**, *1*, 915–922. [CrossRef]
15. Xiao, Y.; Miao, Y.; Hu, S.; Gong, F.; Yu, Q.; Zhou, L.; Chen, S. Structural Stability Boosted in 3D Carbon-Free Iron Selenide through Engineering Heterointerfaces with Se-P Bonds for Appealing Na^+-Storage. *Adv. Funct. Mater.* **2022**, *33*, 2210042. [CrossRef]
16. Sun, Z.; Shi, W.; Chen, J.; Gu, Z.; Lai, S.; Gan, S.; Xie, J.; Li, Q.; Qu, D.; Wu, X.L.; et al. Engineering Honeycomb-like Carbon Nanosheets Encapsulated Iron Chalcogenides: Superior Cyclability and Rate Capability for Sodium Ion Half/Full Batteries. *Electrochim. Acta* **2022**, *431*, 141084. [CrossRef]
17. Sun, Z.; Wu, X.; Gu, Z.; Han, P.; Zhao, B.; Qu, D.; Gao, L.; Liu, Z.; Han, D.; Niu, L. Rationally Designed Nitrogen-Doped Yolk-Shell Fe_7Se_8/Carbon Nanoboxes with Enhanced Sodium Storage in Half/Full Cells. *Carbon* **2020**, *166*, 175–182. [CrossRef]
18. Liu, Y.; Deng, Q.; Li, Y.; Li, Y.; Zhong, W.; Hu, Y.; Ji, X.; Yang, C.; Lin, Z.; Huang, K. CoSe@N-Doped Carbon Nanotubes as a Potassium-Ion Battery Anode with High Initial Coulombic Efficiency and Superior Capacity Retention. *ACS Nano* **2021**, *15*, 1121–1132. [CrossRef]
19. Li, X.; Han, Z.; Yang, W.; Li, Q.; Li, H.; Xu, J.; Li, H.; Liu, B.; Zhao, H.; Li, S.; et al. 3D Ordered Porous Hybrid of ZnSe/N-Doped Carbon with Anomalously High Na^+ Mobility and Ultrathin Solid Electrolyte Interphase for Sodium-Ion Batteries. *Adv. Funct. Mater.* **2021**, *31*, 2106194. [CrossRef]
20. Ge, J.M.; Fan, L.; Wang, J.; Zhang, Q.; Liu, Z.; Zhang, E.; Liu, Q.; Yu, X.; Lu, B. $MoSe_2$/N-Doped Carbon as Anodes for Potassium-Ion Batteries. *Adv. Energy Mater.* **2018**, *8*, 1801477. [CrossRef]
21. Liu, Y.; Yang, C.; Li, Y.; Zheng, F.; Li, Y.; Deng, Q.; Zhong, W.; Wang, G.; Liu, T. $FeSe_2$/Nitrogen-Doped Carbon as Anode Material for Potassium-Ion Batteries. *Chem. Eng. J.* **2020**, *393*, 124590. [CrossRef]

22. Sun, D.; Liu, S.; Zhang, G.; Zhou, J. NiTe$_2$/N-Doped Graphitic Carbon Nanosheets Derived from Ni-Hexamine Coordination Frameworks for Na-Ion Storage. *Chem. Eng. J.* **2019**, *359*, 1659–1667. [CrossRef]
23. Park, G.D.; Kang, Y.C. Conversion Reaction Mechanism for Yolk-Shell-Structured Iron Telluride-C Nanospheres and Exploration of Their Electrochemical Performance as an Anode Material for Potassium-Ion Batteries. *Small Methods* **2020**, *4*, 2000556. [CrossRef]
24. Bao, S.J.; Li, Y.; Li, C.M.; Bao, Q.; Lu, Q.; Guo, J. Shape Evolution and Magnetic Properties of Cobalt Sulfide. *Cryst. Growth Des.* **2008**, *8*, 3745–3749. [CrossRef]
25. Wang, L.H.; Teng, X.L.; Qin, Y.F.; Li, Q. High Electrochemical Performance and Structural Stability of CoO Nanosheets/CoO Film as Self-Supported Anodes for Lithium-Ion Batteries. *Ceram. Int.* **2021**, *47*, 5739–5746. [CrossRef]
26. Ren, L.L.; Wang, L.H.; Qin, Y.F.; Li, Q. One-Pot Synthesized Amorphous Cobalt Sulfide with Enhanced Electrochemical Performance as Anodes for Lithium-Ion Batteries. *Front. Chem.* **2022**, *9*, 818255. [CrossRef] [PubMed]
27. Lyu, Y.; Wu, X.; Wang, K.; Feng, Z.; Cheng, T.; Liu, Y.; Wang, M.; Chen, R.; Xu, L.; Zhou, J.; et al. An Overview on the Advances of LiCoO$_2$ Cathodes for Lithium-Ion Batteries. *Adv. Energy Mater.* **2021**, *11*, 2000982. [CrossRef]
28. Xiang, X.; Zhang, K.; Chen, J. Recent Advances and Prospects of Cathode Materials for Sodium-Ion Batteries. *Adv. Mater.* **2015**, *27*, 5343–5364. [CrossRef]
29. Kim, W.; Shin, D.; Seo, B.; Chae, S.; Jo, E.; Choi, W. Precisely Tunable Synthesis of Binder-Free Cobalt Oxide-Based Li-Ion Battery Anode Using Scalable Electrothermal Waves. *ACS Nano* **2022**, *16*, 17313–17325. [CrossRef]
30. Xiao, Q.; Song, Q.; Zheng, K.; Zhang, L.; Zhu, Y.; Chen, Z. CoSe$_2$ Nanodots Confined in Multidimensional Porous Nanoarchitecture towards Efficient Sodium Ion Storage. *Nano Energy* **2022**, *98*, 107326. [CrossRef]
31. Zhou, Y.; Han, Y.; Zhang, H.; Sui, D.; Sun, Z.; Xiao, P.; Wang, X.; Ma, Y.; Chen, Y. A Carbon Cloth-Based Lithium Composite Anode for High-Performance Lithium Metal Batteries. *Energy Storage Mater.* **2018**, *14*, 222–229. [CrossRef]
32. Liu, J.; Liang, J.; Wang, C.; Ma, J. Electrospun CoSe@N-Doped Carbon Nanofibers with Highly Capacitive Li Storage. *J. Energy Chem.* **2019**, *33*, 160–166. [CrossRef]
33. Wang, C.; Zhang, B.; Xia, H.; Cao, L.; Luo, B.; Fan, X.; Zhang, J.; Ou, X. Composition and Architecture Design of Double-Shelled Co$_{0.85}$Se$_{1-x}$S$_x$@Carbon/Graphene Hollow Polyhedron with Superior Alkali (Li, Na, K)-Ion Storage. *Small* **2020**, *16*, e1905853. [CrossRef] [PubMed]
34. Yu, Q.; Jiang, B.; Hu, J.; Lao, C.Y.; Gao, Y.; Li, P.; Liu, Z.; Suo, G.; He, D.; Wang, W.; et al. Metallic Octahedral CoSe$_2$ Threaded by N-Doped Carbon Nanotubes: A Flexible Framework for High-Performance Potassium-Ion Batteries. *Adv. Sci.* **2018**, *5*, 1800782. [CrossRef]
35. Yang, E.; Ji, H.; Jung, Y. Two-Dimensional Transition Metal Dichalcogenide Monolayers as Promising Sodium Ion Battery Anodes. *J. Phys. Chem. C* **2015**, *119*, 26374–26380. [CrossRef]
36. Ni, W.; Li, X.; Shi, L.Y.; Ma, J. Research Progress on ZnSe and ZnTe Anodes for Rechargeable Batteries. *Nanoscale* **2022**, *14*, 9609–9635. [CrossRef]
37. Yang, S.H.; Park, S.K.; Park, G.D.; Lee, J.H.; Kang, Y.C. Conversion Reaction Mechanism of Ultrafine Bimetallic Co–Fe Selenides Embedded in Hollow Mesoporous Carbon Nanospheres and Their Excellent K-Ion Storage Performance. *Small* **2020**, *16*, e2002345. [CrossRef]
38. Xu, X.; Zhang, Y.; Sun, H.; Zhou, J.; Liu, Z.; Qiu, Z.; Wang, D.; Yang, C.; Zeng, Q.; Peng, Z.; et al. Orthorhombic Cobalt Ditelluride with Te Vacancy Defects Anchoring on Elastic MXene Enables Efficient Potassium-Ion Storage. *Adv. Mater.* **2021**, *33*, 2100272. [CrossRef]
39. Xiao, S.; Li, X.; Zhang, W.; Xiang, Y.; Li, T.; Niu, X.; Chen, J.S.; Yan, Q. Bilateral Interfaces in In$_2$Se$_3$–CoIn$_2$–CoSe$_2$ Heterostructures for High-Rate Reversible Sodium Storage. *ACS Nano* **2021**, *15*, 13307–13318. [CrossRef]
40. Zhang, C.; Li, H.; Zeng, X.; Xi, S.; Wang, R.; Zhang, L.; Liang, G.; Davey, K.; Liu, Y.; Zhang, L.; et al. Accelerated Diffusion Kinetics in ZnTe/CoTe$_2$ Heterojunctions for High Rate Potassium Storage. *Adv. Energy Mater.* **2022**, *12*, 2202577. [CrossRef]
41. Li, Q.; Peng, J.; Zhang, W.; Wang, L.; Liang, Z.; Wang, G.; Wu, J.; Fan, W.; Li, H.; Wang, J.; et al. Manipulating the Polytellurides of Metallic Telluride for Ultra-Stable Potassium-Ion Storage: A Case Study of Carbon-Confined CoTe$_2$ Nanofibers. *Adv. Energy Mater.* **2023**, *13*, 2300150. [CrossRef]
42. Zhang, G.; Liu, K.; Zhou, J. Cobalt Telluride/Graphene Composite Nanosheets for Excellent Gravimetric and Volumetric Na-Ion Storage. *J. Mater. Chem. A Mater.* **2018**, *6*, 6335–6343. [CrossRef]
43. Cho, J.S.; Won, J.M.; Lee, J.K.; Kang, Y.C. Design and Synthesis of Multiroom-Structured Metal Compounds-Carbon Hybrid Microspheres as Anode Materials for Rechargeable Batteries. *Nano Energy* **2016**, *26*, 466–478. [CrossRef]
44. Li, J.; Yan, D.; Lu, T.; Yao, Y.; Pan, L. An Advanced CoSe Embedded within Porous Carbon Polyhedra Hybrid for High Performance Lithium-Ion and Sodium-Ion Batteries. *Chem. Eng. J.* **2017**, *325*, 14–24. [CrossRef]
45. Yang, X.; Wang, S.; Yu, D.Y.W.; Rogach, A.L. Direct Conversion of Metal-Organic Frameworks into Selenium/Selenide/Carbon Composites with High Sodium Storage Capacity. *Nano Energy* **2019**, *58*, 392–398. [CrossRef]
46. Sun, Z.; Gu, Z.; Shi, W.; Sun, Z.; Gan, S.; Xu, L.; Liang, H.; Ma, Y.; Qu, D.; Zhong, L.; et al. Mesoporous N-Doped Carbon-Coated CoSe Nanocrystals Encapsulated in S-Doped Carbon Nanosheets as Advanced Anode with Ultrathin Solid Electrolyte Interphase for High-Performance Sodium-Ion Half/Full Batteries. *J. Mater. Chem. A Mater.* **2022**, *10*, 2113–2121. [CrossRef]
47. Zhang, W.; Wang, X.; Wong, K.W.; Zhang, W.; Chen, T.; Zhao, W.; Huang, S. Rational Design of Embedded CoTe$_2$ nanoparticles in Freestanding N-Doped Multichannel Carbon Fibers for Sodium-Ion Batteries with Ultralong Cycle Lifespan. *ACS Appl. Mater. Interfaces* **2021**, *13*, 34134–34144. [CrossRef]

48. Chen, Z.; Chen, M.; Yan, X.; Jia, H.; Fei, B.; Ha, Y.; Qing, H.; Yang, H.; Liu, M.; Wu, R. Vacancy Occupation-Driven Polymorphic Transformation in Cobalt Ditelluride for Boosted Oxygen Evolution Reaction. *ACS Nano* **2020**, *14*, 6968–6979. [CrossRef]
49. Fan, H.; Mao, P.; Sun, H.; Wang, Y.; Mofarah, S.S.; Koshy, P.; Arandiyan, H.; Wang, Z.; Liu, Y.; Shao, Z. Recent Advances of Metal Telluride Anodes for High-Performance Lithium/Sodium-Ion Batteries. *Mater. Horiz.* **2022**, *9*, 524–546. [CrossRef]
50. Jiang, Y.; Zou, G.; Hou, H.; Li, J.; Liu, C.; Qiu, X.; Ji, X. Composition Engineering Boosts Voltage Windows for Advanced Sodium-Ion Batteries. *ACS Nano* **2019**, *13*, 10787–10797. [CrossRef]
51. Jiang, Y.; Xie, M.; Wu, F.; Ye, Z.; Zhang, Y.; Wang, Z.; Zhou, Y.; Li, L.; Chen, R. Cobalt Selenide Hollow Polyhedron Encapsulated in Graphene for High-Performance Lithium/Sodium Storage. *Small* **2021**, *17*, 2102893. [CrossRef] [PubMed]
52. Ganesan, V.; Nam, K.H.; Park, C.M. Robust Polyhedral CoTe$_2$–C Nanocomposites as High-Performance Li-A Nd Na-Ion Battery Anodes. *ACS Appl. Energy Mater.* **2020**, *3*, 4877–4887. [CrossRef]
53. Park, S.; Park, S.; Mathew, V.; Sambandam, B.; Hwang, J.Y.; Kim, J. A New Tellurium-Based Ni$_3$TeO$_6$-Carbon Nanotubes Composite Anode for Na-Ion Battery. *Int. J. Energy Res.* **2022**, *46*, 16041–16049. [CrossRef]
54. Ge, P.; Li, S.; Xu, L.; Zou, K.; Gao, X.; Cao, X.; Zou, G.; Hou, H.; Ji, X. Hierarchical Hollow-Microsphere Metal–Selenide@Carbon Composites with Rational Surface Engineering for Advanced Sodium Storage. *Adv. Energy Mater.* **2019**, *9*, 1803035. [CrossRef]
55. Miao, Y.; Xiao, Y.; Hu, S.; Chen, S. Chalcogenides Metal-Based Heterostructure Anode Materials toward Na$^+$-Storage Application. *Nano Res.* **2022**, *16*, 2347–2365. [CrossRef]
56. Banhart, F.; Kotakoski, J.; Krasheninnikov, A.V. Structural Defects in Graphene. *ACS Nano* **2011**, *5*, 26–41. [CrossRef]
57. Dong, C.; Zhang, X.; Dong, W.; Lin, X.; Cheng, Y.; Tang, Y.; Zhao, S.; Li, G.; Huang, F. ZnO/ZnS Heterostructure with Enhanced Interfacial Lithium Absorption for Robust and Large-Capacity Energy Storage. *Energy Environ. Sci.* **2022**, *15*, 4738–4747. [CrossRef]
58. Arandiyan, H.; Mofarah, S.S.; Sorrell, C.C.; Doustkhah, E.; Sajjadi, B.; Hao, D.; Wang, Y.; Sun, H.; Ni, B.J.; Rezaei, M.; et al. Defect Engineering of Oxide Perovskites for Catalysis and Energy Storage: Synthesis of Chemistry and Materials Science. *Chem. Soc. Rev.* **2021**, *50*, 10116–10211. [CrossRef]
59. Wang, Y.; Arandiyan, H.; Chen, X.; Zhao, T.; Bo, X.; Su, Z.; Zhao, C. Microwave-Induced Plasma Synthesis of Defect-Rich, Highly Ordered Porous Phosphorus-Doped Cobalt Oxides for Overall Water Electrolysis. *J. Phys. Chem. C* **2020**, *124*, 9971–9978. [CrossRef]
60. Wang, X.; Li, F.; Li, W.; Gao, W.; Tang, Y.; Li, R. Hollow Bimetallic Cobalt-Based Selenide Polyhedrons Derived from Metal-Organic Framework: An Efficient Bifunctional Electrocatalyst for Overall Water Splitting. *J. Mater. Chem. A Mater.* **2017**, *5*, 17982–17989. [CrossRef]
61. Fang, C.; Huang, Y.; Zhang, W.; Han, J.; Deng, Z.; Cao, Y.; Yang, H. Routes to High Energy Cathodes of Sodium-Ion Batteries. *Adv. Energy Mater.* **2016**, *6*, 1501727. [CrossRef]
62. Billaud, J.; Singh, G.; Armstrong, A.R.; Gonzalo, E.; Roddatis, V.; Armand, M.; Rojo, T.; Bruce, P.G. Na$_{0.67}$Mn$_{1−x}$Mg$_x$O$_2$ ($0 \leq x \leq 0.2$): A High Capacity Cathode for Sodium-Ion Batteries. *Energy Environ. Sci.* **2014**, *7*, 1387–1391. [CrossRef]
63. Narsimulu, D.; Kakarla, A.K.; Vamsi Krishna, B.N.; Shanthappa, R.; Yu, J.S. Nitrogen-Doped Reduced Graphene Oxide Incorporated Ni$_2$O$_3$–Co$_3$O$_4$@MoS$_2$ Hollow Nanocubes for High-Performance Energy Storage Devices. *J. Alloys Compd.* **2022**, *922*, 166131. [CrossRef]
64. Jana, J.; Sharma, T.S.K.; Chung, J.S.; Choi, W.M.; Hur, S.H. The Role of Surface Carbide/Oxide Heterojunction in Electrocatalytic Behavior of 3D-Nanonest Supported FeIMoj Composites. *J. Alloys Compd.* **2023**, *946*, 169395. [CrossRef]
65. Pellow, M.A.; Emmott, C.J.M.; Barnhart, C.J.; Benson, S.M. Hydrogen or Batteries for Grid Storage? A Net Energy Analysis. *Energy Environ. Sci.* **2015**, *8*, 1938–1952. [CrossRef]
66. Ge, P.; Hou, H.; Li, S.; Huang, L.; Ji, X. Three-Dimensional Hierarchical Framework Assembled by Cobblestone-Like CoSe$_2$@C Nanospheres for Ultrastable Sodium-Ion Storage. *ACS Appl. Mater. Interfaces* **2018**, *10*, 14716–14726. [CrossRef]
67. Ge, P.; Hou, H.; Li, S.; Yang, L.; Ji, X. Tailoring Rod-Like FeSe$_2$ Coated with Nitrogen-Doped Carbon for High-Performance Sodium Storage. *Adv. Funct. Mater.* **2018**, *28*, 1801765. [CrossRef]
68. Lu, M.; Liu, C.; Li, X.; Jiang, S.; Yao, Z.; Liu, T.; Yang, Y. Mixed Phase Mo-Doped CoSe$_2$ Nanosheets Encapsulated in N-Doped Carbon Shell with Boosted Sodium Storage Performance. *J. Alloys Compd.* **2022**, *922*, 166265. [CrossRef]
69. Wang, M.Y.; Wang, X.L.; Yao, Z.J.; Xie, D.; Xia, X.H.; Gu, C.D.; Tu, J.P. Molybdenum-Doped Tin Oxide Nanoflake Arrays Anchored on Carbon Foam as Flexible Anodes for Sodium-Ion Batteries. *J. Colloid Interface Sci.* **2020**, *560*, 169–176. [CrossRef]
70. Wang, P.; Huang, J.; Zhang, J.; Wang, L.; Sun, P.; Yang, Y.; Yao, Z. Coupling Hierarchical Iron Cobalt Selenide Arrays with N-Doped Carbon as Advanced Anodes for Sodium Ion Storage. *J. Mater. Chem. A Mater.* **2021**, *9*, 7248–7256. [CrossRef]
71. Hu, L.; Li, L.; Zhang, Y.; Tan, X.; Yang, H.; Lin, X.; Tong, Y. Construction of Cobalt Vacancies in Cobalt Telluride to Induce Fast Ionic/Electronic Diffusion Kinetics for Lithium-Ion Half/Full Batteries. *J. Mater. Sci. Technol.* **2022**, *127*, 124–132. [CrossRef]
72. Fang, G.; Wang, Q.; Zhou, J.; Lei, Y.; Chen, Z.; Wang, Z.; Pan, A.; Liang, S. Metal Organic Framework-Templated Synthesis of Bimetallic Selenides with Rich Phase Boundaries for Sodium-Ion Storage and Oxygen Evolution Reaction. *ACS Nano* **2019**, *13*, 5635–5645. [CrossRef] [PubMed]
73. Ye, J.; Li, X.; Xia, G.; Gong, G.; Zheng, Z.; Chen, C.; Hu, C. P-Doped CoSe$_2$ Nanoparticles Embedded in 3D Honeycomb-like Carbon Network for Long Cycle-Life Na-Ion Batteries. *J. Mater. Sci. Technol.* **2021**, *77*, 100–107. [CrossRef]
74. Shin, I.; Cho, W.J.; An, E.S.; Park, S.; Jeong, H.W.; Jang, S.; Baek, W.J.; Park, S.Y.; Yang, D.H.; Seo, J.H.; et al. Spin–Orbit Torque Switching in an All-Van Der Waals Heterostructure. *Adv. Mater.* **2022**, *34*, e2101730. [CrossRef]

75. Pan, L.; Grutter, A.; Zhang, P.; Che, X.; Nozaki, T.; Stern, A.; Street, M.; Zhang, B.; Casas, B.; He, Q.L.; et al. Observation of Quantum Anomalous Hall Effect and Exchange Interaction in Topological Insulator/Antiferromagnet Heterostructure. *Adv. Mater.* **2020**, *32*, 2001460. [CrossRef] [PubMed]
76. Wang, H.; Niu, R.; Liu, J.; Guo, S.; Yang, Y.; Liu, Z.; Li, J. Electrostatic Self-Assembly of 2D/2D $CoWO_4$/g–C_3N_4 p—n Heterojunction for Improved Photocatalytic Hydrogen Evolution: Built-in Electric Field Modulated Charge Separation and Mechanism Unveiling. *Nano Res.* **2022**, *15*, 6987–6998. [CrossRef]
77. Huang, S.; Wang, Z.; von Lim, Y.; Wang, Y.; Li, Y.; Zhang, D.; Yang, H.Y. Recent Advances in Heterostructure Engineering for Lithium–Sulfur Batteries. *Adv. Energy Mater.* **2021**, *11*, 2003689. [CrossRef]
78. Li, Y.; Zhang, J.; Chen, Q.; Xia, X.; Chen, M. Emerging of Heterostructure Materials in Energy Storage: A Review. *Adv. Mater.* **2021**, *33*, 2100855. [CrossRef]
79. Liang, L.; Gu, W.; Wu, Y.; Zhang, B.; Wang, G.; Yang, Y.; Ji, G. Heterointerface Engineering in Electromagnetic Absorbers: New Insights and Opportunities. *Adv. Mater.* **2022**, *34*, 2106195. [CrossRef]
80. Wang, S.; Yang, Y.; Quan, W.; Hong, Y.; Zhang, Z.; Tang, Z.; Li, J. Ti^{3+}-Free Three-Phase $Li_4Ti_5O_{12}$/TiO_2 for High-Rate Lithium Ion Batteries: Capacity and Conductivity Enhancement by Phase Boundaries. *Nano Energy* **2017**, *32*, 294–301. [CrossRef]
81. Luo, D.; Ma, C.; Hou, J.; Zhang, Z.; Feng, R.; Yang, L.; Zhang, X.; Lu, H.; Liu, J.; Li, Y.; et al. Integrating Nanoreactor with O–Nb–C Heterointerface Design and Defects Engineering Toward High-Efficiency and Longevous Sodium Ion Battery. *Adv. Energy Mater.* **2022**, *12*, 2103716. [CrossRef]
82. Yang, C.; Liang, X.; Ou, X.; Zhang, Q.; Zheng, H.S.; Zheng, F.; Wang, J.H.; Huang, K.; Liu, M. Heterostructured Nanocube-Shaped Binary Sulfide (SnCo)S_2 Interlaced with S-Doped Graphene as a High-Performance Anode for Advanced Na^+ Batteries. *Adv. Funct. Mater.* **2019**, *29*, 1807971. [CrossRef]
83. Xiao, Y.; Miao, Y.; Wan, S.; Sun, Y.K.; Chen, S. Synergistic Engineering of Se Vacancies and Heterointerfaces in Zinc-Cobalt Selenide Anode for Highly Efficient Na-Ion Batteries. *Small* **2022**, *18*, 2202582. [CrossRef] [PubMed]
84. Saliba, D.; Ammar, M.; Rammal, M.; Al-Ghoul, M.; Hmadeh, M. Crystal Growth of ZIF-8, ZIF-67, and Their Mixed-Metal Derivatives. *J. Am. Chem. Soc.* **2018**, *140*, 1812–1823. [CrossRef] [PubMed]
85. Venna, S.R.; Jasinski, J.B.; Carreon, M.A. Structural Evolution of Zeolitic Imidazolate Framework-8. *J. Am. Chem. Soc.* **2010**, *132*, 18030–18033. [CrossRef]
86. He, Y.; Luo, M.; Dong, C.; Ding, X.; Yin, C.; Nie, A.; Chen, Y.; Qian, Y.; Xu, L. Coral-like $Ni_xCo_{1-x}Se_2$ for Na-Ion Battery with Ultralong Cycle Life and Ultrahigh Rate Capability. *J. Mater. Chem. A Mater.* **2019**, *7*, 3933–3940. [CrossRef]
87. Zhu, H.; Li, Z.; Xu, F.; Qin, Z.; Sun, R.; Wang, C.; Lu, S.; Zhang, Y.; Fan, H. Ni_3Se_4@$CoSe_2$ Hetero-Nanocrystals Encapsulated into CNT-Porous Carbon Interpenetrating Frameworks for High-Performance Sodium Ion Battery. *J. Colloid Interface Sci.* **2022**, *611*, 718–725. [CrossRef]
88. Zhang, K.; Park, M.; Zhou, L.; Lee, G.H.; Li, W.; Kang, Y.M.; Chen, J. Urchin-Like $CoSe_2$ as a High-Performance Anode Material for Sodium-Ion Batteries. *Adv. Funct. Mater.* **2016**, *26*, 6728–6735. [CrossRef]
89. Ko, Y.N.; Choi, S.H.; Kang, Y.C. Hollow Cobalt Selenide Microspheres: Synthesis and Application as Anode Materials for Na-Ion Batteries. *ACS Appl. Mater. Interfaces* **2016**, *8*, 6449–6456. [CrossRef]
90. Hu, J.-Z.; Liu, W.-J.; Zheng, J.-H.; Li, G.-C.; Bu, Y.-F.; Qiao, F.; Lian, J.-B.; Zhao, Y. Coral-like Cobalt Selenide/Carbon Nanosheet Arrays Attached on Carbon Nanofibers for High-Rate Sodium-Ion Storage. *Rare Met.* **2023**, *42*, 916–928. [CrossRef]
91. Hong, L.; Ju, S.; Yang, Y.; Zheng, J.; Xia, G.; Huang, Z.; Liu, X.; Yu, X. Hollow-Shell Structured Porous $CoSe_2$ Microspheres Encapsulated by MXene Nanosheets for Advanced Lithium Storage. *Sustain. Energy Fuels* **2020**, *4*, 2352–2362. [CrossRef]
92. Gu, X.; Zhang, L.; Zhang, W.; Liu, S.; Wen, S.; Mao, X.; Dai, P.; Li, L.; Liu, D.; Zhao, X.; et al. A CoSe-C@C Core-Shell Structure with Stable Potassium Storage Performance Realized by an Effective Solid Electrolyte Interphase Layer. *J. Mater. Chem. A Mater.* **2021**, *9*, 11397–11404. [CrossRef]
93. Fan, H.; Liu, C.; Lan, G.; Mao, P.; Zheng, R.; Wang, Z.; Liu, Y.; Sun, H. Uniform Carbon Coating Mediated Multiphase Interface in Submicron Sized Rodlike Cobalt Ditelluride Anodes for High-Capacity and Fast Lithium Storage. *Electrochim. Acta* **2023**, *439*, 141614. [CrossRef]
94. Nie, P.; Le, Z.; Chen, G.; Liu, D.; Liu, X.; Wu, H.B.; Xu, P.; Li, X.; Liu, F.; Chang, L.; et al. Graphene Caging Silicon Particles for High-Performance Lithium-Ion Batteries. *Small* **2018**, *14*, 1800635. [CrossRef]
95. Xu, H.; Cao, J.; Shan, C.; Wang, B.; Xi, P.; Liu, W.; Tang, Y. MOF-Derived Hollow CoS Decorated with CeO_x Nanoparticles for Boosting Oxygen Evolution Reaction Electrocatalysis. *Angew. Chem.* **2018**, *130*, 8790–8794. [CrossRef]
96. Huang, P.; Ying, H.; Zhang, S.; Zhang, Z.; Han, W.Q. In Situ Fabrication of MXene/CuS Hybrids with Interfacial Covalent Bonding via Lewis Acidic Etching Route for Efficient Sodium Storage. *J. Mater. Chem. A Mater.* **2022**, *10*, 22135–22144. [CrossRef]
97. Jin, T.; Han, Q.; Wang, Y.; Jiao, L. 1D Nanomaterials: Design, Synthesis, and Applications in Sodium–Ion Batteries. *Small* **2018**, *14*, 1703086. [CrossRef]
98. Nie, S.; Liu, L.; Liu, J.; Xie, J.; Zhang, Y.; Xia, J.; Yan, H.; Yuan, Y.; Wang, X. Nitrogen-Doped TiO_2–C Composite Nanofibers with High-Capacity and Long-Cycle Life as Anode Materials for Sodium-Ion Batteries. *Nanomicro Lett.* **2018**, *10*, 71. [CrossRef]
99. La Monaca, A.; Paolella, A.; Guerfi, A.; Rosei, F.; Zaghib, K. Electrospun Ceramic Nanofibers as 1D Solid Electrolytes for Lithium Batteries. *Electrochem. Commun.* **2019**, *104*, 106483. [CrossRef]
100. Park, J.S.; Park, G.D.; Kang, Y.C. Exploration of Cobalt Selenite–Carbon Composite Porous Nanofibers as Anode for Sodium-Ion Batteries and Unveiling Their Conversion Reaction Mechanism. *J. Mater. Sci. Technol.* **2021**, *89*, 24–35. [CrossRef]

101. Li, F.; Li, L.; Yao, T.; Liu, T.; Zhu, L.; Li, Y.; Lu, H.; Qian, R.; Liu, Y.; Wang, H. Electrospinning Synthesis of Porous Carbon Nanofiber Supported CoSe$_2$ Nanoparticles towards Enhanced Sodium Ion Storage. *Mater. Chem. Phys.* **2021**, *262*, 124314. [CrossRef]
102. Lin, J.; Zhong, Z.; Wang, H.; Zheng, X.; Wang, Y.; Qi, J.; Cao, J.; Fei, W.; Huang, Y.; Feng, J. Rational Constructing Free-Standing Se Doped Nickel-Cobalt Sulfides Nanotubes as Battery-Type Electrode for High-Performance Supercapattery. *J. Power Sources* **2018**, *407*, 6–13. [CrossRef]
103. Xia, Z.; Sun, H.; He, X.; Sun, Z.; Lu, C.; Li, J.; Peng, Y.; Dou, S.; Sun, J.; Liu, Z. In Situ Construction of CoSe$_2$@vertical-Oriented Graphene Arrays as Self-Supporting Electrodes for Sodium-Ion Capacitors and Electrocatalytic Oxygen Evolution. *Nano Energy* **2019**, *60*, 385–393. [CrossRef]
104. Li, Z.; Zhang, J.; Lou, X.W.D. Hollow Carbon Nanofibers Filled with MnO$_2$ Nanosheets as Efficient Sulfur Hosts for Lithium-Sulfur Batteries. *Angew. Chem.* **2015**, *127*, 13078–13082. [CrossRef]
105. Xin, D.; He, S.; Han, X.; Zhang, X.; Cheng, Z.; Xia, M. Co-Doped 1T-MoS$_2$ Nanosheets Anchored on Carbon Cloth as Self-Supporting Anode for High-Performance Lithium Storage. *J. Alloys Compd.* **2022**, *921*, 166099. [CrossRef]
106. Wang, G.; Chen, J.; Cai, P.; Jia, J.; Wen, Z. A Self-Supported Ni–Co Perselenide Nanorod Array as a High-Activity Bifunctional Electrode for a Hydrogen-Producing Hydrazine Fuel Cell. *J. Mater. Chem. A Mater.* **2018**, *6*, 17763–17770. [CrossRef]
107. Liu, B.; Li, H.; Cao, B.; Jiang, J.; Gao, R.; Zhang, J. Few Layered N, P Dual-Doped Carbon-Encapsulated Ultrafine MoP Nanocrystal/MoP Cluster Hybrids on Carbon Cloth: An Ultrahigh Active and Durable 3D Self-Supported Integrated Electrode for Hydrogen Evolution Reaction in a Wide PH Range. *Adv. Funct. Mater.* **2018**, *28*, 1801527. [CrossRef]
108. Sun, T.; Huang, C.; Shu, H.; Luo, L.; Liang, Q.; Chen, M.; Su, J.; Wang, X. Porous NiCo$_2$S$_4$ Nanoneedle Arrays with Highly Efficient Electrocatalysis Anchored on Carbon Cloths as Self-Supported Hosts for High-Loading Li–S Batteries. *ACS Appl. Mater. Interfaces* **2020**, *12*, 57975–57986. [CrossRef]
109. Zeng, M.; Wang, M.; Zheng, L.; Gao, W.; Liu, R.; Pan, J.; Zhang, H.; Yang, Z.; Li, X. In Situ Enhance Lithium Polysulfides Redox Kinetics by Carbon Cloth/MoO$_3$ Self-Standing Electrode for Lithium–Sulfur Battery. *J. Mater. Sci.* **2022**, *57*, 10003–10016. [CrossRef]
110. Kim, I.; Park, S.W.; Kim, D.W. Carbon-Coated Tungsten Diselenide Nanosheets Uniformly Assembled on Porous Carbon Cloth as Flexible Binder-Free Anodes for Sodium-Ion Batteries with Improved Electrochemical Performance. *J. Alloys Compd.* **2020**, *827*, 154348. [CrossRef]
111. Li, T.; Bai, Y.; Wang, Y.; Xu, H.; Jin, H. Advances in Transition-Metal (Zn, Mn, Cu)-Based MOFs and Their Derivatives for Anode of Lithium-Ion Batteries. *Coord Chem. Rev.* **2020**, *410*, 213221. [CrossRef]
112. Wang, H.; Zhu, Q.L.; Zou, R.; Xu, Q. Metal-Organic Frameworks for Energy Applications. *Chem* **2017**, *2*, 52–80. [CrossRef]
113. Miao, W.; Zhang, Y.; Li, H.; Zhang, Z.; Li, L.; Yu, Z.; Zhang, W. ZIF-8/ZIF-67-Derived 3D Amorphous Carbon-Encapsulated CoS/NCNTs Supported on CoS-Coated Carbon Nanofibers as an Advanced Potassium-Ion Battery Anode. *J. Mater. Chem. A Mater.* **2019**, *7*, 5504–5512. [CrossRef]
114. Cheng, Z.; Du, T.; Gong, H.; Zhou, L. In-Situ Synthesis of Fe$_7$S$_8$ on Metal Sites of MOFs as High-Capacity and Fast-Kinetics Anodes for Sodium Ion Batteries. *J. Alloys Compd.* **2023**, *940*, 168854. [CrossRef]
115. Guo, W.; Sun, W.; Lv, L.P.; Kong, S.; Wang, Y. Microwave-Assisted Morphology Evolution of Fe-Based Metal-Organic Frameworks and Their Derived Fe$_2$O$_3$ Nanostructures for Li-Ion Storage. *ACS Nano* **2017**, *11*, 4198–4205. [CrossRef] [PubMed]
116. Zhang, Y.; Pan, A.; Ding, L.; Zhou, Z.; Wang, Y.; Niu, S.; Liang, S.; Cao, G. Nitrogen-Doped Yolk-Shell-Structured CoSe/C Dodecahedra for High-Performance Sodium Ion Batteries. *ACS Appl. Mater. Interfaces* **2017**, *9*, 3624–3633. [CrossRef]
117. Feng, J.; Luo, S.H.; Yan, S.X.; Zhan, Y.; Wang, Q.; Zhang, Y.H.; Liu, X.; Chang, L.J. Rational Design of Yolk–Shell Zn-Co-Se@N-Doped Dual Carbon Architectures as Long-Life and High-Rate Anodes for Half/Full Na-Ion Batteries. *Small* **2021**, *17*, e2101887. [CrossRef]
118. Yin, Y.; Rioux, R.M.; Erdonmez, C.K.; Hughes, S.; Somorjai, G.A.; Alivisatos, A.P. Formation of Hollow Nanocrystals through the Nanoscale Kirkendall Effect. *Science* **2004**, *304*, 711–714. [CrossRef]
119. Wang, B.; Miao, X.; Dong, H.; Ma, X.; Wu, J.; Cheng, Y.; Geng, H.; Li, C.C. In Situ construction of Active Interfaces towards Improved High-Rate Performance of CoSe$_2$. *J. Mater. Chem. A Mater.* **2021**, *9*, 14582–14592. [CrossRef]
120. Pang, J.; Mendes, R.G.; Bachmatiuk, A.; Zhao, L.; Ta, H.Q.; Gemming, T.; Liu, H.; Liu, Z.; Rummeli, M.H. Applications of 2D MXenes in Energy Conversion and Storage Systems. *Chem. Soc. Rev.* **2019**, *48*, 72–133. [CrossRef]
121. Yu, H.; Wang, Y.; Jing, Y.; Ma, J.; Du, C.F.; Yan, Q. Surface Modified MXene-Based Nanocomposites for Electrochemical Energy Conversion and Storage. *Small* **2019**, *15*, e1901503. [CrossRef] [PubMed]
122. Sun, H.; Zhang, Y.; Xu, X.; Zhou, J.; Yang, F.; Li, H.; Chen, H.; Chen, Y.; Liu, Z.; Qiu, Z.; et al. Strongly Coupled Te-SnS$_2$/MXene Superstructure with Self-Autoadjustable Function for Fast and Stable Potassium Ion Storage. *J. Energy Chem.* **2021**, *61*, 416–424. [CrossRef]
123. Kannan, K.; Sadasivuni, K.K.; Abdullah, A.M.; Kumar, B. Current Trends in MXene-Based Nanomaterials for Energy Storage and Conversion System: A Mini Review. *Catalysts* **2020**, *10*, 495. [CrossRef]
124. Nguyen, V.H.; Nguyen, B.S.; Hu, C.; Nguyen, C.C.; Nguyen, D.L.T.; Dinh, M.T.N.; Vo, D.V.N.; Trinh, Q.T.; Shokouhimehr, M.; Hasani, A.; et al. Novel Architecture Titanium Carbide (Ti$_3$C$_2$T$_x$) Mxene Cocatalysts toward Photocatalytic Hydrogen Production: A Mini-Review. *Nanomaterials* **2020**, *10*, 602. [CrossRef] [PubMed]
125. Ran, J.; Gao, G.; Li, F.T.; Ma, T.Y.; Du, A.; Qiao, S.Z. Ti$_3$C$_2$ MXene Co-Catalyst on Metal Sulfide Photo-Absorbers for Enhanced Visible-Light Photocatalytic Hydrogen Production. *Nat. Commun.* **2017**, *8*, 13907. [CrossRef] [PubMed]

126. Peng, Q.; Guo, J.; Zhang, Q.; Xiang, J.; Liu, B.; Zhou, A.; Liu, R.; Tian, Y. Unique Lead Adsorption Behavior of Activated Hydroxyl Group in Two-Dimensional Titanium Carbide. *J. Am. Chem. Soc.* **2014**, *136*, 4113–4116. [CrossRef]
127. Wang, M.; Peng, A.; Zeng, M.; Chen, L.; Li, X.; Yang, Z.; Chen, J.; Guo, B.; Ma, Z.; Li, X. Facilitating the Redox Conversion of CoSe$_2$ nanorods by Ti$_3$C$_2$T$_x$ to Improve the Electrode Durability as Anodes for Sodium-Ion Batteries. *Sustain. Energy Fuels* **2021**, *5*, 6381–6391. [CrossRef]
128. Acharya, J.; Ojha, G.P.; Pant, B.; Park, M. Construction of Self-Supported Bimetallic MOF-Mediated Hollow and Porous Tri-Metallic Selenide Nanosheet Arrays as Battery-Type Electrodes for High-Performance Asymmetric Supercapacitors. *J. Mater. Chem. A Mater.* **2021**, *9*, 23977–23993. [CrossRef]
129. Sarkar, D.; Das, D.; Nagarajan, S.; Mitlin, D. Thermally Fabricated Cobalt Telluride in Nitrogen-Rich Carbon Dodecahedra as High-Rate Potassium and Sodium Ion Battery Anodes. *Sustain. Energy Fuels* **2022**, *6*, 3582–3590. [CrossRef]
130. Zhao, H.; Qi, Y.; Liang, K.; Li, J.; Zhou, L.; Chen, J.; Huang, X.; Ren, Y. Interface-Driven Pseudocapacitance Endowing Sandwiched CoSe$_2$/N-Doped Carbon/TiO$_2$ Microcubes with Ultra-Stable Sodium Storage and Long-Term Cycling Stability. *ACS Appl. Mater. Interfaces* **2021**, *13*, 61555–61564. [CrossRef]
131. Lei, Y.X.; Miao, N.X.; Zhou, J.P.; Hassan, Q.U.; Wang, J.Z. Novel Magnetic Properties of CoTe Nanorods and Diversified CoTe$_2$ Nanostructures Obtained at Different NaOH Concentrations. *Sci. Technol. Adv. Mater.* **2017**, *18*, 325–333. [CrossRef] [PubMed]
132. Keles, O.; Karahan, B.D.; Eryilmaz, L.; Amine, R.; Abouimrane, A.; Chen, Z.; Zuo, X.; Zhu, Z.; Al-Hallaj, S.; Amine, K. Superlattice-Structured Films by Magnetron Sputtering as New Era Electrodes for Advanced Lithium-Ion Batteries. *Nano Energy* **2020**, *76*, 105094. [CrossRef]
133. Sun, C.; Gao, L.; Yang, Y.; Yan, Z.; Zhang, D.; Bian, X. Ultrafast Microwave-Induced Synthesis of Lithiophilic Oxides Modified 3D Porous Mesh Skeleton for High-Stability Li-Metal Anode. *Chem. Eng. J.* **2023**, *452*, 139407. [CrossRef]
134. Huang, P.; Ying, H.; Zhang, S.; Zhang, Z.; Han, W.Q. Molten Salts Etching Route Driven Universal Construction of MXene/Transition Metal Sulfides Heterostructures with Interfacial Electronic Coupling for Superior Sodium Storage. *Adv. Energy Mater.* **2022**, *12*, 2202052. [CrossRef]
135. Hu, H.; Zhang, J.; Guan, B.; Lou, X.W.D. Unusual Formation of CoSe@carbon Nanoboxes, Which Have an Inhomogeneous Shell, for Efficient Lithium Storage. *Angew. Chem.* **2016**, *128*, 9666–9670. [CrossRef]
136. Sui, Y.; Guo, J.; Chen, X.; Guan, J.; Chen, X.; Wei, H.; Liu, Q.; Wei, B.; Geng, H. Highly Dispersive CoSe$_2$ Nanoparticles Encapsulated in Carbon Nanotube-Grafted Multichannel Carbon Fibers as Advanced Anodes for Sodium-Ion Half/Full Batteries. *Inorg. Chem. Front.* **2022**, *9*, 5217–5225. [CrossRef]
137. He, Y.; Li, H.; Yu, J.; Wang, Q.; Dai, Y.; Shan, R.; Zhou, G. Hierarchical Mesoporous Binary Nickel Cobalt Selenides Nanospheres for Enhanced Sodium-Ion Storage. *J. Alloys Compd.* **2023**, *938*, 168608. [CrossRef]
138. Zhao, J.; Wu, H.; Li, L.; Lu, S.; Mao, H.; Ding, S. A CoSe$_2$-Based 3D Conductive Network for High-Performance Potassium Storage: Enhancing Charge Transportation by Encapsulation and Restriction Strategy. *Mater. Chem. Front.* **2021**, *5*, 5351–5360. [CrossRef]
139. Jiang, Y.; Wu, F.; Ye, Z.; Zhou, Y.; Chen, Y.; Zhang, Y.; Lv, Z.; Li, L.; Xie, M.; Chen, R. Superimposed Effect of Hollow Carbon Polyhedron and Interconnected Graphene Network to Achieve CoTe$_2$ Anode for Fast and Ultralong Sodium Storage. *J. Power Sources* **2023**, *554*, 232174. [CrossRef]
140. Zhao, W.; Zhang, W.; Lei, Y.; Wang, L.; Wang, G.; Wu, J.; Fan, W.; Huang, S. Dual-Type Carbon Confinement Strategy: Improving the Stability of CoTe$_2$ Nanocrystals for Sodium-Ion Batteries with a Long Lifespan. *ACS Appl. Mater. Interfaces* **2022**, *14*, 6801–6809. [CrossRef]
141. Ding, Y.; Wang, W.; Bi, M.; Guo, J.; Fang, Z. CoTe Nanorods/RGO Composites as a Potential Anode Material for Sodium-Ion Storage. *Electrochim. Acta* **2019**, *313*, 331–340. [CrossRef]
142. Yang, S.; Park, G.D.; Kang, Y.C. Conversion Reaction Mechanism of Cobalt Telluride-Carbon Composite Microspheres Synthesized by Spray Pyrolysis Process for K-Ion Storage. *Appl. Surf. Sci.* **2020**, *529*, 147140. [CrossRef]

Disclaimer/Publisher's Note: The statements, opinions and data contained in all publications are solely those of the individual author(s) and contributor(s) and not of MDPI and/or the editor(s). MDPI and/or the editor(s) disclaim responsibility for any injury to people or property resulting from any ideas, methods, instructions or products referred to in the content.

Review

Annealing and Doping Effects on Transition Metal Dichalcogenides—Based Devices: A Review

Raksan Ko, Dong Hyun Lee and Hocheon Yoo *

Department of Electronic Engineering, Gachon University, 1342 Seongnam-daero, Seongnam 13120, Republic of Korea
* Correspondence: hyoo@gachon.ac.kr

Abstract: Transition metal dichalcogenides (TMDC) have been considered promising electronic materials in recent years. Annealing and chemical doping are two core processes used in manufacturing electronic devices to modify properties and improve device performance, where annealing enhances crystal quality, reduces defects, and enhances carrier mobility, while chemical doping modifies conductivity and introduces new energy levels within the bandgap. In this study, we investigate the annealing effects of various types of dopants, time, and ambient conditions on the diverse material properties of TMDCs, including crystal structure quality, defect density, carrier mobility, electronic properties, and energy levels within the bandgap.

Keywords: two-dimensional transition metal dichalcogenides; field effect transistor; annealing process; chemical doping

Citation: Ko, R.; Lee, D.H.; Yoo, H. Annealing and Doping Effects on Transition Metal Dichalcogenides—Based Devices: A Review. *Coatings* **2023**, *13*, 1364. https://doi.org/10.3390/coatings13081364

Academic Editor: Ning Sun

Received: 2 July 2023
Revised: 25 July 2023
Accepted: 26 July 2023
Published: 3 August 2023

Copyright: © 2023 by the authors. Licensee MDPI, Basel, Switzerland. This article is an open access article distributed under the terms and conditions of the Creative Commons Attribution (CC BY) license (https://creativecommons.org/licenses/by/4.0/).

1. Introduction

Transition metal dichalcogenides (TMDCs) have received considerable attention over the years. TMDC is a type of two-dimensional (2D) material similar to graphene [1–4], but rather than a single layer of carbon atoms in graphene, it consists of a single layer of transition metal atoms including molybdenum (Mo) [5–8], tungsten (W) [9–11], and platinum (Pt) [12–15] sandwiched between two layers of chalcogen. An atom includes sulfur (S) [16–19], selenium (Se) [20–23], and tellurium (Te) [24–26]. TMDCs offer various interesting properties, such as high electron mobility, strong optical-matter interactions, and mechanical flexibility. These properties make them promising candidates for a variety of applications in fields such as thin-film transistors [27–30], photosensors [31–33], gas detectors [34–36], neuromorphic devices [37–40], biosensors [41–44] and energy harvesting devices [45–47]. In recent years, significant efforts have been made to develop new methods for synthesizing TMDC and to explore its properties in more detail. This allowed a better understanding of the fundamental physics of TMDCs and opened new avenues for their use in a variety of applications.

Annealing is a simple yet effective method for controlling the properties of TMDCs [48–51]. This process involves heating the material to a specific temperature for a defined duration and slowly cooling it in a controlled environment. The outcome of annealing on TMDCs can be influenced by various factors, such as the annealing temperature [52–55], duration [56–58], and atmosphere [59–62]. One of the most significant advantages of annealing on TMDCs is the elimination of defects and impurities that can degrade their electronic and optical properties, thus enhancing device performance [63–65].

In addition, annealing can also affect the electronic and optical properties of TMDCs [66–69]. For instance, it can cause shifts in the bandgap energy [70,71], making it useful in optoelectronic devices [72–74]. Furthermore, annealing can improve electron mobility [75–77] and carrier concentration [78,79] and enhance electrical conductivity [80–82]. The effect of annealing on TMDC is also dependent on the atmosphere in which the process is carried out. For example, annealing in a controlled atmosphere can create sulfur or selenium voids that can alter the electronic and optical properties of the material [83,84]. Annealing in an oxidizing atmosphere can also generate an oxide layer on the material surface, which can impact electronic and optical properties [85–87].

In addition, the chemical doping method is considered to improve electrical properties of TMDC while minimizing lattice structure damage. Chemical dopant could modulate the distance between the Fermi level and the conduction band or valence band [88]. Therefore, doping the TMDC with an appropriate dopant reduces the height of the Schottky barrier and enhances the carrier density [89].

Here, we present the effects of annealing and chemical doping on the various TMDC—Based transistors. This review provides details on the TMDC transistor's annealing process in different environments, including air, nitrogen, and vacuum. Additionally, we discuss changes in the electrical characteristics of TMDC transistors resulting from organic dopants such as dielectrics and SAM. Finally, we examine the potential for performance enhancement in electronic engineering based on TMDC materials and suggest future directions for research.

2. Synthesis of Transition Metal Dichalcogenides

TMDC materials possess unique electronic and optical properties, valuable for electronic and optoelectronic devices. High-quality TMDC monolayers or few-layer structures are crucial for these applications. Mechanical exfoliation, using Scotch tape or metal assistance, provides a simple and cost-effective method for obtaining single layers from bulk crystals [90–92]. Meanwhile, chemical vapor deposition (CVD) allows controlled growth of large-area monolayers with desired properties on different substrates [93,94]. Liquid-phase exfoliation is a low-cost, scalable method for mass production but faces challenges in obtaining single atomic layers with specific structures [95,96]. Ongoing research aims to optimize exfoliation processes for addressing this issue.

3. The Fundamental Processes Occurring during Annealing in the TMDC

Annealing is utilized in shaping the crystal structure, crystallinity, and defects of TMDC, exerting crucial control over the material's electrical and optical properties [97–101]. Different from graphene, monolayer TMDC consists of sandwich structures with chalcogen atoms forming upper and lower layers and a transition metal ion plane in between. These TMDCs can be classified into 2H and 1T phases, each with distinct properties. The 2H phase exhibits excellent optical emission characteristics but limited electrical catalytic properties, while the 1T phase demonstrates superior catalytic properties but lacks light-absorbing capabilities due to its metallic nature. Through annealing, transitions between these crystal structures can be induced, enabling precise control to achieve desired characteristics.

Moreover, annealing influences the growth of crystalline domains, impacting crystal boundaries and thereby regulating charge transport and electrical properties. Additionally, the annealing process reduces defects and imperfections, leading to enhanced electrical and optical properties. In intentionally doped TMDC, annealing facilitates the diffusion of dopants into the crystal lattice, resulting in improved electronic properties and the ability to tailor specific functionalities. The outcomes of these annealing effects depend on various conditions such as temperature, time, and environment, allowing for fine tuning of TMDC properties for diverse applications.

4. Annealing Effects of TMDC—Based Transistors

4.1. Annealing Effects of MoS_2—Based Transistors

The n-doping effect of molybdenum disulfide (MoS_2) FET was demonstrated through a simple annealing process. This study performed annealing for 2 h in a nitrogen atmosphere [102]. Figure 1a shows the transfer curve of a MoS_2 FET with the effect of vacuum annealing, where the pristine MoS_2 FET exhibits a low on/off-current ratio of 100 A/A due to its high off current of 10^{-5} A. The off current of the MoS_2 FET annealed at 200 °C was below 10^{-11} A at negative gate voltage, indicating the n-type doping effect. The off current of the MoS$_2$ FET annealed at 300 °C shows a slight increase compared to that at 200 °C. The decrease in the off current is due to the rearranged surface of the MoS$_2$ channel and the removal of residual material through heat. On the other hand, increasing the annealing temperature to 400 °C resulted in a decrease in the on current and an increase in the off current, with a drain current of 10^{-5} A and an ambiguous boundary between the on/off currents. The high off current of the MoS$_2$ FET is attributed to the phase transformation to Mo$_2$S$_5$ caused by high-temperature energy. Moreover, when the electrical characteristics are measured in atmospheric environments, the adsorption of water and oxygen molecules on the annealing-treated MoS$_2$ channel region leads to an increase in the off current. Figure 1b shows the on/off current of the MoS$_2$ FET as a function of annealing temperature. The on current continues to decrease as the annealing temperature current becomes the initial drain current value. Figure 1c shows the extracted charge carrier mobility of the MoS$_2$ FET with annealing temperature. The electron mobility of a pristine MoS$_2$ FET is 8.5 cm^2 V^{-1} s^{-1}, but the highest electron mobility of 20.7 cm^2 V^{-1} s^{-1} appears at the lowest off-state current under annealing conditions of 200 °C, and as the annealing temperature increases above 200 °C, the electron mobility decreases.

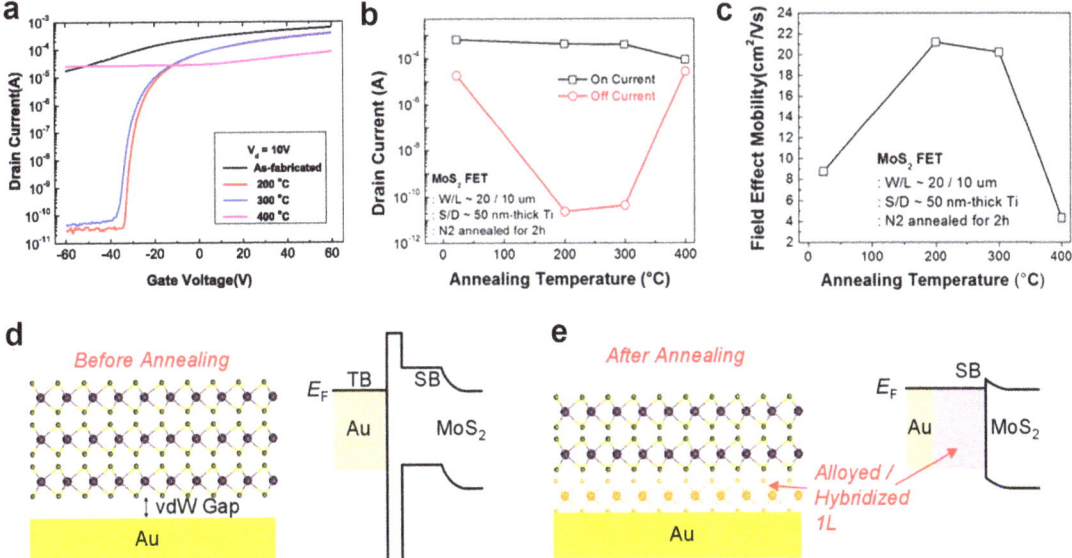

Figure 1. (a) Transfer curve of MoS$_2$ FET according to annealing temperature; (b) The plotted on- and off-current values of MoS$_2$ FET; (c) The extracted field effect mobility of MoS$_2$ FET according to annealing temperature (adapted from [102] with permission from the Springer Science and Business Media); Band structure of the Au contact electrode and MoS$_2$ channel (d) before annealing and, (e) after annealing (adapted from [103] with permission from the AIP Publishing LLC).

As another example, the induced n-doping effect in MoS$_2$ FET through a vacuum annealing process was reported by Islam et al. [103]. The pristine MoS$_2$ FET was mounted on a furnace and annealed at 250 °C under a vacuum of 15 mTorr for 1 h. The contact and channel resistances of the pristine MoS$_2$ FET were 4 MΩ and 112 kΩ, respectively, while the annealed MoS$_2$ FET had a contact resistance of 2 kΩ and a channel resistance of 171 kΩ. Figure 1d shows the band structure of the pristine MoS$_2$ FET. A vacuum annealing process reduced the contact and channel resistances of the device, despite the presence of a high contact resistance caused by the tunneling barrier between the Au contact electrode and the MoS$_2$ channel due to the Van der Waals gap. The vacuum-annealed interface between the Au contact electrode and MoS$_2$ channel is alloyed, eliminating the Van der

Waals gap and tunneling barrier (Figure 1e). Additionally, the alloyed Au contact electrode and MoS$_2$ channel had reduced work function, leading to a decreased Schottky barrier, and consequently lowering the contact resistance. This vacuum annealing process increases the mobility of MoS$_2$ FET from 0.1 to 8 cm^2 V^{-1} s^{-1} and enhances the on/off ratio 10 fold.

4.2. Annealing Effects of MoTe$_2$—Based Transistors

To change the electrical characteristics of n-type molybdenum ditelluride (MoTe$_2$) FETs, a p-type doping method through an annealing process was reported [104]. This annealing process led to a MoTe$_2$—Based complementary metal-oxide semiconductor (CMOS) inverter using a p-type MoTe$_2$ FET with an annealing process and an n-type MoTe$_2$ FET with electron beam irradiation. To manufacture a p-type MoTe$_2$ FET, a rapid thermal annealing (RTA) process was performed at 250 °C for 3 h. Figure 2a shows the molecular structure of oxidized MoTe$_2$ resulting from the annealing process. MoTe$_2$ has bonds such as Te vacancy and antisites. Upon high-temperature annealing treatment, the Te vacancy was increased and induced the absorption of oxygen molecules. As a result, oxygen molecules with high electronegativity are adsorbed onto the Te vacancy in oxidized MoTe$_2$, acting as electron acceptors and inducing a p-type doping effect on MoTe$_2$ FET. Figure 2b shows the electrical properties of the pristine MoTe$_2$ FET and the annealed MoTe$_2$ FET. After the annealing process, the MoTe$_2$ FET enhanced p-type behavior, and the hysteresis was reduced. Also, the doping effect of p-type MoTe$_2$ FET induced by high-temperature annealing lasted for 2 weeks. Figure 2c shows the voltage transfer characteristics (VTC) of a CMOS inverter implemented using an annealed treated p-type MoTe$_2$ FET and an n-type MoTe$_2$ FET fabricated by electron beam irradiation. The CMOS inverter exhibits full-swing operation over a drain voltage range of 1 V to 5 V.

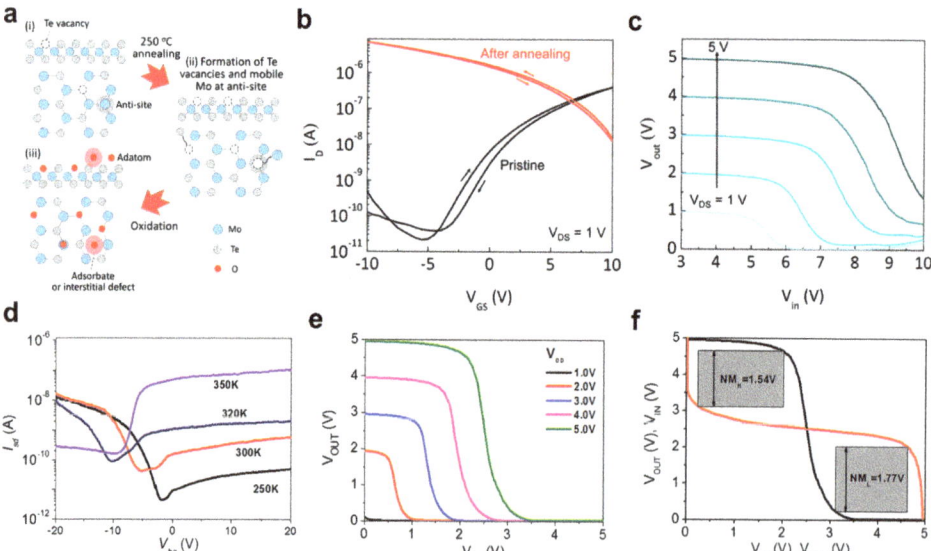

Figure 2. (a) Schematic of MoTe$_2$ structure with annealing and oxidation; (b) Transfer curve of pristine MoTe$_2$ FET and annealed MoTe$_2$ FET; (c) Voltage transfer characteristics of MoTe$_2$ based CMOS inverter (V$_{DS}$ was applied from 1 V to 5 V at 1 V interval). (adapted from [104] with permission from the John Wiley and Sons); (d) Transfer curve of MoTe$_2$ FET according to vacuum annealing temperature; (e) Voltage transfer characteristics of MoTe$_2$ inverter; (f) Voltage transfer characteristics (black lines) and their mirrors (red lines) of MoTe$_2$ inverter at V$_{DD}$ = 5 V (adapted from [105] with permission from the Springer Science and Business Media).

When oxygen or water molecules in the air are adsorbed onto the MoTe$_2$ channel, it operates as a *p*-type transistor. However, the vacuum annealing process removes the adsorbed molecules from the MoTe$_2$ channel. Consequently, Te vacancies are generated in the MoTe$_2$ channel due to the removed adsorbates, resulting in the acquisition of n-type conductivity [105]. Figure 2d shows the transfer curves of the n-type MoTe$_2$ FET through the vacuum annealing process. The pristine MoTe$_2$ FET operates as a *p*-type due to the incomplete removal of adsorbates from the MoTe$_2$ channel. In contrast, when the temperature is increased to 76.85 °C, the drain current at both positive and negative gate voltages decreases and increases, respectively. The n-type MoTe$_2$ FET has on/off ratio of 3.8×10^2 A/A, an S.S of 1.1 V/dec, and electron mobility of 2 cm^2 V^{-1} s^{-1}. In addition, the inverter was demonstrated using unipolar *p*-type and n-type MoTe$_2$ FETs. Figure 2e shows the VTC characteristics of the MoTe$_2$ inverter. The transition voltage was half of V$_{DD}$ due to the transfer curve of *p*- and n-type of MoTe$_2$ FET being symmetrical. Figure 2f shows the noise margin and VTC curve of the MoTe$_2$ inverter when V$_{DD}$ is 5 V. The low-level noise margin (NM$_L$) is 1.54 V, while the high-level noise margin (NM$_H$) is 1.77 V.

4.3. Annealing Effects of WSe$_2$—Based Transistors

The work function of the electrode and the subsequent annealing process changed the electrical characteristics of the tungsten diselenide (WSe$_2$) FET. Bandyopadhyay et al. fabricated WSe$_2$ FET using various metal electrodes with different work functions, such as gold (Au, ϕ_m = ~5.4 eV), molybdenum (Mo, ϕ_m = ~4.53 eV), and aluminum (Al, ϕ_m = ~4.08 eV) [106]. The WSe$_2$ FET with metal electrodes of different work functions operates as both n-type and *p*-type transistors. Furthermore, the field effect mobility and on current were increased by 300 °C vacuum, annealing for 3 h. Figure 3a shows the transfer curve of a WSe$_2$ FET with Au/Ti electrodes. The hole mobility of the pristine Au/Ti/WSe$_2$ FET was 432 cm^2 V^{-1} s^{-1}, but it increased by 150% to 625 cm^2 V^{-1} s^{-1} after the annealing process with enhancing *p*-type operation. This indicates the interface properties between the WSe$_2$ channel and gate dielectric are enhanced by annealing effects. Figure 3b shows the transfer curve of WSe$_2$ FETs with Mo electrodes. The pristine Mo/WSe$_2$ FET had ambipolar behavior with *p*-type dominance, and the hole and electron mobilities were 210 and 30 cm^2 V^{-1} s^{-1}, respectively. The additional annealing process further improved the ambipolar conduction of the Mo/WSe$_2$ FET. The annealed Mo/WSe$_2$ FET had a hole mobility of 410 cm^2 V^{-1} s^{-1} and an electron mobility of 70 cm^2 V^{-1} s^{-1}. On the other hand, WSe$_2$ FET with Al electrodes resulted in an n-type operation (Figure 3c). The electron mobility of the Al/WSe$_2$ FET before and after annealing was 242 and 366 cm^2 V^{-1} s^{-1}, respectively. In summary, the conductivity of WSe$_2$ FETs with dissimilar electrode work functions can be controlled as either n-type or *p*-type, and additional vacuum annealing shows the annealing effects of the device, inducing high mobility.

The annealing process performed in ambient air causes the formation of an oxide layer (WO$_3$) due to the interaction between the WSe$_2$ surface and oxygen molecules. The *p*-doping effect on ambipolar WSe$_2$ FET using an annealing process in ambient environment was observed [107]. The work function of WO$_3$ (~6.7 eV) formed on the WSe$_2$ surface under this condition is lower than that of WSe$_2$ (~4.4 eV), resulting in a WSe$_2$ FET operating as a *p*-type transistor. Figure 3d shows the modified transfer curves of the ambipolar WSe$_2$ FET after annealing in air ambient. The WSe$_2$ FET was annealed at 200 °C on a hot plate for 1 h. After annealing, the n-type current (V$_G$ = 70 V) of the ambipolar WSe$_2$ FET decreased at all drain voltages (V$_D$ = 4 V, 6 V), and the *p*-type current increased (V$_G$ = −70 V). Also, the hole and electron mobility before annealing are 0.13 and 5.5 cm^2 V^{-1} s^{-1}, respectively. In contrast, when WSe$_2$ FET was annealed, the extracted hole mobility increased up to 1.3 cm^2 V^{-1} s^{-1}, while the electron mobility decreased to 0.69 cm^2 V^{-1} s^{-1}. Figure 3e,f show the photo-switching characteristics of the pristine WSe$_2$ FET and the annealed WSe$_2$ FET. A 405 nm laser with a power density of 11 mW/cm^2 was applied. In the switching operation, the rise time and decay time of the pristine WSe$_2$ FET were 92.3 s and 57.6 s, respectively. In contrast, the rise time of the annealed WSe$_2$ FET decreased significantly by 610 times to 0.15 s. Also, the decay time decreased to less than 0.33 s with a 170 times difference. The improved optical switching behavior is attributed to the lattice mismatch between WSe$_2$ and WO$_3$ generated by the proposed annealing process, and the recombination of photo-generated electron-hole pairs is facilitated by the trap regions resulting from the lattice mismatch.

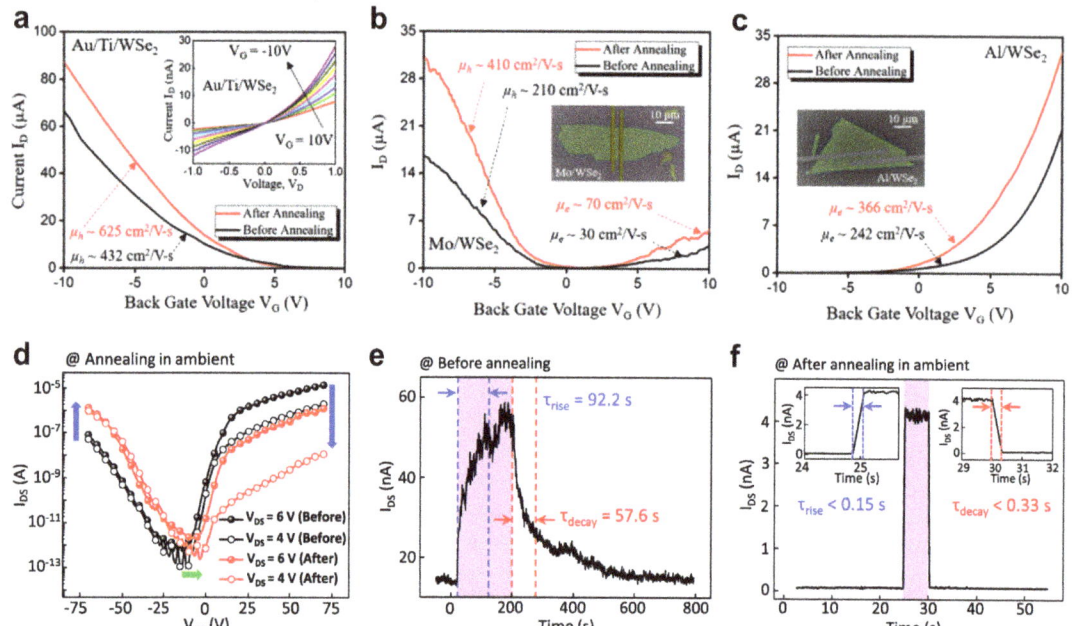

Figure 3. Transfer curve of WSe$_2$ FETs with various metal electrodes such as (**a**) gold/titanium, (**b**) molybdenum, and (**c**) aluminum (adapted from [106] with permission from the Elsevier B.V.); (**d**) Transfer curve of pristine WSe$_2$ FET and annealed WSe$_2$ FET under air ambient; Photo-switching behavior of (**e**) pristine WSe$_2$ FET and, (**f**) WSe$_2$ FET annealing at air ambient (adapted from [107] with permission from the Springer Science and Business Media).

4.4. Annealing Effects of WS$_2$—Based Transistors

The n-type behavior of multilayer tungsten disulfide (WS$_2$) FET through a double annealing process, performed both before and after electrode deposition, was demonstrated by Ji et al. in 2022 [108]. The first annealing process was carried out for 2 h at 200 °C on a WS$_2$ transferred substrate mounted in a vacuum tube furnace. The first annealing process removed the organic residue and improved the WS$_2$/electrode interface by desorbing the surface adsorbent. The second annealing process was also performed in a vacuum tube furnace at 200 °C after electrode deposition. Figure 4a shows the transfer curve of pristine WS$_2$ FET (Group A) and annealed after electrode deposition WS$_2$ FET (Group B). The on/off ratio and field effect mobility of pristine WS$_2$ FET were 1.9×10^5 A/A and 11.1 cm^2 V^{-1} s^{-1}, respectively. In contrast, the WS$_2$ FET performed by annealing process after electrode deposition had 1.6×10^6 A/A of on/off ratio and 20.8 cm^2 V^{-1} s^{-1} of field effect mobility. Also, the output curves of WS$_2$ FET from groups A and B showed a decrease in contact resistance from ~6.7×10^2 kΩ·μm to ~3.8×10^2 kΩ·μm (Figure 4b). Statistical analysis of the field effect mobility for 50 WS$_2$ FET in each group was also performed (Figure 4c). Group C performed with an annealing process after the WS$_2$ flake transfer, while Group D comprised WS$_2$ FETs that carried out a double annealing process. The average field effect mobility of pristine WS$_2$ FET was the lowest at 5.6 cm^2 V^{-1} s^{-1}.

The average field effect mobility with the first and second annealing processes was 14.8 and 16.0 cm^2 V^{-1} s^{-1}, respectively. In contrast, the average field effect mobility of WS$_2$ FET performed double annealing process was the highest at 23.8 cm^2 V^{-1} s^{-1}. Thus, the mobility and contact resistance of the WS$_2$ FET were improved through double annealing processing.

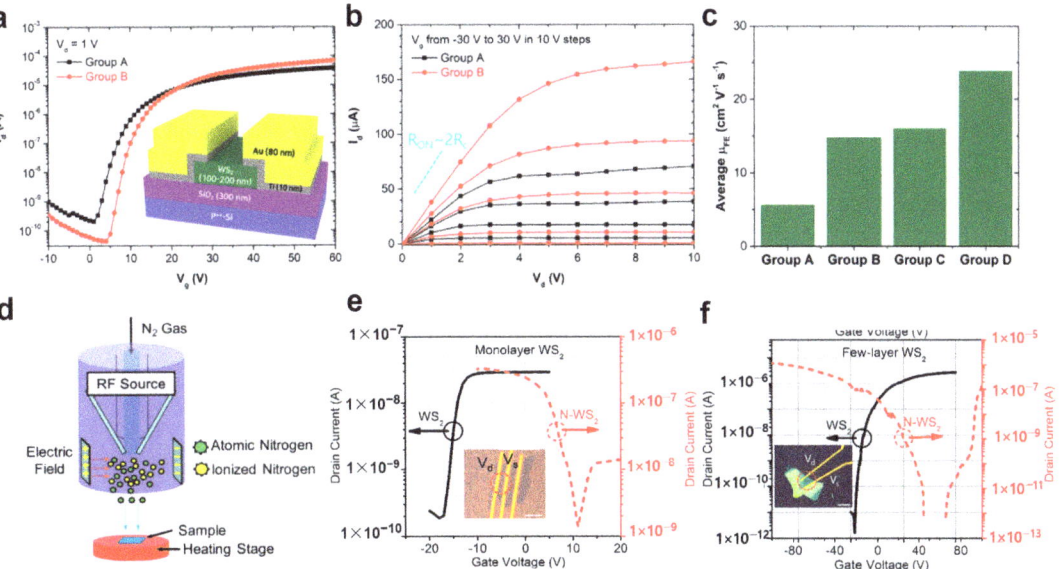

Figure 4. Electrical characteristics comparison between the pristine WS$_2$ FET and the annealed after electrode deposition WS$_2$ FET using (**a**) Transfer curves and, (**b**) Output curves; (**c**) Statistical analysis of 50 WS$_2$ FET with A to D groups (adapted from [108] with permission from the Elsevier B.V.); (**d**) 3D schematic of the WS$_2$ nitrogen substitutional doping and annealing process; Transfer curves of pristine WS$_2$ FET and nitrogen doping WS$_2$ FET with (**e**) monolayer WS$_2$ and, (**f**) few layer WS$_2$ (adapted from [109] with permission from the American Chemical Society).

WS$_2$ typically exhibits n-type conductivity in both exfoliated flakes and chemical vapor deposition (CVD) growth. Nitrogen substitution doping and annealing manifest *p*-doping effects on WS$_2$ FETs through Fermi-level pinning, which is a technological achievement for implementing TMDC—Based CMOS FETs [109]. The WS$_2$ films were grown using radio frequency (RF) magnetron sputtering, followed by nitrogen annealing at 300 °C under a pressure of 7.0 × 10^{-5} mbar. The nitrogen radicals composed of ionized nitrogen and atomic nitrogen were effused into WS$_2$ samples mounted on the main chamber and created W–N bonds (Figure 4d). The W–N bonds created by substituting S atoms in WS$_2$ with N atoms result in an acceptor level 0.24 eV lower than the conduction band edge of WS$_2$, enabling *p*-type doping. Figure 4e shows the transfer curves of a monolayer n-type WS$_2$ and a *p*-type WS$_2$ FET after nitrogen treatment. The hole mobility and the threshold voltage of the n-type WS$_2$ FET are 0.53 cm^2 V^{-1} s^{-1} and 9 V, respectively. However, a *p*-type WS$_2$ FET with nitrogen substitutional doping had 1.70 cm^2 V^{-1} s^{-1} of hole mobility and −12 V of the threshold voltage. Similarly, nitrogen annealing also enables the *p*-type operation of multilayer WS$_2$ FET (Figure 4f). The *p*-type doping effect of WS$_2$ FET by nitrogen substitutional doping and the annealing process is due to Fermi-level pinning.

The electrical properties of TMDCs can be changed depending on the annealing temperature and the environment (Table 1). Thus, depending on the intended application, the appropriate type of TMDC should be selected, and the required electrical properties should be achieved with suitable processing methods. Table 1 summarizes the characteristics of TMDC—Based transistors depending on the annealing conditions.

Table 1. The influence of annealing on the characteristics of TMDC—Based transistors.

Materials	Annealing Temperature	Annealing Time	Annealing Ambient	Doping Effect	Mobility	Ref.
MoS_2	200 °C	2 h	Nitrogen	n-type doping	20.7 cm^2 V^{-1} s^{-1}	[102]
MoS_2	250 °C	1 h	Vacuum	n-type doping	8 cm^2 V^{-1} s^{-1}	[103]
$MoTe_2$	250 °C	3 h	Vacuum	p-type doping	N/A	[104]
$MoTe_2$	76.85 °C	N/A	Vacuum	n-type doping	2 cm^2 V^{-1} s^{-1}	[105]
WSe_2	300 °C	3 h	Vacuum	p-type doping	625 cm^2 V^{-1} s^{-1}	[106]
WSe_2	200 °C	1 h	air	p-type doping	1.3 cm^2 V^{-1} s^{-1}	[107]
WS_2	200 °C (double)	2 h	Vacuum	n-type doping	23.8 cm^2 V^{-1} s^{-1}	[108]
WS_2	300 °C	N/A	Nitrogen	p-type doping	1.7 cm^2 V^{-1} s^{-1}	[109]

5. Annealing and Chemical Doping Effects in TMDC—Based Transistors

5.1. Annealing and Chemical Doping Effects in MoS$_2$—Based Transistors

The formation of a triphenylphosphine (PPh$_3$) layer can be controlled by annealing temperature. The use of PPh$_3$ provided the n-doping effect of MoS$_2$ FET through continuous annealing at three different temperatures (150, 250, and 350 °C) after spin coating a 7.5 wt% PPh$_3$ solution on the surface of transistor [110]. The coated phosphorus atoms of PPh$_3$ transfer electrons to the MoS$_2$ surface, inducing an n-doping effect. Also, the PPh$_3$ layers increase the number of electrons moving to the MoS$_2$ channel as the annealing temperature increases. Raman spectroscopy was performed to analyze the n-doping effect of MoS$_2$. Figure 5a shows the redshift values of the E_{2g}^1 peak and the A_{1g} peak of MoS$_2$ according to the annealing temperature. At 150 °C annealing conditions, both the E_{2g}^1 peak and the A_{1g} peak are redshifted by −0.78 cm^1. As the annealing temperature increases to 350 °C, the E_{2g}^{-1} peak and the A_{1g} peak are redshifted by −2.16 cm^{-1} and −2.75 cm^{-1}, respectively. This indicates that PPh$_3$ formation at higher annealing temperatures results in an enhanced n-doping effect. Figure 5b shows the transfer curve of pristine MoS$_2$ and PPh$_3$-doped MoS$_2$ FET with annealed from 150 °C to 350 °C. As the annealing temperature increases, the on current of the PPh$_3$-doped MoS$_2$ FET increases, and the threshold voltage shifts towards negative gate voltages. Furthermore, the energy barrier height of PPh$_3$ decreases, resulting in a decrease in contact resistance from 2.82 kΩ to 0.24 kΩ (Figure 5c). In addition, the on/off-current ratio and carrier mobility of PPh$_3$-doped MoS$_2$ FET improved to 8.70 × 10^5 A/A and 241 cm^2 V^{-1} s^{-1}, respectively, from their initial values of 8.72 × 10^4 A/A and 12.1 cm^2 V^{-1} s^{-1}.

In the demonstration, an ambipolar MoS$_2$ FET with strong p-type conductivity was achieved through chemical doping and annealing effects. In 2021, Lee et al. spin-coated a 10 mg mL^{-1} solution of poly(9,9-di-n-octylfluorenyl-2,7-diyl) (PFO) onto the device and annealed it on a hotplate at various temperatures of 60, 300, and 350 °C for 10 min [111]. Figure 5d illustrates the 3D schematic diagram of the doping process of a PFO-doped MoS$_2$ FET. Additionally, the electrical characteristics of a PFO-doped MoS$_2$ FET were measured at various annealing temperatures (Figure 5e). The pristine MoS$_2$ FET shows conventional n-type behavior and has 0.93 µA of on current at V_G = 30 V and 10^2 A/A of on/off ratio. However, as the annealing temperature was increased, the PFO-doped MoS$_2$ FET exhibited ambipolar behavior due to the enhanced p-type conductivity. Also, at an annealing temperature of 350 °C, the reduced n-type current and increased p-type current suggest ambipolar behavior of MoS$_2$ FETs by PFO doping. Figure 5f shows the variation of n-type and p-type currents with pristine MoS$_2$ FET and PFO-doped MoS$_2$ FET. As PFO doping was performed and the annealing temperature increases, the p-type current gradually increases. Also, the p-type current of 350 °C annealed PFO-doped MoS$_2$ FET was 0.20 µA with 24 times increase compared to the pristine MoS$_2$ FET. On the other hand, the on current decreased 14 times from 0.93 µA to 0.06 µA.

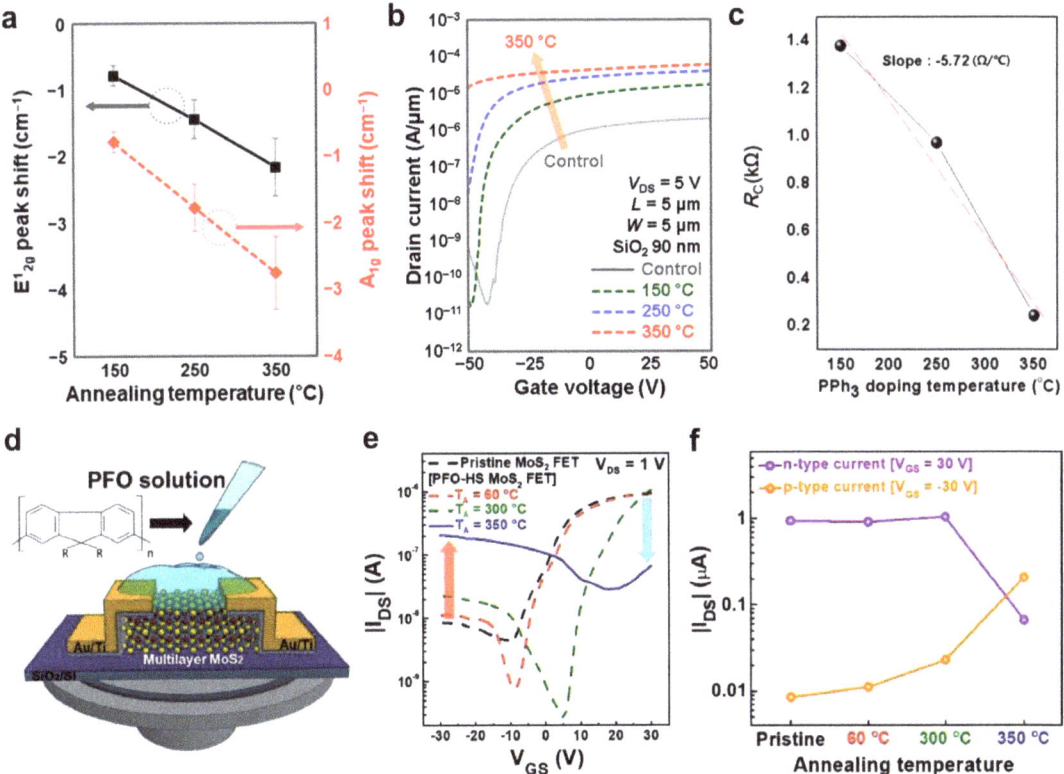

Figure 5. (a) Raman peak shift values against the formation temperatures of PPh$_3$ (black line: E$_{2g}^1$ peak, red line: A$_{1g}$ peak); (b) Transfer curve of pristine MoS$_2$ FET and PPh$_3$-doped MoS$_2$ FET annealing at 3 different temperatures (150, 250, and 350 °C); (c) Contact resistance of PPh$_3$-doped MoS$_2$ FET measured at various annealing temperatures (black line: 150, 250, and 350 °C, red line: line profile of contact resistance); (adapted from [110] with permission from the American Chemical Society); (d) 3D schematic of PFO-doped MoS$_2$ FET, where red spheres represent Mo atoms, and yellow spheres represent S atoms. The inset shows the chemical structure of poly(9,9-di-n-octylfluorenyl-2,7-diyl) (PFO); Electrical characteristics of the pristine MoS$_2$ FET and PFO-doped MoS$_2$ FET annealed at various temperature (60, 300, and 350 °C); (e) Transfer curve; (f) Extracted n- and p-type current variation (adapted from [111] with permission from the Elsevier B.V.).

5.2. Annealing and Chemical Doping Effects in MoTe$_2$—Based Transistors

MoTe$_2$ exhibits instability in the surrounding environment due to oxidation. The oxidation achieved p-type doping of the device and improved stability in the surrounding environment through bis(trifluoromethane)sulfonamide (TFSI) doping and PMMA encapsulation [112]. The mechanically exfoliated MoTe$_2$ flakes were transferred onto a substrate and coated with PMMA, followed by annealing at 100 °C. The flakes and half of the electrodes were etched to expose the PMMA-coated device and the uncoated device. Subsequently, both samples were immersed in a nonreactive atmosphere for a 5 min period in a chlorobenzene (CB) solvent solution containing TFSI. Figure 6a illustrates a schematic diagram of the device. Chlorobenzene (CB), a chlorine—Based solvent containing lone electron pairs, can act as an electron donor to MoTe$_2$, resulting in an n-type doping effect. However, the electron-donating effect is nullified by the electron-withdrawing effect of TFSI. Figure 6b illustrates the transfer characteristics of TSFI-MoTe$_2$ FETs with and without a PMMA layer, compared to the pristine MoTe$_2$ FET. After TSFI doping, the device without a PMMA layer exhibited unipolar *p*-type behavior with an on/off ratio exceeding 10^3 A/A and a maximum mobility of 0.58 cm^2 V^{-1} s^{-1}. The decrease in current in the n-branch indicates that TSFI induces *p*-type doping. The significant shift of the threshold voltage to 0 V after PMMA coating reflects the improved efficiency of TSFI doping for achieving *p*-type operation in the device. Conclusively, the device doped with TSFI along with a PMMA layer exhibited unipolar *p*-type behavior with an on/off ratio exceeding 10^6 A/A and a maximum mobility of 30 cm^2 V^{-1} s^{-1}, demonstrating a 250-fold improvement compared to the pristine device. This demonstrates superior on/off ratio and mobility enhancement characteristics compared to other doped TMD FETs.

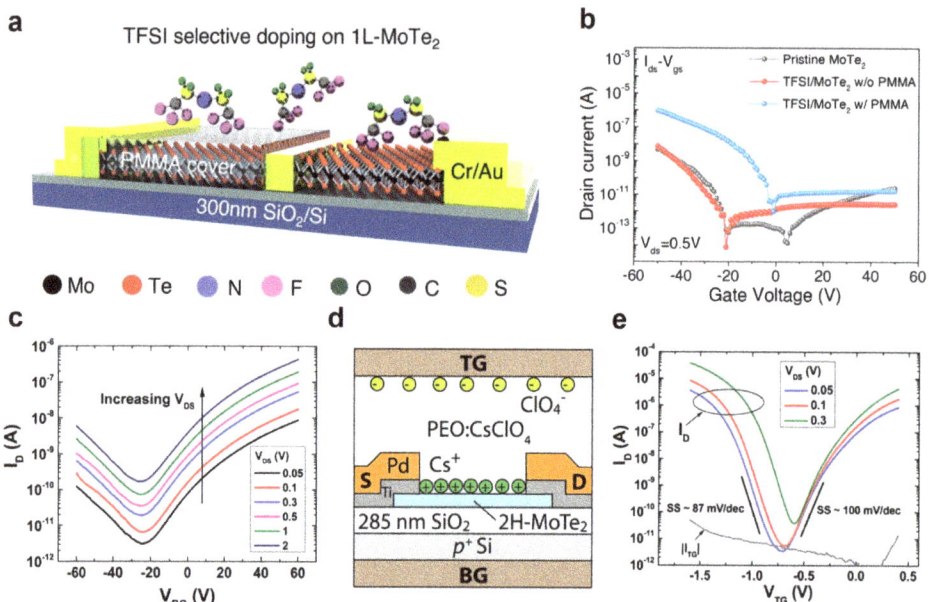

Figure 6. (a) Structure of TFSI-MoTe$_2$ with areas encapsulated by PMMA and those without encapsulation; (b) Transfer characteristics of pristine MoTe$_2$ device compared to TFSI-MoTe$_2$ with areas encapsulated by PMMA and those without encapsulation (adapted from [112] with permission from the Applied Surface Science); (c) Transfer curve of pristine MoTe$_2$ FET with back-gate structure; (d) Cross section of ion gate MoTe$_2$ FET; (e) Transfer curve of ion gate MoTe$_2$ FET with various supply voltage (adapted from [113] with permission from the American Chemical Society).

The ambipolar MoTe$_2$ FET doped with poly(ethylene oxide) (PEO) and CsClO$_4$ as the top-gate dielectric exhibited both n-doping and *p*-doping effects [113]. The PEO:CsClO$_4$ solution was drop cast on the pristine MoTe$_2$ FET with back-gated structure and was annealed at 90 °C for 3 min. Figure 6c shows the transfer curve of a pristine MoTe$_2$ FET. The pristine MoTe$_2$ FET shows ambipolar behavior at various drain-source voltages (0.05 V to 2 V). Figure 6d shows a cross-section of a MoTe$_2$ FET doped with a top-gate dielectric of PEO:CsClO$_4$. To construct the ion gate, the PEO:CsClO$_4$ with gate dielectric and palladium (Pd) top gate electrodes were sequentially deposited on the MoTe$_2$ channel. A positive or negative gate voltage applied to the top gate electrode induces n-type and *p*-type doping, respectively. The positive gate voltage applied to the top gate electrode induces electrons in the channel due to Cs$^+$ ions, resulting in n-type doping effect. On the other hand, a negative gate voltage results in *p*-type doping of MoTe$_2$ due to ClO$_4^-$ ions. Figure 6e shows the transfer curve of an ion-gate MoTe$_2$ FET. Compared to pristine MoTe$_2$ FETs, the on/off ratio and on current of ion-gate MoTe$_2$ FET increased 20 times and 40 times in the n-branch and *p*-branch, respectively (V_{DS} = 0.05 V).

5.3. Annealing and Chemical Doping Effects in WSe$_2$—Based Transistors

Using self-assembled monolayers (SAM) and a double annealing process, the changes in the characteristics of WSe$_2$ FET was observed. The device was immersed in an octadecyl-trichlorosilane (OTS) solution, which served as the material for SAM, and annealed at 120 °C for 20 min [114]. Figure 7a shows the 3D schematic and energy band diagram of the OTS-doped WSe$_2$ FET. The methyl groups of OTS doping on the WSe$_2$ surface possess positive poles, and the electrons accumulate on the WSe$_2$ surface due to the dipolar effect of OTS, resulting in the occurrence of a p-type doping effect. Also, as the concentration of OTS solution was increased, the p-type doping effect on WSe$_2$ FET was enhanced. The OTS doping effect on WSe$_2$ FETs reduces the Schottky barrier height. The upward valence band of the OTS-doped WSe$_2$ FET enhances the hole injection by the narrowed tunneling barrier. Figure 7b shows the transfer curves of a pristine WSe$_2$ FET and a WSe$_2$ FET doped with OTS at a concentration of 1.2%. The pristine WSe$_2$ FET has a threshold voltage of −9.1 V and on current of 1.64×10^{-6} A/μm. On the other hand, threshold voltage and on current of WSe$_2$ FET doped with 1.2% OTS were −0.45 V and 1.42×10^{-5} A/μm, respectively. In addition, the mobility increased from 30 ± 4 cm^2 V^{-1} s^{-1} (before doping) to 192 cm^2 V^{-1} s^{-1} for the device doped with 0.024% OTS (~105 cm^2 V^{-1} s^{-1} in the 1.2% OTS-doped device). Figure 7c shows the extracted threshold voltage of OTS-doped WSe$_2$ FETs according to time in an air ambient. The threshold voltage of WSe$_2$ FETs doped with various OTS concentrations (0.024, 0.24, 0.12, 1.2%) is shifted towards a negative gate voltage due to moisture in the air. However, an additional annealing process at 120 °C confirmed that shifts in the threshold voltage back to positive values restores the p-doping effect. The additional annealing process reduces the Si–O–Si bonds of OTS formed in an air exposure and increases the Si–OH bonds that enhance the p-doping effect.

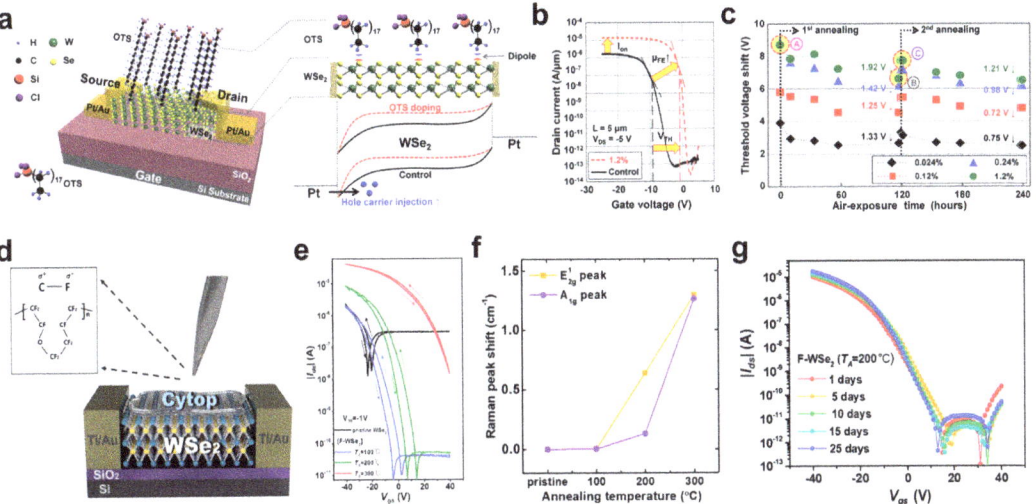

Figure 7. (a) 3D structure and energy band diagram of OTS-doped WSe$_2$ FET; (b) Transfer curve of pristine WSe$_2$ FET and OTS-doped WSe$_2$ FET; (c) Extracted threshold voltage of OTS-doped WSe$_2$ FET according to air exposure time (adapted from [114] with permission from the American Chemical Society); (d) 3D schematic of CYTOP-doped WSe$_2$ FET and inset shows the chemical structure of CYTOP; (e) Transfer curve of pristine WSe$_2$ FET and CYTOP-doped WSe$_2$ FET with various annealing temperature (100, 200, and 300 °C); (f) Raman peak shift of pristine WSe$_2$ FET and CYTOP-doped WSe$_2$ FET; (g) Electrical stability of CYTOP-doped WSe$_2$ FET over 25 days in air condition (adapted from [115] with permission from the MDPI).

Meanwhile, In 2021, the p-doping effect of WSe$_2$ FET was reported by means of the fluoropolymer CYTOP as a p-type dopant [115]. The pristine WSe$_2$ FETs were coated with CYTOP solution and annealed at various temperatures (100, 200, and 300 °C). Figure 7d shows a 3D schematic of the CYTOP-doped WSe$_2$ FET. The C–F bond of CYTOP coated on the WSe$_2$ surface enhances hole accumulation, resulting in the increased p-type current. Figure 7e shows the transfer curves of pristine WSe$_2$ FET and CYTOP-doped WSe$_2$ FETs. The on and off currents of the pristine WSe$_2$ FET are 2.30×10^{-6} A and 1.26×10^{-8} A, respectively. However, as CYTOP is doped and the annealing temperature increases, the p-type current increases and the n-type current significantly decreases. The off current of the CYTOP-doped WSe$_2$ FET annealed at 100 °C is reduced to 8.46×10^{-12} A. Subsequently, at the annealing temperature of 200 °C, the on current increases to 8.52×10^{-6} A. In particular, the CYTOP-doped WSe$_2$ FET with 300 °C annealing processed has an on current of 4.10×10^{-5} A, which is 6 times higher than the initial value, and off current of 3.15×10^{-6} A. In addition, CYTOP coating and additional annealing treatments remove impurities on the surface of WSe$_2$ and reduce hysteresis due to reduced traps. Also, the p-doping effect of CYTOP was investigated by Raman analysis (Figure 7f). The A_{1g} and E_{2g}^1 peaks of CYTOP-coated WSe$_2$ become blue shifted, and their values increase as the annealing temperature increases. At last, the electrical characteristics of a CYTOP-doped WSe$_2$ FET annealed at 100 °C were stable for 25 days in air exposure (Figure 7g).

5.4. Annealing and Chemical Doping Effects in MoSe$_2$ and WS$_2$—Based Transistor

The molecular arrangement of poly-(diketopyrrolopyrrole terthiophene) (PDPP3T) can be controlled through the adjustment of annealing temperature. Yoo et al. demonstrated the n-doping effect in multilayer molybdenum diselenide (MoSe$_2$) FET by doping PDPP3T [116]. In particular, at annealing temperatures above 200 °C, the molecular structure of PDPP3T changed edge-on state with molecules oriented vertically. As a result, the edge-on PDPP3T induced an enhanced charge transport effect to MoSe$_2$ FET due to the superimposed molecular dipole moment. Figure 8a shows the 3D schematic of the PDPP3T-doped MoSe$_2$ FET. The coated PDPP3T was annealed at various temperatures (100, 200, and 300 °C). Figure 8b shows the transfer curves of pristine MoSe$_2$ FET and PDPP3T-doped MoSe$_2$ FET according to the annealing temperature. The pristine MoSe$_2$ FET had ambipolar behavior with enhanced p-type conductivity with 10^4 A/A of on/off ratio. Annealing temperatures of 100 °C and 200 °C slightly increase the n-type current and p-type current of the PDPP3T-doped MoSe$_2$ FET. However, the annealing process at 300 °C increased the n-type current rapidly and decreased the p-type current. In particular, the n-type current of 1.5×10^{-6} A has improved 2000 times compared to the initial value of 7×10^{-10} A. Also, the PDPP3T-doped MoSe$_2$ FET has enhanced photoresponsivity compared to the pristine device at different wavelengths of light (832, 638, and 405 nm) (Figure 8c). Also, a maximum photoresponsivity of 91.2 AW^{-1} was obtained when light with a wavelength of 638 nm is applied at an intensity of 20 mW/cm^2.

The n-doping effect in WS$_2$ FETs has been reported using lithium fluoride (LiF) as a dopant [117]. The pristine WS$_2$ FET is immersed in 0.01 M of dopant solution and then annealed at 80 °C for 2 min to achieve the n-doping effect. Figure 8d shows the transfer curve of the n-type doped WS$_2$ FET using LiF dopant. The on/off ratio and the threshold voltage of the pristine WS$_2$ FET are 4.83×10^5 A/A and 14 V, respectively. However, the LiF-doped WS$_2$ FET has an on/off ratio of 1.05×10^6 A/A and a threshold voltage of -6 V due to the negative shifted gate voltage. Also, the contact resistance was extracted using the transfer length method (TLM). Figure 8e shows the TLM resistance of the WS$_2$ FET before and after doping according to the channel length. The doping effect of the WS$_2$ FET on the LiF significantly reduces the channel and contact resistance, especially leading to the lowest contact resistance of 0.9 kΩ·μm. Figure 8f shows the changed mobility and threshold voltage according to LiF immersing time. As the immersion time in the LiF solution increases, the mobility increases, and the threshold voltage shifts towards the negative gate voltage. Also, the LiF immersing time of 120 min improves the mobility of the WS$_2$ FET from 13.2 to 34.7 cm^2 V^{-1} s^{-1}.

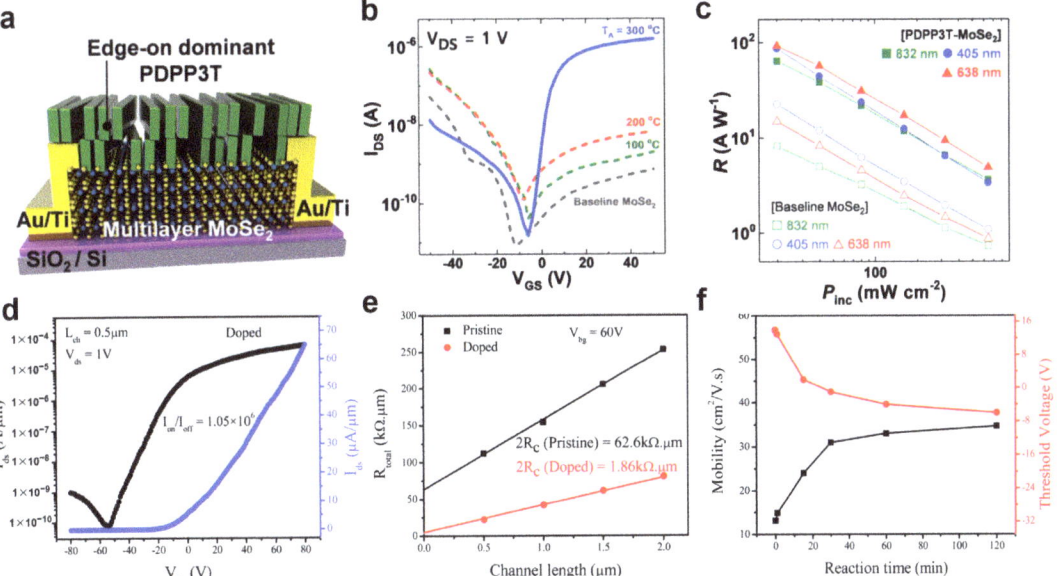

Figure 8. (**a**) 3D schematic of PDPP3T-doped MoSe$_2$ FET; (**b**) Transfer curve of pristine MoSe$_2$ FET and PDPP3T-doped MoSe$_2$ FET with various annealing temperatures (100, 200, and 300 °C); (**c**) comparison photoresponsivity under various light irradiation (832, 638, and 405 nm) (adapted from [116] with permission from the John Wiley and Sons); (**d**) Transfer curve of LiF-doped WS$_2$ FET with log and linear scale; (**e**) Channel and contact resistance of pristine WS$_2$ FET and LiF-doped WS$_2$ FET according to channel length; (**f**) Extracted mobility and threshold voltage of LiF-doped WS$_2$ FET with LiF immersing time (adapted from [117] with permission from the American Chemical Society).

Chemically doped TMDC FETs exhibit changes in electrical properties as both the TMDC and the chemical dopant are affected by the annealing process. Hence, to achieve the target electrical properties, optimized annealing conditions and appropriate chemical doping should be employed. Table 2 summarizes the post-annealing characteristics of chemically doped TMDC—Based transistors.

Table 2. The annealing effect on the characteristics of TMDC—Based transistors doped with chemical dopants.

Materials	Dopant	Doping Method	Annealing Temperature	Doping Effect	Mobility	Ref.
MoS$_2$	PPh$_3$	Spin-coating	350 °C	n-type doping	241 cm^2 V^{-1} s^{-1}	[110]
MoS$_2$	PFO	Spin-coating	350 °C	p-type doping	0.24 cm^2 V^{-1} s^{-1}	[111]
MoTe$_2$	TFSI	Dipping	100 °C	p-type doping	30 cm^2 V^{-1} s^{-1}	[112]
MoTe$_2$	PEO:CsClO$_4$	Drop-casting	90 °C	n- and p-type doping	7 cm^2 V^{-1} s^{-1} (electron) 26 cm^2 V^{-1} s^{-1} (hole)	[113]
WSe$_2$	OTS	Dipping	120 °C	p-type doping	192 cm^2 V^{-1} s^{-1}	[114]
WSe$_2$	CYTOP	Spin-coating	200 °C	p-type doping	85 cm^2 V^{-1} s^{-1}	[115]
MoSe$_2$	PDPP3T	Spin-coating	300 °C	n-type doping	75.6 cm^2 V^{-1} s^{-1}	[116]
WS$_2$	LiF	Dipping	80 °C	n-type doping	34.7 cm^2 V^{-1} s^{-1}	[117]

6. Conclusions and Future Aspect

This review discusses diverse techniques aimed at enhancing the electrical characteristics of transistors using TMDCs. The methods explored include annealing processes and chemical doping, which serve as effective means for engineering the electrical properties of TMDC transistors. Studies have examined the impact of annealing processes performed in various atmospheres, such as air, vacuum, and nitrogen, on the performance of TMDC—Based transistors. Additionally, the bandgap structure and Schottky barrier of TMDC can be controlled through functionalization with chemical dopants, such as polymers, SAMs, organic molecules, and inorganic compounds. Optimized annealing processes can further enhance the doping effect of TMDC—Based transistors.

However, several technological barriers remain to be addressed for the development and industrialization of annealing processing and chemical doping techniques for TMDC. Precise control of temperature, humidity, and ambient conditions during the TMDC annealing process is essential to achieve consistent experimental results. Ensuring uniform functionalization of dopants on the TMDC surface is crucial for accurately identifying the doping mechanism. Moreover, the thermal and chemical stability of the dopants should be considered to improve compatibility with subsequent processes. Efforts are ongoing to overcome these barriers through various technological advancements and research initiatives. Continuous optimization of research has significantly expanded the applicability of TMDC—Based devices in diverse fields, including photodetectors, neuromorphic sensors, and logic circuits. Innovative approaches, such as advanced annealing processes, chemical doping methods, integration with other materials, and the development of new device architectures, are actively being explored. As a result of these efforts, TMDC—Based FETs demonstrate great potential as semiconductor devices and are poised for success in industrialization. With the ongoing improvements and innovations, TMDC—Based FETs are expected to play a crucial role in various technological applications.

Author Contributions: Conceptualization: H.Y.; Literature survey: R.K. and D.H.L.; Writing draft: R.K. and D.H.L.; review and editing, R.K., D.H.L. and H.Y. All authors have read and agreed to the published version of the manuscript.

Funding: This work was supported by the Technology Innovation Program (00144300, Interface Technology of 3D Stacked Heterogeous System for SCM—Based Process-in-Memory) funded by the Ministry of Trade, Industry & Energy (MOTIE, Korea). This work was supported in part by the Gachon University Research Fund of 2021 (GCU-202106380001).

Institutional Review Board Statement: Not applicable.

Informed Consent Statement: Not applicable.

Data Availability Statement: Not applicable.

Conflicts of Interest: The authors declare no conflict of interest.

References

1. Stankovich, S.; Dikin, D.A.; Dommett, G.H.; Kohlhaas, K.M.; Zimney, E.J.; Stach, E.A.; Piner, R.D.; Nguyen, S.T.; Ruoff, R.S. Graphene—Based composite materials. *Nature* **2006**, *442*, 282–286. [CrossRef] [PubMed]
2. Tiwari, S.K.; Mishra, R.K.; Ha, S.K.; Huczko, A. Evolution of graphene oxide and graphene: From imagination to industrialization. *ChemNanoMat* **2018**, *4*, 598–620. [CrossRef]
3. Hughes, Z.E.; Walsh, T.R. Computational chemistry for graphene—Based energy applications: Progress and challenges. *Nanoscale* **2015**, *7*, 6883–6908. [CrossRef] [PubMed]
4. Tiwari, S.K.; Sahoo, S.; Wang, N.; Huczko, A. Graphene research and their outputs: Status and prospect. *J. Sci. Adv. Mater. Dev.* **2020**, *5*, 10–29.
5. Chen, Y.; Wang, X.; Wang, P.; Huang, H.; Wu, G.; Tian, B.; Hong, Z.; Wang, Y.; Sun, S.; Shen, H. Optoelectronic properties of few-layer MoS_2 FET gated by ferroelectric relaxor polymer. *ACS Appl. Mater. Interfaces* **2016**, *8*, 32083–32088. [CrossRef]
6. Ghatak, S.; Pal, A.N.; Ghosh, A. Nature of electronic states in atomically thin MoS_2 field-effect transistors. *ACS Nano* **2011**, *5*, 7707–7712. [CrossRef] [PubMed]
7. Wang, Q.H.; Kalantar-Zadeh, K.; Kis, A.; Coleman, J.N.; Strano, M.S. Electronics and optoelectronics of two-dimensional transition metal dichalcogenides. *Nat. Nanotechnol.* **2012**, *7*, 699–712. [CrossRef]

8. Cao, L. Two-dimensional transition-metal dichalcogenide materials: Toward an age of atomic-scale photonics. *MRS Bull.* **2015**, *40*, 592–599. [CrossRef]
9. Movva, H.C.; Rai, A.; Kang, S.; Kim, K.; Fallahazad, B.; Taniguchi, T.; Watanabe, K.; Tutuc, E.; Banerjee, S.K. High-mobility holes in dual-gated WSe$_2$ field-effect transistors. *ACS Nano* **2015**, *9*, 10402–10410. [CrossRef]
10. Tang, H.; Shi, B.; Pan, Y.; Li, J.; Zhang, X.; Yan, J.; Liu, S.; Yang, J.; Xu, L.; Yang, J. Schottky contact in monolayer WS$_2$ field-effect transistors. *Adv. Theory Simul.* **2019**, *2*, 1900001. [CrossRef]
11. Iqbal, M.W.; Iqbal, M.Z.; Khan, M.F.; Shehzad, M.A.; Seo, Y.; Park, J.H.; Hwang, C.; Eom, J. High-mobility and air-stable single-layer WS$_2$ field-effect transistors sandwiched between chemical vapor deposition-grown hexagonal BN films. *Sci. Rep.* **2015**, *5*, 10699. [CrossRef]
12. Sajjad, M.; Singh, N.; Schwingenschlögl, U. Strongly bound excitons in monolayer PtS$_2$ and PtSe$_2$. *Appl. Phys. Lett.* **2018**, *112*, 043101. [CrossRef]
13. Lu, J.; Zhang, X.; Su, G.; Yang, W.; Han, K.; Yu, X.; Wan, Y.; Wang, X.; Yang, P. Large-area uniform few-layer PtS$_2$: Synthesis, structure and physical properties. *Mater. Today Phys.* **2021**, *18*, 100376. [CrossRef]
14. Sato, Y.; Nishimura, T.; Duanfei, D.; Ueno, K.; Shinokita, K.; Matsuda, K.; Nagashio, K. Intrinsic electronic transport properties and carrier densities in PtS$_2$ and SnSe$_2$: Exploration of n+-Source for 2D tunnel FETs. *Adv. Electron. Mater.* **2021**, *7*, 2100292. [CrossRef]
15. Yin, S.; Luo, Q.; Wei, D.; Guo, G.; Sun, X.; Tang, Y.; Dai, X. A type-II PtS$_2$/MoTe$_2$ van der Waals heterostructure with adjustable electronic and optical properties. *Results Phys.* **2022**, *33*, 105172. [CrossRef]
16. Zhang, Y.; Feng, Q.; Hao, R.; Zhang, M. Fabrication of Large-Area Short-Wave Infrared Array Photodetectors under High Operating Temperature by High Quality PtS$_2$ Continuous Films. *Electronics* **2022**, *11*, 838. [CrossRef]
17. Bertolazzi, S.; Brivio, J.; Kis, A. Stretching and breaking of ultrathin MoS$_2$. *ACS Nano* **2011**, *5*, 9703–9709. [CrossRef]
18. Park, J.; Woo, H.; Jeon, S. Impact of fast transient charging and ambient on mobility of WS$_2$ field-effect transistor. *J. Vac. Sci. Technol. B* **2017**, *35*, 050601. [CrossRef]
19. Zeng, H.; Liu, G.-B.; Dai, J.; Yan, Y.; Zhu, B.; He, R.; Xie, L.; Xu, S.; Chen, X.; Yao, W.. Optical signature of symmetry variations and spin-valley coupling in atomically thin tungsten dichalcogenides. *Sci. Rep.* **2013**, *3*, 1608. [CrossRef]
20. Wang, Z.; Li, Q.; Besenbacher, F.; Dong, M. Facile synthesis of single crystal PtSe$_2$ nanosheets for nanoscale electronics. *Adv. Mater.* **2016**, *28*, 10224–10229. [CrossRef]
21. AlMutairi, A.; Yin, D.; Yoon, Y. PtSe$_2$ field-effect transistors: New opportunities for electronic devices. *IEEE Electron Device Lett.* **2017**, *39*, 151–154. [CrossRef]
22. Pudasaini, P.R.; Oyedele, A.; Zhang, C.; Stanford, M.G.; Cross, N.; Wong, A.T.; Hoffman, A.N.; Xiao, K.; Duscher, G.; Mandrus, D.G. High-performance multilayer WSe$_2$ field-effect transistors with carrier type control. *Nano Res.* **2018**, *11*, 722–730. [CrossRef]
23. Lu, X.; Utama, M.I.B.; Lin, J.; Gong, X.; Zhang, J.; Zhao, Y.; Pantelides, S.T.; Wang, J.; Dong, Z.; Liu, Z. Large-area synthesis of monolayer and few-layer MoSe$_2$ films on SiO$_2$ substrates. *Nano Lett.* **2014**, *14*, 2419–2425. [CrossRef] [PubMed]
24. Liu, X.; Yang, Y.; Hu, T.; Zhao, G.; Chen, C.; Ren, W. Vertical ferroelectric switching by in-plane sliding of two-dimensional bilayer WTe$_2$. *Nanoscale* **2019**, *11*, 18575–18581. [CrossRef] [PubMed]
25. Zhang, E.; Chen, R.; Huang, C.; Yu, J.; Zhang, K.; Wang, W.; Liu, S.; Ling, J.; Wan, X.; Lu, H.-Z. Tunable positive to negative magnetoresistance in atomically thin WTe$_2$. *Nano Lett.* **2017**, *17*, 878–885. [CrossRef]
26. Qu, D.; Liu, X.; Huang, M.; Lee, C.; Ahmed, F.; Kim, H.; Ruoff, R.S.; Hone, J.; Yoo, W.J. Carrier-type modulation and mobility improvement of thin MoTe$_2$. *Adv. Mater.* **2017**, *29*, 1606433. [CrossRef]
27. Choi, W.; Yin, D.; Choo, S.; Jeong, S.-H.; Kwon, H.-J.; Yoon, Y.; Kim, S. Low-temperature behaviors of multilayer MoS$_2$ transistors with ohmic and Schottky contacts. *Appl. Phys. Lett.* **2019**, *115*, 033501. [CrossRef]
28. Zou, T.; Kim, H.J.; Kim, S.; Liu, A.; Choi, M.Y.; Jung, H.; Zhu, H.; You, I.; Reo, Y.; Lee, W.J. High-Performance Solution-Processed 2D P-Type WSe$_2$ Transistors and Circuits through Molecular Doping. *Adv. Mater.* **2023**, *35*, 2208934. [CrossRef] [PubMed]
29. Fathipour, S.; Ma, N.; Hwang, W.; Protasenko, V.; Vishwanath, S.; Xing, H.; Xu, H.; Jena, D.; Appenzeller, J.; Seabaugh, A. Exfoliated multilayer MoTe$_2$ field-effect transistors. *Appl. Phys. Lett.* **2014**, *105*, 192101. [CrossRef]
30. Park, G.H.; Nielsch, K.; Thomas, A. 2D transition metal dichalcogenide thin films obtained by chemical gas phase deposition techniques. *Adv. Mater. Interfaces* **2019**, *6*, 1800688. [CrossRef]
31. Patel, A.B.; Chauhan, P.; Machhi, H.K.; Narayan, S.; Sumesh, C.; Patel, K.; Soni, S.S.; Jha, P.; Solanki, G.; Pathak, V. Transferrable thin film of ultrasonically exfoliated MoSe$_2$ nanocrystals for efficient visible-light photodetector. *Phys. E Low Dimens. Syst. Nanostruct.* **2020**, *119*, 114019. [CrossRef]
32. Seo, S.G.; Ryu, J.H.; Lee, W.Y.; Jin, S.H. Visible Light Illumination Effects on Instability of MoS$_2$ Thin-Film Transistors for Optical Sensor Application. *Phys. Status Solidi A* **2022**, *219*, 2200052. [CrossRef]
33. Yang, Y.; Li, J.; Choi, S.; Jeon, S.; Cho, J.H.; Lee, B.H.; Lee, S. High-responsivity PtSe$_2$ photodetector enhanced by photogating effect. *Appl. Phys. Lett.* **2021**, *118*, 013103. [CrossRef]
34. Pham, T.; Li, G.; Bekyarova, E.; Itkis, M.E.; Mulchandani, A. MoS$_2$—Based optoelectronic gas sensor with sub-parts-per-billion limit of NO$_2$ gas detection. *ACS Nano* **2019**, *13*, 3196–3205. [CrossRef]
35. Zong, B.; Li, Q.; Chen, X.; Liu, C.; Li, L.; Ruan, J.; Mao, S. Highly enhanced gas sensing performance using a 1T/2H Heterophase MoS$_2$ field-effect transistor at room temperature. *ACS Appl. Mater. Interfaces* **2020**, *12*, 50610–50618. [CrossRef]

36. Zheng, W.; Liu, X.; Xie, J.; Lu, G.; Zhang, J. Emerging van der Waals junctions based on TMDs materials for advanced gas sensors. *Coord. Chem. Rev.* **2021**, *447*, 214151. [CrossRef]
37. Li, D. Nanofabrication Technologies for Making Neuromorphic Devices Based on Two-Dimensional MoS_2. Ph.D. Thesis, University of Michigan, Ann Arbor, MI, USA, 2020.
38. Sangwan, V.K.; Jariwala, D.; Kim, I.S.; Chen, K.-S.; Marks, T.J.; Lauhon, L.J.; Hersam, M.C. Gate-tunable memristive phenomena mediated by grain boundaries in single-layer MoS_2. *Nat. Nanotechnol.* **2015**, *10*, 403–406. [CrossRef]
39. Chen, S.; Mahmoodi, M.R.; Shi, Y.; Mahata, C.; Yuan, B.; Liang, X.; Wen, C.; Hui, F.; Akinwande, D.; Strukov, D.B. Wafer-scale integration of two-dimensional materials in high-density memristive crossbar arrays for artificial neural networks. *Nat. Electron.* **2020**, *3*, 638–645. [CrossRef]
40. Bian, H.; Goh, Y.Y.; Liu, Y.; Ling, H.; Xie, L.; Liu, X. Stimuli-Responsive Memristive Materials for Artificial Synapses and Neuromorphic Computing. *Adv. Mater.* **2021**, *33*, 2006469. [CrossRef]
41. Lee, H.W.; Kang, D.-H.; Cho, J.H.; Lee, S.; Jun, D.-H.; Park, J.-H. Highly sensitive and reusable membraneless field-effect transistor (FET)-type tungsten diselenide (WSe_2) biosensors. *ACS Appl. Mater. Interfaces* **2018**, *10*, 17639–17645. [CrossRef]
42. Park, H.; Han, G.; Lee, S.W.; Lee, H.; Jeong, S.H.; Naqi, M.; AlMutairi, A.; Kim, Y.J.; Lee, J.; Kim, W.-J. Label-free and recalibrated multilayer MoS_2 biosensor for point-of-care diagnostics. *ACS Appl. Mater. Interfaces* **2017**, *9*, 43490–43497. [CrossRef] [PubMed]
43. Ryu, B.; Nam, H.; Oh, B.-R.; Song, Y.; Chen, P.; Park, Y.; Wan, W.; Kurabayashi, K.; Liang, X. Cyclewise operation of printed MoS_2 transistor biosensors for rapid biomolecule quantification at femtomolar levels. *ACS Sens.* **2017**, *2*, 274–281. [CrossRef] [PubMed]
44. Fathi-Hafshejani, P.; Azam, N.; Wang, L.; Kuroda, M.A.; Hamilton, M.C.; Hasim, S.; Mahjouri-Samani, M. Two-dimensional-material—Based field-effect transistor biosensor for detecting COVID-19 virus (SARS-CoV-2). *ACS Nano* **2021**, *15*, 11461–11469. [CrossRef]
45. Jeong, Y.; Shin, D.; Park, J.H.; Park, J.; Yi, Y.; Im, S. Integrated advantages from perovskite photovoltaic cell and 2D $MoTe_2$ transistor towards self-power energy harvesting and photosensing. *Nano Energy* **2019**, *63*, 103833. [CrossRef]
46. Lee, M.H.; Wu, W. 2D Materials for Wearable Energy Harvesting. *Adv. Mater. Technol.* **2022**, *7*, 2101623. [CrossRef]
47. Tahir, M.B.; Fatima, U. Recent trends and emerging challenges in two-dimensional materials for energy harvesting and storage applications. *Energy Storage* **2022**, *4*, e244. [CrossRef]
48. Taube, A.; Judek, J.; Łapińska, A.; Zdrojek, M. Temperature-dependent thermal properties of supported MoS_2 monolayers. *ACS Appl. Mater. Interfaces* **2015**, *7*, 5061–5065. [CrossRef]
49. Ahmed, S.; Viboon, P.; Ding, X.; Bao, N.; Du, Y.; Herng, T.; Ding, J.; Yi, J. Annealing effect on the ferromagnetism of MoS_2 nanoparticles. *J. Alloys Compd.* **2018**, *746*, 399–404. [CrossRef]
50. Kim, H.J.; Kim, D.; Jung, S.; Bae, M.H.; Yun, Y.J.; Yi, S.N.; Yu, J.S.; Kim, J.H.; Ha, D.H. Changes in the Raman spectra of monolayer MoS_2 upon thermal annealing. *J. Raman Spectrosc.* **2018**, *49*, 1938–1944. [CrossRef]
51. Lin, J.; Pantelides, S.T.; Zhou, W. Vacancy-induced formation and growth of inversion domains in transition-metal dichalcogenide monolayer. *ACS Nano* **2015**, *9*, 5189–5197. [CrossRef]
52. Kim, S.; Hong, S.; Yoo, H. Control of Charge Transport Properties in Molybdenum Diselenide Field-Effect Transistors for Enhanced Noise-Margin and Inverter Characteristics. *IEEE Trans. Nanotechnol.* **2022**, *21*, 266–270. [CrossRef]
53. Khan, M.A.; Mehmood, M.Q.; Massoud, Y. High-Temperature Annealing Effects on Atomically Thin Tungsten Diselenide Field-Effect Transistor. *Appl. Sci.* **2022**, *12*, 8119. [CrossRef]
54. Xu, S.; Wu, Z.; Lu, H.; Han, Y.; Long, G.; Chen, X.; Han, T.; Ye, W.; Wu, Y.; Lin, J. Universal low-temperature Ohmic contacts for quantum transport in transition metal dichalcogenides. *2D Mater.* **2016**, *3*, 021007. [CrossRef]
55. Choi, Y.; Park, H.; Lee, N.; Kim, B.; Lee, J.; Lee, G.; Jeon, H. Deposition of the tin sulfide thin films using ALD and a vacuum annealing process for tuning the phase transition. *J. Alloys Compd.* **2022**, *896*, 162806. [CrossRef]
56. Liu, X.; Islam, A.; Guo, J.; Feng, P.X.-L. Controlling polarity of $MoTe_2$ transistors for monolithic complementary logic via Schottky contact engineering. *ACS Nano* **2020**, *14*, 1457–1467. [CrossRef]
57. Wang, X.; Feng, H.; Wu, Y.; Jiao, L. Controlled synthesis of highly crystalline MoS_2 flakes by chemical vapor deposition. *J. Am. Chem. Soc.* **2013**, *135*, 5304–5307. [CrossRef]
58. Tongay, S.; Zhou, J.; Ataca, C.; Liu, J.; Kang, J.S.; Matthews, T.S.; You, L.; Li, J.; Grossman, J.C.; Wu, J. Broad-range modulation of light emission in two-dimensional semiconductors by molecular physisorption gating. *Nano Lett.* **2013**, *13*, 2831–2836. [CrossRef]
59. Ueno, K.; Fukushima, K. Changes in structure and chemical composition of α-$MoTe_2$ and β-$MoTe_2$ during heating in vacuum conditions. *Appl. Phys. Express* **2015**, *8*, 095201. [CrossRef]
60. Kim, Y.J.; Park, W.; Yang, J.H.; Kim, Y.; Lee, B.H. Contact resistance reduction of WS_2 FETs using high-pressure hydrogen annealing. *IEEE J. Electron Devices Soc.* **2017**, *6*, 164–168. [CrossRef]
61. Park, H.; Son, J.; Kim, J. Reducing the contact and channel resistances of black phosphorus via low-temperature vacuum annealing. *J. Mater. Chem. C* **2018**, *6*, 1567–1572. [CrossRef]
62. Chow, W.L.; Yu, P.; Liu, F.; Hong, J.; Wang, X.; Zeng, Q.; Hsu, C.H.; Zhu, C.; Zhou, J.; Wang, X. High mobility 2D palladium diselenide field-effect transistors with tunable ambipolar characteristics. *Adv. Mater.* **2017**, *29*, 1602969. [CrossRef]
63. Yang, C.-M.; Chen, T.-C.; Yang, Y.-C.; Meyyappan, M. Annealing effect on UV-illuminated recovery in gas response of graphene—Based NO_2 sensors. *RSC Adv.* **2019**, *9*, 23343–23351. [CrossRef] [PubMed]

64. Chen, W.-H.; Kawakami, N.; Hsueh, J.-W.; Kuo, L.-H.; Chen, J.-Y.; Liao, T.-W.; Kuo, C.-N.; Lue, C.-S.; Lai, Y.-L.; Hsu, Y.-J. Toward Perfect Surfaces of Transition Metal Dichalcogenides with Ion Bombardment and Annealing Treatment. *ACS Appl. Mater. Interfaces* **2023**, *15*, 16153–16161. [CrossRef] [PubMed]
65. Merve, A.; ERTUGRUL, M. Investigation of the dependence of ambipolarity on channel thickness for TMDC based field effect transistors. *Erzincan Univ. J. Sci. Technol.* **2021**, *14*, 825–836.
66. Kim, I.S.; Sangwan, V.K.; Jariwala, D.; Wood, J.D.; Park, S.; Chen, K.-S.; Shi, F.; Ruiz-Zepeda, F.; Ponce, A.; Jose-Yacaman, M. Influence of stoichiometry on the optical and electrical properties of chemical vapor deposition derived MoS_2. *ACS Nano* **2014**, *8*, 10551–10558. [CrossRef]
67. Golovynskyi, S.; Irfan, I.; Bosi, M.; Seravalli, L.; Datsenko, O.I.; Golovynska, I.; Li, B.; Lin, D.; Qu, J. Exciton and trion in few-layer MoS_2: Thickness-and temperature-dependent photoluminescence. *Appl. Surf. Sci.* **2020**, *515*, 146033. [CrossRef]
68. Si, K.; Ma, J.; Guo, Y.; Zhou, Y.; Lu, C.; Xu, X.; Xu, X. Improving photoelectric performance of MoS_2 photoelectrodes by annealing. *Ceram. Int.* **2018**, *44*, 21153–21158. [CrossRef]
69. Choi, J.; Zhang, H.; Choi, J.H. Modulating optoelectronic properties of two-dimensional transition metal dichalcogenide semiconductors by photoinduced charge transfer. *ACS Nano* **2016**, *10*, 1671–1680. [CrossRef]
70. Kufer, D.; Konstantatos, G. Highly sensitive, encapsulated MoS_2 photodetector with gate controllable gain and speed. *Nano Lett.* **2015**, *15*, 7307–7313. [CrossRef]
71. Chaves, A.; Azadani, J.G.; Alsalman, H.; Da Costa, D.; Frisenda, R.; Chaves, A.; Song, S.H.; Kim, Y.D.; He, D.; Zhou, J. Bandgap engineering of two-dimensional semiconductor materials. *NPJ 2D Mater. Appl.* **2020**, *4*, 29. [CrossRef]
72. Iacovella, F.; Koroleva, A.; Rybkin, A.G.; Fouskaki, M.; Chaniotakis, N.; Savvidis, P.; Deligeorgis, G. Impact of thermal annealing in forming gas on the optical and electrical properties of MoS_2 monolayer. *J. Phys. Condens. Matter* **2020**, *33*, 035001. [CrossRef] [PubMed]
73. Frisenda, R.; Drüppel, M.; Schmidt, R.; Michaelis de Vasconcellos, S.; Perez de Lara, D.; Bratschitsch, R.; Rohlfing, M.; Castellanos-Gomez, A. Biaxial strain tuning of the optical properties of single-layer transition metal dichalcogenides. *NPJ 2D Mater. Appl.* **2017**, *1*, 10. [CrossRef]
74. Chee, S.-S.; Oh, C.; Son, M.; Son, G.-C.; Jang, H.; Yoo, T.J.; Lee, S.; Lee, W.; Hwang, J.Y.; Choi, H. Sulfur vacancy-induced reversible doping of transition metal disulfides via hydrazine treatment. *Nanoscale* **2017**, *9*, 9333–9339. [CrossRef] [PubMed]
75. Eda, G.; Yamaguchi, H.; Voiry, D.; Fujita, T.; Chen, M.; Chhowalla, M. Photoluminescence from chemically exfoliated MoS_2. *Nano Lett.* **2011**, *11*, 5111–5116. [CrossRef] [PubMed]
76. Allain, A.; Kis, A. Electron and hole mobilities in single-layer WSe_2. *ACS Nano* **2014**, *8*, 7180–7185. [CrossRef] [PubMed]
77. Baugher, B.W.; Churchill, H.O.; Yang, Y.; Jarillo-Herrero, P. Intrinsic electronic transport properties of high-quality monolayer and bilayer MoS_2. *Nano Lett.* **2013**, *13*, 4212–4216. [CrossRef] [PubMed]
78. Shahbazi, M.; Khanlary, M.R. Study of optical, electrochemical, and morphological properties of MoS_2 thin films prepared by thermal evaporation. *Braz. J. Phys.* **2021**, *51*, 1182–1190. [CrossRef]
79. Yang, R.; Zheng, X.; Wang, Z.; Miller, C.J.; Feng, P.X.-L. Multilayer MoS_2 transistors enabled by a facile dry-transfer technique and thermal annealing. *J. Vac. Sci. Technol. B* **2014**, *32*, 061203. [CrossRef]
80. Klots, A.; Newaz, A.; Wang, B.; Prasai, D.; Krzyzanowska, H.; Lin, J.; Caudel, D.; Ghimire, N.; Yan, J.; Ivanov, B. Probing excitonic states in suspended two-dimensional semiconductors by photocurrent spectroscopy. *Sci. Rep.* **2014**, *4*, 6608. [CrossRef]
81. Islam, Z.; Kozhakhmetov, A.; Robinson, J.; Haque, A. Enhancement of WSe_2 FET Performance Using Low-Temperature Annealing. *J. Electron. Mater.* **2020**, *49*, 3770–3779. [CrossRef]
82. Zhang, S.; Li, R.; Yao, Z.; Liao, P.; Li, Y.; Tian, H.; Wang, J.; Liu, P.; Guo, J.; Liu, K. Laser annealing towards high-performance monolayer MoS_2 and WSe_2 field effect transistors. *Nanotechnology* **2020**, *31*, 30LT02. [CrossRef] [PubMed]
83. Wang, X.; Gong, Y.; Shi, G.; Chow, W.L.; Keyshar, K.; Ye, G.; Vajtai, R.; Lou, J.; Liu, Z.; Ringe, E. Chemical vapor deposition growth of crystalline monolayer $MoSe_2$. *ACS Nano* **2014**, *8*, 5125–5131. [CrossRef]
84. Shi, Y.; Zhou, W.; Lu, A.-Y.; Fang, W.; Lee, Y.-H.; Hsu, A.L.; Kim, S.M.; Kim, K.K.; Yang, H.Y.; Li, L.-J. Van der Waals epitaxy of MoS_2 layers using graphene as growth templates. *Nano Lett.* **2012**, *12*, 2784–2791. [CrossRef] [PubMed]
85. Mirabelli, G.; Walsh, L.A.; Gity, F.; Bhattacharjee, S.; Cullen, C.P.; Coileáin, C.Ó.; Monaghan, S.; McEvoy, N.; Nagle, R.; Hurley, P.K. Effects of annealing temperature and ambient on metal/$PtSe_2$ contact alloy formation. *ACS Omega* **2019**, *4*, 17487–17493. [CrossRef]
86. Lee, D.; Jang, J.H.; Song, W.; Moon, J.; Kim, Y.; Lee, J.; Jeong, B.; Park, S. In situ work-function measurement during chemical transformation of MoS_2 to MoO_3 by ambient-pressure x-ray photoelectron spectroscopy. *2D Mater.* **2020**, *7*, 025014. [CrossRef]
87. Wu, J.; Li, H.; Yin, Z.; Li, H.; Liu, J.; Cao, X.; Zhang, Q.; Zhang, H. Layer thinning and etching of mechanically exfoliated MoS_2 nanosheets by thermal annealing in air. *Small* **2013**, *9*, 3314–3319.
88. Liu, X.; Choi, M.S.; Hwang, E.; Yoo, W.J.; Sun, J. Fermi level pinning dependent 2D semiconductor devices: Challenges and prospects. *Adv. Mater.* **2022**, *34*, 2108425. [CrossRef]
89. Wang, Y.; Liu, S.; Li, Q.; Quhe, R.; Yang, C.; Guo, Y.; Zhang, X.; Pan, Y.; Li, J.; Zhang, H. Schottky barrier heights in two-dimensional field-effect transistors: From theory to experiment. *Rep. Prog. Phys.* **2021**, *84*, 056501. [CrossRef]

90. Malavika, C.; Roshini, R.A.; Kanthi, R.S.; Kannan, E. Single crystal flake parameters of MoS_2 and $MoSe_2$ exfoliated using anodic bonding technique and its potential in rapid prototyping. *J. Phys. Commun.* **2020**, *4*, 105015. [CrossRef]
91. Heyl, M.; List-Kratochvil, E.J. Only gold can pull this off: Mechanical exfoliations of transition metal dichalcogenides beyond scotch tape. *Appl. Phys. A* **2023**, *129*, 16. [CrossRef]
92. Novoselov, K.S.; Geim, A.K.; Morozov, S.V.; Jiang, D.-E.; Zhang, Y.; Dubonos, S.V.; Grigorieva, I.V.; Firsov, A.A. Electric field effect in atomically thin carbon films. *Science* **2004**, *306*, 666–669. [CrossRef] [PubMed]
93. Hoang, A.T.; Qu, K.; Chen, X.; Ahn, J.-H. Large-area synthesis of transition metal dichalcogenides via CVD and solution—Based approaches and their device applications. *Nanoscale* **2021**, *13*, 615–633. [CrossRef] [PubMed]
94. Wang, J.; Li, T.; Wang, Q.; Wang, W.; Shi, R.; Wang, N.; Amini, A.; Cheng, C. Controlled growth of atomically thin transition metal dichalcogenides via chemical vapor deposition method. *Mater. Today Adv.* **2020**, *8*, 100098. [CrossRef]
95. Bernal, M.; Álvarez, L.; Giovanelli, E.; Arnáiz, A.; Ruiz-González, L.; Casado, S.; Granados, D.; Pizarro, A.; Castellanos-Gomez, A.; Pérez, E. Luminescent Transition Metal Dichalcogenide Nanosheets through One-Step Liquid Phase Exfoliation. *2D Mater.* **2016**, *3*, 035014. [CrossRef]
96. Coleman, J.N.; Lotya, M.; O'Neill, A.; Bergin, S.D.; King, P.J.; Khan, U.; Young, K.; Gaucher, A.; De, S.; Smith, R.J. Two-dimensional nanosheets produced by liquid exfoliation of layered materials. *Science* **2011**, *331*, 568–571. [CrossRef]
97. Yang, G.; Fang, X.; Gu, Y.; Danner, A.; Xie, F.; Zhang, X.; Lu, N.; Wang, Y.; Hua, B.; Gu, X. Insights on the enhanced Raman scattering of monolayer TMDCs (Mo, W)(S, Se)$_2$ with Ag nanoparticles via rapid thermal annealing. *Appl. Surf. Sci.* **2020**, *520*, 146367. [CrossRef]
98. Hu, W.; Wang, Y.; He, K.; He, X.; Bai, Y.; Liu, C.; Zhou, N.; Wang, H.; Li, P.; Ma, X. Straining of atomically thin WSe_2 crystals: Suppressing slippage by thermal annealing. *J. Appl. Phys.* **2022**, *132*, 085104. [CrossRef]
99. Jin, Z.; Shin, S.; Han, S.-J.; Min, Y.-S. Novel chemical route for atomic layer deposition of MoS_2 thin film on SiO_2/Si substrate. *Nanoscale* **2014**, *6*, 14453–14458. [CrossRef]
100. Etzkorn, J.; Therese, H.A.; Rocker, F.; Zink, N.; Kolb, U.; Tremel, W. Metal-Organic Chemical Vapor Deposition Synthesis of Hollow Inorganic-Fullerene-Type MoS_2 and $MoSe_2$ Nanoparticles. *Adv. Mater.* **2005**, *17*, 2372–2375. [CrossRef]
101. Kosmala, T.; Palczynski, P.; Amati, M.; Gregoratti, L.; Sezen, H.; Mattevi, C.; Agnoli, S.; Granozzi, G. Strain Induced Phase Transition of WS_2 by Local Dewetting of Au/Mica Film upon Annealing. *Surfaces* **2020**, *4*, 1–8. [CrossRef]
102. Namgung, S.D.; Yang, S.; Park, K.; Cho, A.-J.; Kim, H.; Kwon, J.-Y. Influence of post-annealing on the off current of MoS_2 field-effect transistors. *Nanoscale Res. Lett.* **2015**, *10*, 62. [CrossRef] [PubMed]
103. Islam, A.; Lee, J.; Feng, P.X.-L. All-dry transferred single-and few-layer MoS_2 field effect transistor with enhanced performance by thermal annealing. *J. Appl. Phys.* **2018**, *123*, 025701. [CrossRef]
104. Choi, M.S.; Lee, M.; Ngo, T.D.; Hone, J.; Yoo, W.J. Chemical Dopant-Free Doping by Annealing and Electron Beam Irradiation on 2D Materials. *Adv. Electron. Mater.* **2021**, *7*, 2100449. [CrossRef]
105. Liu, J.; Wang, Y.; Xiao, X.; Zhang, K.; Guo, N.; Jia, Y.; Zhou, S.; Wu, Y.; Li, Q.; Xiao, L. Conversion of multi-layered $MoTe_2$ transistor between P-type and N-type and their use in inverter. *Nanoscale Res. Lett.* **2018**, *13*, 291. [CrossRef]
106. Bandyopadhyay, A.S.; Saenz, G.A.; Kaul, A.B. Role of metal contacts and effect of annealing in high performance 2D WSe_2 field-effect transistors. *Surf. Coat. Technol.* **2020**, *381*, 125084. [CrossRef]
107. Seo, J.; Cho, K.; Lee, W.; Shin, J.; Kim, J.-K.; Kim, J.; Pak, J.; Lee, T. Effect of Facile p-Doping on Electrical and Optoelectronic Characteristics of Ambipolar WSe_2 Field-Effect Transistors. *Nanoscale Res. Lett.* **2019**, *14*, 313. [CrossRef]
108. Ji, M.; Choi, W. Performance enhancement of WS_2 transistors via double annealing. *Microelectron. Eng.* **2022**, *255*, 111709. [CrossRef]
109. Tang, B.; Yu, Z.G.; Huang, L.; Chai, J.; Wong, S.L.; Deng, J.; Yang, W.; Gong, H.; Wang, S.; Ang, K.-W. Direct n- to p-type channel conversion in monolayer/few-layer WS_2 field-effect transistors by atomic nitrogen treatment. *ACS Nano* **2018**, *12*, 2506–2513. [CrossRef]
110. Heo, K.; Jo, S.-H.; Shim, J.; Kang, D.-H.; Kim, J.-H.; Park, J.-H. Stable and reversible triphenylphosphine—Based n-type doping technique for molybdenum disulfide (MoS_2). *ACS Appl. Mater. Interfaces* **2018**, *10*, 32765–32772. [CrossRef]
111. Lee, D.H.; Yun, H.J.; Hong, S.; Yoo, H. Ambipolar conduction and multicolor photosensing behaviors from poly (9,9-di-n-octylfluorenyl-2,7-diyl)-molybdenum disulfide heterointerfaces. *Surf. Interfaces* **2021**, *27*, 101448. [CrossRef]
112. Nguyen, P.H.; Nguyen, D.H.; Kim, H.; Jeong, H.M.; Oh, H.M.; Jeong, M.S. Synergistic hole-doping on ultrathin $MoTe_2$ for highly stable unipolar field-effect transistor. *Appl. Surf. Sci.* **2022**, *596*, 153567. [CrossRef]
113. Xu, H.; Fathipour, S.; Kinder, E.W.; Seabaugh, A.C.; Fullerton-Shirey, S.K. Reconfigurable ion gating of 2H-$MoTe_2$ field-effect transistors using poly (ethylene oxide)-$CsClO_4$ solid polymer electrolyte. *ACS Nano* **2015**, *9*, 4900–4910. [CrossRef] [PubMed]
114. Kang, D.-H.; Shim, J.; Jang, S.K.; Jeon, J.; Jeon, M.H.; Yeom, G.Y.; Jung, W.-S.; Jang, Y.H.; Lee, S.; Park, J.-H. Controllable nondegenerate p-type doping of tungsten diselenide by octadecyltrichlorosilane. *ACS Nano* **2015**, *9*, 1099–1107. [CrossRef] [PubMed]
115. Lee, H.; Hong, S.; Yoo, H. Interfacial doping effects in fluoropolymer-tungsten diselenide composites providing high-performance P-type transistors. *Polymers* **2021**, *13*, 1087. [CrossRef]

116. Yoo, H.; Hong, S.; Moon, H.; On, S.; Ahn, H.; Lee, H.K.; Kim, S.; Hong, Y.K.; Kim, J.J. Chemical doping effects on CVD-grown multilayer MoSe$_2$ transistor. *Adv. Electron. Mater.* **2018**, *4*, 1700639. [CrossRef]
117. Khalil, H.M.; Khan, M.F.; Eom, J.; Noh, H. Highly stable and tunable chemical doping of multilayer WS$_2$ field effect transistor: Reduction in contact resistance. *ACS Appl. Mater. Interfaces* **2015**, *7*, 23589–23596. [CrossRef] [PubMed]

Disclaimer/Publisher's Note: The statements, opinions and data contained in all publications are solely those of the individual author(s) and contributor(s) and not of MDPI and/or the editor(s). MDPI and/or the editor(s) disclaim responsibility for any injury to people or property resulting from any ideas, methods, instructions or products referred to in the content.

Review

Review of Degradation Mechanism and Health Estimation Method of VRLA Battery Used for Standby Power Supply in Power System

Ruxin Yu [1], Gang Liu [1], Linbo Xu [1], Yanqiang Ma [2], Haobin Wang [2,*] and Chen Hu [3,*]

1. Zhejiang Zheneng Jiahua Electric Power Generation Co., Ltd., Jiaxing 314201, China
2. Hebei Chuangke Electronic Technology Co., Ltd., Handan 056107, China
3. Laboratory of Operation and Control of Renewable Energy & Storage Systems, China Electric Power Research Institute, Beijing 100192, China
* Correspondence: whbwag@163.com (H.W.); whhuchen@163.com (C.H.)

Abstract: As the backup power supply of power plants and substations, valve-regulated lead-acid (VRLA) batteries are the last safety guarantee for the safe and reliable operation of power systems, and the batteries' status of health (SOH) directly affects the stability and safety of power system equipment. In recent years, serious safety accidents have often occurred due to aging and failure of VRLA batteries, so it is urgent to accurately evaluate the health status of batteries. Accurate estimation of battery SOH is conducive to real-time monitoring of single-battery health information, providing a reliable guarantee for fault diagnosis and improving the overall life and economic performance of the battery pack. In this paper, first, the floating charging operation characteristics and aging failure mechanism of a VRLA battery are summarized. Then, the definition and estimation methods of battery SOH are reviewed, including an experimental method, model method, data-driven method and fusion method. The advantages and disadvantages of various methods and their application conditions are analyzed. Finally, for a future big data power system backup power application scenario, the existing problems and development prospects of battery health state estimation are summarized and prospected.

Keywords: backup power supply; VRLA batteries; aging failure mechanism; state of health; evaluation methods

1. Introduction

Direct current (DC) power supply systems play a very important role in power plants and substations. VRLA batteries are widely used as backup power to ensure normal operation of power plants and substations. Once alternating current (AC) suddenly loses power, the VRLA batteries immediately start to supply power to important DC load equipment, such as relay control and protection devices, automation devices, opening and closing mechanisms of high-voltage circuit breakers, communication equipment, emergency lighting lamps, etc. [1,2]. The abnormal failure of a VRLA battery as the backup power supply for a DC power supply system, seriously affects the safe operations of the DC power supply system. Therefore, the battery is considered the most core component of the DC power supply system, and it is an important guarantee for the safe and stable operation of power plants and substation systems [3].

In recent years, major accidents have occurred in power plants and substations due to VRLA battery failures, which has caused considerable security risks to the safe and stable operation of power grids. In 2013, AC an circuit was out of power due to a lightning strike at a substation of a power grid company in China. Due to the failure of the battery pack, some circuit breaker switches could not work normally, which eventually led to a serious accident of voltage loss in the whole substation. Cause analysis revealed that some negative plate straps of VRLA batteries in the substation were seriously corroded. When faced with heavy load impact, the output voltage of the battery pack dropped considerably and therefore could not meet the requirement of the minimum voltage of switching action. As a result, some switches could not work normally, and the breakdown could not be isolated in time, which eventually led to total substation voltage loss [4]. In June 2016, a fire broke out due to equipment failure, and the lead-acid battery failed to supply power in time due to failure, resulting in an accident of voltage loss in a 330 kV substation in the western suburb of Xi'an, China [4]. Similar serious accidents due to VRLA failure have also occurred in substations in other areas [5,6].

Therefore, analysis of the causes of battery failure and estimation and prediction of the state of health (SOH) of batteries are helpful to identify malfunctioning batteries in a timely manner and make maintenance plans. This is of great significance for prolonging the service life of the battery, reducing the maintenance cost of the system and ensuring the safe operation of the power system.

Many studies have been conducted on the SOH of lithium-ion power battery [7], but few comprehensive reviews have been conducted on the SOH of batteries for standby power supply [8]. Cuma et al. [9] conducted a comprehensive review of various estimation strategies used in hybrid and battery electric vehicles, focusing on battery fault diagnosis, state of charge (SOC) and state of health (SOH) estimation. However, in this study included few applications involving lead-acid batteries. Waltari et al. [10] introduced fault classification and state-of-health monitoring methods for lead-acid batteries. Battery failures are classified into three categories: high impedance, low impedance (short circuit) and deterioration of capacity. Battery health monitoring methods including string-voltage-based, cell-voltage-based, current-based and impedance-based methods were reviewed. However, the article was published in 1999, and the reviewed methods of battery health monitoring are simple and outdated. Ouyang et al. [11] reviewed capacity forecasting technology for VRLA batteries. They divided the capacity forecasting methods into open-circuit voltage measuring, Coulomb counting and internal resistance methods. However, the study did not highlight the influencing factors of SOH.

In view of the lack of summary of the estimation approaches to estimate SOH for VRLA batteries for standby power supply, a detailed and up-to-date summary is necessary. In this paper, the research progress of the decay mechanism of VRLA batteries and the method of estimating SOH are reviewed. First, we introduce the working mode and failure mechanism of the standby VRLA floating charge mode. Then, we describe the principles of SOH estimation methods, practical application cases and the advantages and disadvantages of these estimation methods. This review can provide a decision-making basis for the operation, maintenance and scientific management of standby power supply.

The remainder of this paper is arranged as follows: Section 2 summarizes the floating charge operation characteristics and failure mechanism of backup VRLA batteries. Section 3 introduces the definition of battery SOH. Section 4 introduces the classification and characteristics of different SOH estimation methods for VRLA batteries. Finally, in Section 5, the conclusions are summarized.

2. Operation Characteristics and Failure Mechanism of VRLA Floating Charge

2.1. Operating Characteristics of Standby VRLA

The operation mode of a valve-regulated lead-acid battery for standby power supply includes initial charging before operation, floating charging in normal operation, balanced charging every three months, emergency power supply in case of AC interruption, constant current and constant voltage charging after AC recovery, etc., as shown in Figure 1. VRLA is in the cycle mode of floating charge and equalizing charge for a long time. Once AC power is lost, VRLA is used as the standby DC emergency power supply. After the AC power returns to normal, the charger charges the battery with constant current and voltage. Then, VRLA switches to floating charge and equalizing charge cycle mode [12].

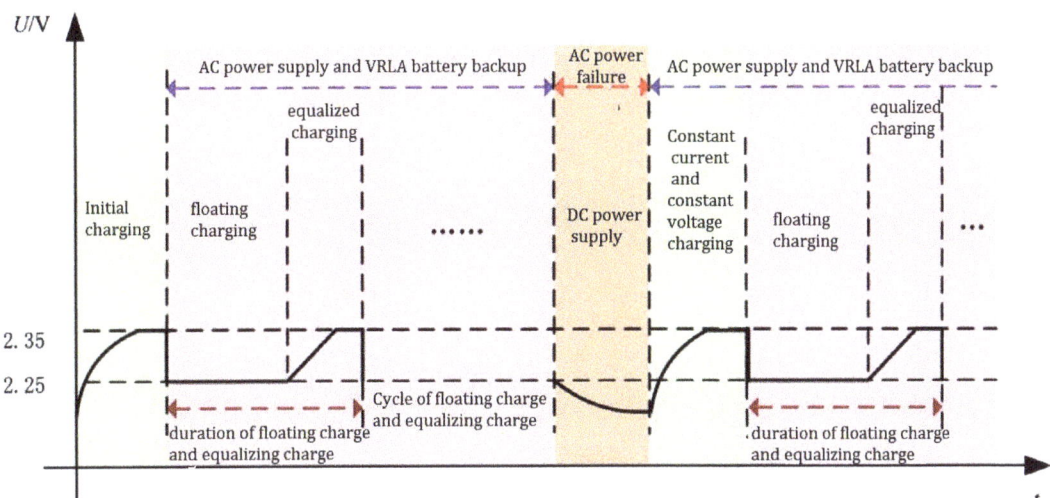

Figure 1. Charging and discharging operation mode diagram of a VRLA battery [12]. Adapted with permission from Ref. [12]. 2022, Chinese Journal of Power Sources magazine.

2.2. Aging Mechanism of VRLA in Floating Charge Operation

VRLA is widely used in the power industry because it has the advantages of low price, mature technology, safety, reliability and easy maintenance. In power plants and substations, VRLA operates in floating charge mode for a long time. Various processes promote the aging of VRLA batteries, such as positive grid corrosion, irreversible sulfation, softening of positive active material, negative plate strap corrosion, water loss, etc., as shown in Table 1 [13,14].

Table 1. VRLA battery failure modes.

Failure Mode	Failure Reason	Phenomena
Positive grid corrosion	In the environment of strong acid, strong oxidation and high potential, the positive grid alloy is thermodynamically unstable, and oxidation corrosion is inevitable	Capacity reduction and ncrease in internal resistance
Irreversible sulfation of negative electrode	When the floating charge voltage is too low, $PbSO_4$ crystals with coarse particles and poor chemical activity are formed on the negative electrode surface.	The battery capacity is significantly reduced; the voltage rises quickly when charging and drops rapidly when discharging
Softening and shedding of positive active material	In the positive active substance, the composite structure of $\alpha\text{-}PbO_2/\beta\text{-}PbO_2$ crystal and $PbO_2\text{-}PbO(OH)_2$ gel is destroyed, which leads to a decrease in the binding force between active substance particles	During the initial stage of use, the battery capacity is reduced
Corrosion of negative plate strap	The metal lead of the negative bus bar is slowly corroded over a long time and transformed into powder $PbSO_4$ crystal	The open-circuit voltage and floating charge voltage are low, and the internal resistance is high

Table 1. Cont.

Failure Mode	Failure Reason	Phenomena
Battery leakage	The battery is not tightly sealed or the shell is broken	Sulfuric acid leakage, pole corrosion and pole temperature rise
Thermal runaway	High voltage and current cause a large amount of heat to accumulate in the battery, causing the battery temperature to rise rapidly	Increase in the floating charge current, temperature rise and battery swelling

(1) Corrosion of positive grid

Corrosion of the positive grid is one of the most common failure modes of VRLA batteries, which refers to the process by which the lead alloy of the positive grid is oxidized to lead dioxide. Under strong oxidation and a high-potential environment, the thermodynamic instability of lead and lead alloys is the fundamental cause of positive grid corrosion. At the end of charging, the positive grid is usually in the potential range of 1.3~1.4 V, which is much higher than the protection potential of lead alloy. The following electrochemical reactions occur:

$$Pb + 2H_2O \rightarrow PbO_2 + 4H^+ + 4e^- \tag{1}$$

In the case of overcharge, the acidity near the positive plate increases due to oxygen evolution reaction [15,16]. The composition of the grid alloy is the main factor affecting the corrosion rate of the positive plate, in addition to environmental temperature, floating charge voltage, casting process and other factors [17]. Corrosion of the positive plate may reduce the mechanical strength of the grid, break the grid, increase the ohmic internal resistance and rapidly increase the voltage during charging [18].

(2) Irreversible sulfation of the negative electrode

The main active substance of a VRLA battery cathode is sponge lead. During discharge, spongy Pb is converted into crystal $PbSO_4$, and $PbSO_4$ is reversibly converted into Pb during charging [19]. When the battery is in the state of deep discharge, undercharge, open-circuit or low-rate discharge for a long time, the $PbSO_4$ crystal of the battery anode cannot be completely converted into spongy Pb [20]. The coarse $PbSO_4$ crystal gradually covers the negative plate surface, and the inert $PbSO_4$ no longer participates in the chemical reaction, that is, irreversible sulfation. Irreversible sulfation affects the recombination of H_2 and O_2 into H_2O in the battery, resulting in the inability of active substances in the electrode plate to participate in the reaction, which increases the battery internal resistance and causes premature battery failure [21].

(3) Negative plate strap corrosion

D. Pavlov [22] et al. thought that the oxygen generated by the positive electrode partially gathered at the upper part of the electrode group, which caused the negative electrode tab and bus bar to lose cathodic protection. If the anode tab and bus bar are farther away from AGM/H_2SO_4 system, the potential of the anode bus bar is higher than the equilibrium potential of $PbSO_4/Pb$, and metallic lead slowly corrodes and transforms into powdered $PbSO_4$ crystal [23]. When the corrosion is serious, the surface and even the inside of the bus bar are seriously pulverized, resulting in the reduction in its mechanical strength. Under the action of stress, the bus bar breaks, resulting in the failure of the battery due to the open circuit inside.

(4) Softening and shedding of positive active material

During charging and discharging, the structure of the positive active material of the battery is damaged, which leads to a reduction in the binding force between the active material and the grid and ultimately leading to the active material falling off. Pavlov believed that in a gel–crystal system, with the charge–discharge cycle, the oxygen evolution reaction in the battery destroys the polymer chain in PAM (positive active material), resulting in an increase in the crystallinity of PAM. D Pavlov thought that the positive active material was a gel–crystal system and that its smallest active material unit was composed of $\alpha\text{-PbO}_2/\beta\text{-PbO}_2$ crystal and $\text{PbO}_2\text{-PbO(OH)}_2\text{(OH)}_2$ gel, which were in a state of mutual balance. With the charge–discharge cycle, the amorphous state in the active material gradually crystallizes, and the crystal area with poor binding force increases. This reduces the binding force between the active material units and ultimately leads to the softening and shedding of the positive active material [24,25].

Zhong et al. [26] classified the main failures of VRLA batteries during floating charging into three categories: positive grid corrosion, negative busbar corrosion and negative sulfation. The positive grid first undergoes electrochemical corrosion, which intensifies the oxygen evolution reaction on the anode. The large amount of oxygen released from the positive electrode increases the oxygen recombination reaction at the negative electrode of the battery and intensifies the corrosion of the negative busbar. At the same time, the process of anode grid corrosion and oxygen release requires water consumption, which increases the oxygen transmission channel. The oxygen recombination reaction is further intensified inside the battery, which also causes the risk of thermal runaway of the battery. In addition, there are other failure modes, such as micro-short circuit, shell rupture, etc. [27]. The aging mechanism of a VRLA battery is often dominated by one failure mode, and the others coexist and interact with each other. Therefore, the degradation of VRLA battery capacity is the result of the interaction of various aging factors [14].

2.3. Aging Mechanism Analysis Method

The methods for analyzing the aging mechanism of batteries can be divided into three categories, namely external characteristic analysis (electrical testing), disassembly analysis and in situ online analysis [28]. External characteristic analysis, such as charge–discharge curve and electrochemical impedance spectroscopy (EIS), incremental capacity analysis (ICA), differential voltage analysis (DVA), etc., are used to extract the aging characteristics of the battery by properly processing the external characteristics (voltage, capacity, internal resistance, etc.) of the battery. The disassembly analysis method is also called material physical and chemical property testing and analysis. First, the aged battery is disassembled in a suitable environment to determine the internal materials of the battery, including grids, separators, positive and negative active substances, electrolytes, etc. Then, these materials are tested by analytical techniques such as scanning electron microscope (SEM), X-ray diffraction technology (XRD) and inductively coupled plasma technology (ICP) to obtain material information such as microscopic morphology, crystal structure and element distribution [29]. The in situ online analysis method involves the use of in situ analysis equipment to monitor the change of the internal physical characteristics of the battery during the cycle and analyze the evolution of the internal material of the battery during the aging process.

The advantages and disadvantages of the three methods for analyzing the aging mechanism of batteries are compared in Table 2, and the main test technologies in each analysis method are listed. The disassembly analysis method and in situ test technology method usually require expensive experimental equipment and cannot analyze the aging behavior of the battery online, so their application is limited. The external characteristic analysis method is based on the battery charge and discharge or impedance spectrum to analyze and extract the aging characteristics of the battery without damage to the battery sample, so it is suitable for online estimation of the aging behavior of the battery.

Table 2. Comparison of three methods for analyzing aging mechanisms of batteries.

Aging Analysis Method	Pros	Cons	Testing Technique
External characteristic analysis	The studied battery is not damaged; the evolution of battery aging at different life stages can be studied	The aging mechanism is analyzed based on speculation and needs to be verified by the disassembly analysis method	ICA, DVA, EIS
Disassembly analysis	The physical and chemical properties of the internal materials of the battery can be directly characterized; internal causes of aging can be determined, and different failure modes can be distinguished	The studied battery is inevitably damaged	SEM, XRD, ICP
In situ online analysis	The studied battery is not damaged; the evolution of the material inside the battery is characterized in situ at different life stages	Requires complicated devices	In situ XRD, neutron diffraction

A VRLA battery is a complex electrochemical system, and its capacity decay is nonlinear. The aging mechanism of a battery is complex and is influenced by many factors. The analysis of battery decay failure mechanisms is helpful to determine the health factors that can best characterize battery SOH. Battery performance is tested by a variety of testing technologies to detect the battery failure mode and obtain aging information. On this basis, combined with various models, the battery SOH can be accurately predicted to ensure the safe operation of the battery.

3. Definition of Battery SOH

SOH represents the ability of a current battery to store electric energy compared with that of a new battery. With the increase in service time, the internal resistance of the battery increases, and the maximum usable capacity decreases. Therefore, capacity and internal resistance parameters are often used to define battery SOH in the industry.

According to the definition in terms of capacity, the SOH can be expressed as

$$\text{SOH} = \frac{C_{curr}}{C_{rated}} \times 100\% \tag{2}$$

where C_{rated} is the nominal capacity, and C_{curr} is the present maximum available capacity, which can be measured by discharging the battery at a fixed current (usually 0.1 C_{rated}) and air temperature (usually 20 °C to 30 °C).

According to the definition in terms of resistance, the SOH can be expressed as

$$\text{SOH} = \frac{R_{EOL} - R_{curr}}{R_{EOL} - R_{new}} \times 100\% \tag{3}$$

where R_{new} represents the initial internal resistance of the new battery, R_{curr} represents the actual internal resistance under the current cycle and R_{EOL} represents the internal resistance at the end of the battery life [30]. Another parameter used to describe the state of a battery is the SOC, which is defined as

$$\text{SOC} = \frac{C_{remain}}{C_{curr}} \times 100\% \tag{4}$$

where C_{remain} is the remaining capacity of the battery. According to Formulas (2) and (4), SOC and SOH are closely linked through C_{curr}. Therefore, the accurate estimation of SOH must be related to SOC.

According to the IEEE 1188-2005 standard, when the actual capacity of a VRLA is less than 80% of the rated capacity, that is, the SOH is less than 80%, the battery must be maintained or replaced [31].

4. SOH Estimation Methods

At present, battery SOH estimation methods include the experimental method, the model method, the data-driven method and the fusion method. In this study, the different SOH estimation methods are classified into four different categories: experimentally based, model-based, data-driven and fusion methods. Figure 2 shows the classification of the different SOH estimation methods.

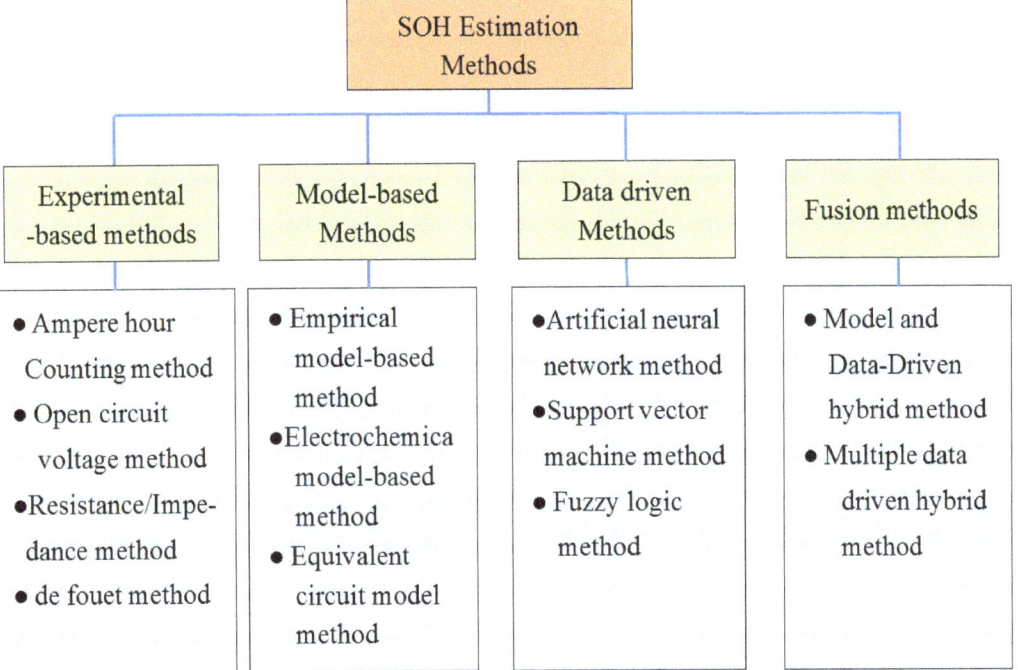

Figure 2. State of health (SOH) estimation methods.

4.1. Experimentally Based Methods

Experimental methods are usually carried out in the laboratory because they require specific equipment and are time-consuming. Experimental methods estimate the SOH by collecting data and measurements that can be used to understand and evaluate the battery aging behavior. The experimental methods usually require less computation and are easy to implement. Therefore, these methods are among the earliest methods used to estimate the SOH of VRLA batteries [23].

4.1.1. Ampere-Hour Counting Method

The ampere-hour counting method is one of the classical methods to estimate the battery SOH [32]. The common procedure of this method is to measure the present maximum capacity of the battery. In order to measure the current maximum capacity of the battery, the battery is first fully charged and then fully discharged; the current of the battery is then recorded. Then, the maximum capacity of the battery can be calculated by integrating the discharge current.

$$C_{curr} = \int_{t_1}^{t_2} I \, dt \tag{5}$$

where I is the discharging current, and t_1 and t_2 are the starting and ending times of the discharge process, respectively.

The initial maximum capacity of a battery ($C_{initial}$) is usually provided by the manufacturer (referred to as nominal capacity); then, the SOH is determined using Equation (5).

The ampere-hour counting method is easy to implement under experimental conditions, and its estimation result is usually regarded as the true value of SOH, which can be used to verify the accuracy of other SOH estimation methods. At present, in power plants and substations, the ampere-hour discharge method is used to check the SOH of the battery pack every 1–2 years. When maintenance staff find a failed battery, they should replace it immediately to maintain the battery in good working condition. The ampere-hour counting method has some disadvantages. For example, it takes too long to test for the battery to be fully charged and discharged, so it is not suitable for online SOH estimation. The full discharge test is also harmful to the battery because deep discharge shortens the service life of the battery [33].

4.1.2. Open-Circuit Voltage Method

The open-circuit voltage (OCV) of the battery has long been known to have a functional relationship with the battery SOH. If the open-circuit voltage of the battery is measured, the battery SOH can be estimated [34].

James H. Aylor et al. [35] proposed a new technology for estimating battery SOH. The technique employs coulometric measurement under loading conditions and open-circuit voltage under no-load conditions in order to predict the change of the battery SOH. This technique was developed to enhance the accuracy and to reduce the required rest period of open-circuit voltage measurement.

Mchrnoosh Shahriari [36] presented an online method for the estimation of the state of health (SOH) of VRLA batteries based on the state of charge (SOC) of the battery. The SOC is estimated using an extended Kalman filter and a neural network model of the battery. Then, the SOH is estimated online based on the relationship between the SOC and the battery open-circuit voltage using fuzzy logic and the recursive least squares method. Experimental results show good estimation of the SOH of VRLA batteries.

The open-circuit voltage of the battery cannot be directly detected in the floating charge mode. In order to accurately measure the open-circuit voltage of the battery, it is necessary to keep the battery offline for a long time to reach a stable state. In addition, in order to improve the estimation accuracy of the open-circuit voltage method, it needs to be used in combination with other methods.

4.1.3. Resistance/Impedance Method

The internal resistance of the battery is considered an important index of SOH because it is seriously affected by the degradation of battery performance. When the SOH of the battery decreases, the internal resistance increases. With the increase in internal resistance, the SOH of the battery decreases. Considering the strong correlation between internal resistance and SOH, internal resistance is regarded as a good tool to estimate SOH [37]. The two main methods used to evaluate battery SOH are the internal resistance method and the electrochemical impedance method [38].

The internal resistance method usually establishes the corresponding relationship between the internal resistance and SOH and then evaluates the battery SOH according to internal resistance. To measure the internal resistance, a sudden current change (ΔI) is exerted on the battery, and the consequent voltage change (ΔU) is measured. The internal resistance can be calculated as $R = \Delta U / \Delta I$. The next step is to perform a regression analysis of the resistance/impedance and SOH. Finally, using the regression function, The SOH of the batteries is estimated [23].

The internal resistance method only needs to obtain the voltage and current, making it suitable application in online estimation of battery SOH. Generally, the internal resistance of the battery has a certain relationship with SOC and SOH, and maintenance personnel can use these relationships to monitor the battery status in real time [39]. However, due to the uncertainty of the relationship between internal resistance and SOC, the error of SOH estimation is slightly larger [40]. In addition, when the capacity of a lead-acid battery is greater than 60%, the internal resistance changes slightly. Therefore, the internal resistance method is only used to roughly judge the battery SOH.

EIS is a kind of electrochemical measurement method whereby a low-amplitude sine wave voltage (or current) disturbance signal is imposed on the battery. EIS has no effect on the internal state of the battery and provides more rich information on electrode process dynamics and electrode interface structure details than other conventional electrochemical methods. Based on the circuit model, the relationship between the EIS curve and SOH can be established to accurately analyze the SOH of the battery. However, EIS measurement requires sophisticated and professional test equipment, which has high requirements for the test environment. As the circuit model itself is a technical difficulty, the process of EIS measurement and SOH calculation is relatively complex, which leads to the time-consuming and high cost of the impact method to estimate SOH. Therefore, a simpler and more general method for obtaining EIS parameters online requires further research.

4.1.4. Coup de Fouet Method

After being fully charged, the battery is discharged with a constant current. In the first few minutes, the discharge voltage reaches the peak voltage and then rises to the discharge platform voltage [41–43]. This phenomenon is referred to as coup de fouet (Figure 3).

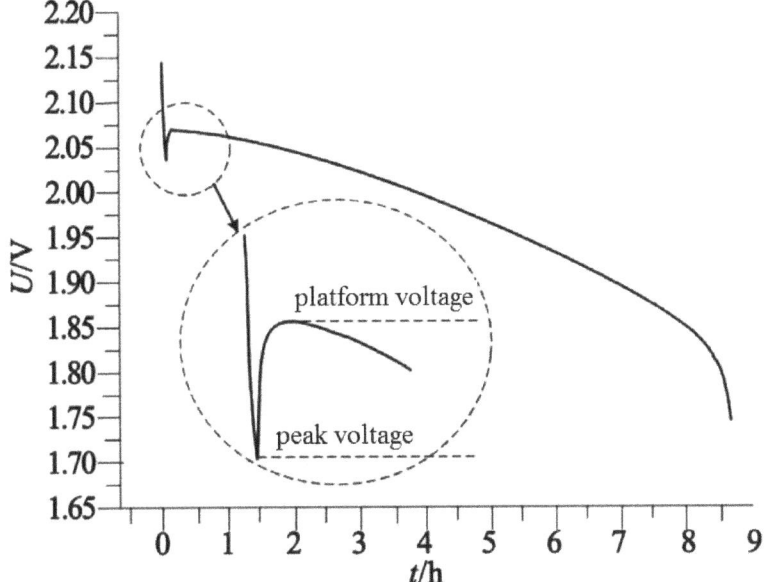

Figure 3. Coup de fouet of a lead-acid battery.

Several studies have applied the "coup de fouet" phenomenon to estimation of battery SOH. Phillip E. Pascoe et al. [44] found that the valley voltage and peak voltage in the coup de fouet phenomenon are linearly related to the actual available capacity of the battery; therefore, the SOH can be estimated according to the peak voltage and the platform voltage. A series of experimental studies revealed that the discharge rate and temperature have effects on the peak voltage and the platform voltage. Yuan et al. [45] assumed that the peak voltage and plateau voltage would be impacted under different discharge conditions (temperature and discharge rate). According to the coup de fouet phenomenon of the battery, the SOH is taken as the output variable, with the peak voltage, plateau voltage, discharge rate and temperature as input variables; accordingly, a battery SOH estimation model based on a BP neural network was built. The results show that the model based on a BP (backpropagation) neural network can effectively predict battery SOH. Due to the short discharge time during the test, the current working state of the battery is not be affected, and SOH can be estimated online. Compared with the traditional discharge test method, the coup de fouet method is more convenient and efficient and is very suitable for online detection of battery SOH as a backup power supply.

4.2. Model-Based Methods

Model-based methods use indirect measurement methods to predict the SOH of the battery. Empirical models, electrochemical models and equivalent circuit models can be applied.

4.2.1. Empirical Model-Based Method

The experience model-based method is used to simulate the aging process of the battery and test the effects of temperature, discharge depth, charge and discharge current on the battery life. The equivalent circuit model or a mathematical model are established with temperature, charging/discharging current and voltage as independent variables and SOH as a dependent variable. Then, the SOH and other battery parameters are calculated based on the model parameters.

Empirical models include the impedance empirical model and the capacity empirical model. The empirical model is used to first test the impedance (or capacity) of batteries in different life stages; then, the change trend of battery impedance (or capacity) over the whole life is obtained. Finally, the SOH of the battery is estimated according to the relationship between impedance and capacity [46].

John Wang et al. comprehensively studied the influence of battery temperature, discharge rate and SOC on battery capacity decay and established an Arrhenius model of capacity decay under the combined influence of temperature and discharge rate.

Compared with the equivalent circuit model, the empirical model is much more complicated, and it can explain many battery phenomena that the equivalent circuit model cannot. However, the empirical model lacks an explanation of the corresponding physical meaning, and its reliability and accuracy of estimating battery SOH often depend on the authenticity of the obtained experimental data. As a result, this method is not very common for SOH estimation.

4.2.2. Electrochemical Model-Based Method

The electrochemical model-based method is also called the battery mechanism model. In the electrochemical model-based method, a series of partial differential equations and algebraic equations is used to describe the physical and chemical processes inside the battery. The parameters in the model can represent the physical and chemical characteristics of the battery and can accurately reflect the changes thereof. Therefore, the electrochemical model-based method can be applied to the analysis of the VRLA performance decay mechanism and SOH estimation and provide theoretical support for prediction of the remaining life of batteries [47–49].

Chao Lyu et al. [50] presented a new method for battery SOH prediction by incorporating an electrochemical model into the particle filtering framework. A simplified electrochemical model of a lead-acid battery was introduced based on the theory of porous electrodes and the theory of diluted solution, which involve the charge conservation, electrode dynamics, liquid phase diffusion, liquid phase equilibrium and potential equilibrium of the solid phase. Figure 4 shows the schematic diagram of lead acid battery. The experimental results show that the model has the advantages of fast calculation speed, minimal damage to the battery and short detection time, making it suitable for SOH estimation and residual life prediction of backup VRLA battery packs in power systems.

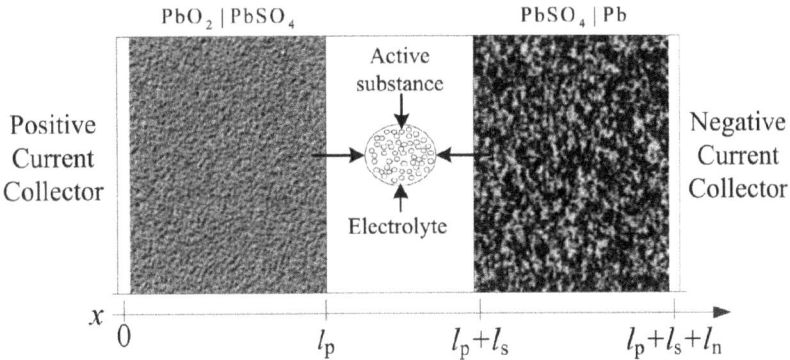

Figure 4. Schematic diagram of a lead-acid battery [50]. Reprinted with permission from Ref. [50]. 2016, Elsevier.

Evaluation of battery SOH using an electrochemical model has the advantages of clear physical meaning, high accuracy, universality, etc.; however, an electrochemical model includes many internal parameters of the battery, the control equation is complex, and many calculations are required, so its practical application is difficult.

4.2.3. Equivalent Circuit Model Method

The equivalent circuit model (ECM) is widely used in battery management systems because it comprises few parameters and is a simple mathematical model [51]. The ECM is composed of a voltage source, inductance, resistance, capacitor and other circuit components, which describe the charging and discharging characteristics of the battery through different combinations and simulate the dynamic characteristics of the battery. The general equivalent circuit model is shown in Figure 5. Generally, the basic equivalent circuit models of lead-acid batteries include the Rint model, the Thevenin model, the second-order RC (resistor–capacitor) model, etc. [52–55]. In the equivalent circuit model, a battery equivalent circuit is first established, and the model parameters are identified to estimate the SOH using algorithms, such as the least square method, Kalman filtering and artificial neural network.

According to the chemical principle of VRLA, Zhang et al. [56] established an equivalent model of a second-order reactive RC circuit to simulate the charging and discharging process. Based on the model, the open-circuit voltage and internal resistance of VRLA could be derived. The parameters of the model are estimated by the least recursive double algorithm with forgetting factor through system identification. Analysis and comparison show that the error of battery parameters identified by this algorithm is very small, and the second-order RC model simulation of the VRLA battery charging and discharging process is very accurate and efficient. Zhang et al. [57] built a second-order RC circuit model and used the terminal voltage comparison method to verify the rationality of the circuit model and the accuracy of identification parameters. In the model, the relationship between the ohmic internal resistance (R_0) and SOH is established, and the SOH of the battery is estimated. Experimental verification shows that the relative error between the SOH of a VRLA battery estimated by the model and the SOH determined by the definition method is about 3%.

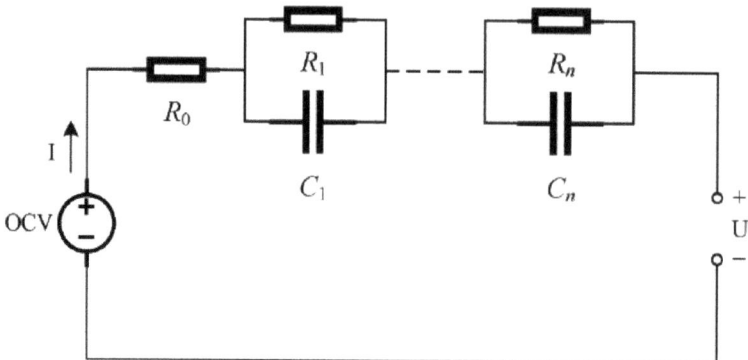

Figure 5. General equivalent circuit model [23]. Reprinted with permission from Ref. [23]. 2021, Elsevier.

The ECM is a model for the external characterization of VRLA batteries. The ECM has advantages such as concise structure, easy parameter identification, simple calculation processes and clear physical meanings and is therefore widely used in SOC and SOH estimation [58]. Because the equivalent circuit model often needs approximate equivalent treatment, when the key parameters of the battery cannot be obtained, some model prediction errors are expanded.

4.3. Data-Driven Methods

With the rapid development of big data and machine learning technology, data-driven technology has broken through the shackles of complex nonlinear systems that are difficult to model and has become the main research direction of battery health. Data-driven methods include artificial neural networks, support vector machines and Gaussian process regression. The general flow of the data-driven prediction method is shown in Figure 6 [59]. First, a large amount of battery information (such as voltage, current, temperature and impedance) is collected, which may come from past historical data or real-time data. Secondly, the battery degradation characteristics are extracted. The third step is to train a machine learning model to showing the relationship between the extracted characteristics and the SOH of the battery. Finally, once the machine learning model is determined, it is applied to evaluate the battery SOH.

4.3.1. Artificial Neural Network Method

An artificial neural network (ANN) is a network formed by the interconnection of neurons in a certain way, and a prediction model is obtained by training the thresholds and ratios of neurons through a large amount of data. Its typical structure consists of an input layer, a hidden layer and an output layer, as shown in Figure 7. Common neural networks for predicting battery SOH include BP neural networks, Elman neural networks and RBF (radical basis function) neural networks [60,61].

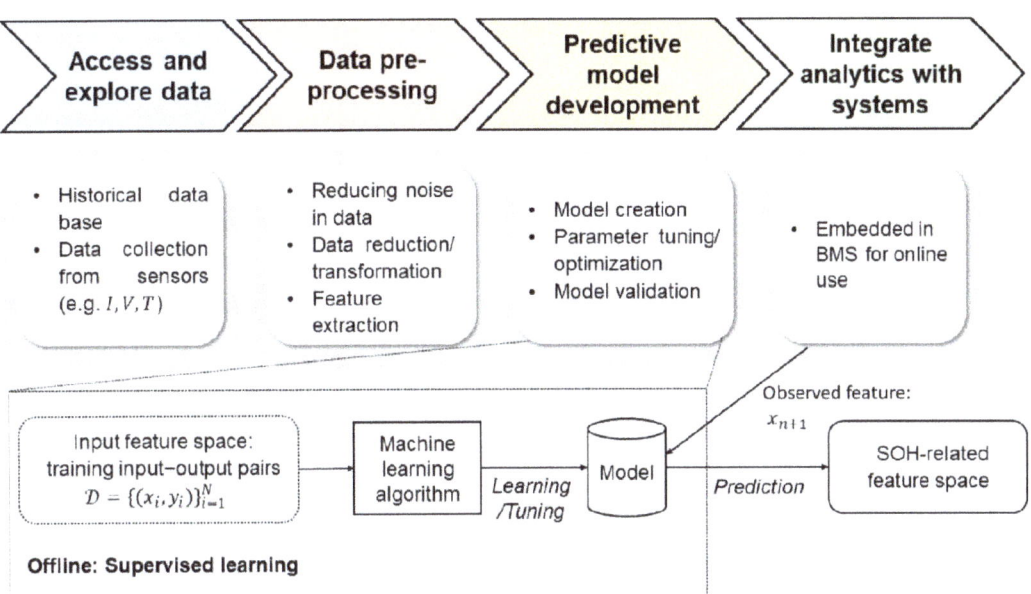

Figure 6. Generic workflow for data-driven SOH estimation methods [59]. Reprinted with permission from Ref. [59]. 2019, Elsevier.

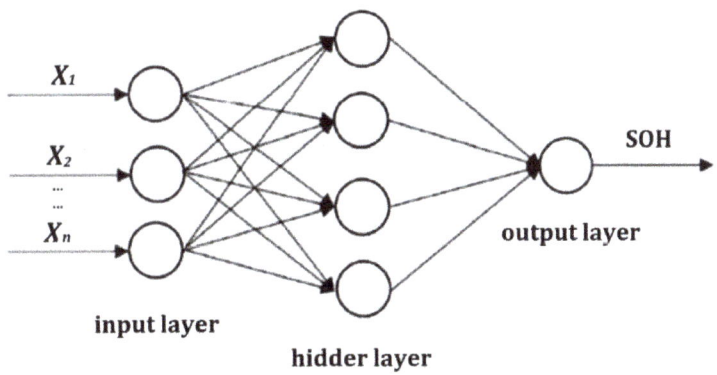

Figure 7. Structure of the BP neural network for model.

Talha et al. [62] proposed a simplified method to estimate the SOC and SOH of VRLA battery online using NN. Firstly, Terminal voltage (V_t) and open-circuit voltage (OCV) were measured for different charging currents (I_{ch}) and discharging currents (I_{dch}). I_{ch}/I_{dch} and V_t were used as NN input variables and OCV was output. Then, the SOC was calculated using an empirical function that described the relationship between SOC and OCV. Finally, the slope of SOC and current were used as training inputs, and the SOH was estimated by the NN. Mei et al. [63] built a wavelet neutral network (WNN) model for prediction of the operating life of a substation VRLA battery. First, a WNN model of battery operating life was established. Then, the experimental data were trained to obtain a WNN model for battery operating life prediction. The predicted results of the WNN model and actual data were compared. The experimental results show that the average relative error of prediction is only 1.49%. The WNN model can quickly and accurately predict the working life of batteries in substations.

An ANN has the advantages of self-organization, self-adaptation, fault tolerance, self-learning evolution, high prediction accuracy and the ability to be applied to a variety of nonlinear prediction fields. However, the prediction ability of an ANN algorithm for a small dataset is poor, and the estimation error is affected by the training data.

4.3.2. Support Vector Machine Model

As a supervised machine learning algorithm, support vector machine (SVM) is used to solve classification and regression problems. Support vector machine constructs a hyperplane or a group of hyperplanes in high-dimensional or infinite dimensional space, which maximizes the isolation edge between positive and negative examples. Its principle diagram is shown in Figure 8. Similar to ANN, SVM is usually used to determine the relationship between input characteristics and SOH [23].

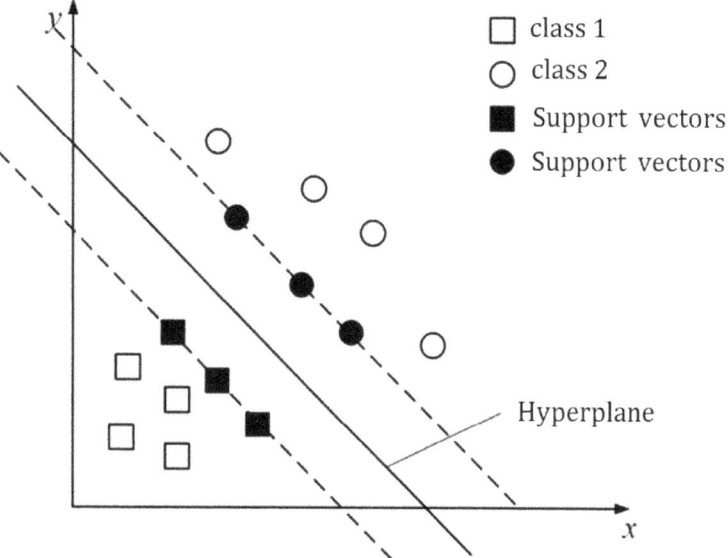

Figure 8. Principle diagram of SVM.

In view of the battery damage caused by checking discharge in substations, Cao [64] proposed a support vector machine algorithm to assess the health of substation batteries. In the study, the voltage, internal resistance, cycle times and activation time of a substation battery pack were taken as the feature vectors of the evaluation model. The experimental results show that the classification accuracy of the model is as high as 97.45%. Chang et al. [65] presented an approach using a Kalman filter (KF) based on SVM to estimate the battery SOH. The basic idea of a Kalman filter based on SVM is that the SVM time series model is first formed based on the measured data, and Kalman filter detection is combined with SVM prediction to estimate the SOH. Experimental results show that the estimation error of SOH was below 3%.

4.3.3. Fuzzy Logic Method

A complete fuzzy logic system (FLS) includes four main components, namely fuzzification, a knowledge database, a rules processor and defuzzification. The structure of a FLS is illustrated in Figure 9. The FIS involves four steps. First, fuzzification: the input values are fuzzified into the fuzzy language variables through the definition of a membership function. Second, knowledge database: a knowledge database is established to describe the membership functions for the input and output variables. Third, rules processor: according to the constraint conditions formulated by the rule base, a reasoning mechanism executes the inference procedure. Finally, defuzzification: the fuzzy output sets are converted to real output values [66].

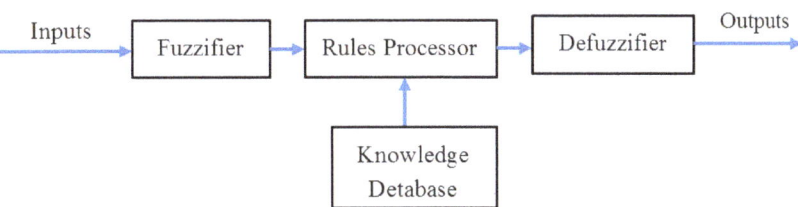

Figure 9. The structure of a fuzzy logic system.

When the fuzzy logic method is implemented in SOH estimation, the output of the fuzzy logic model is the battery SOH, and the inputs are extracted features that are related to the battery SOH. To implement this method, a rule base that describes how each extracted feature contributes to the SOH should first be built based on the training dataset. The rule base may be described by an expert or generated using neutral network algorithms [67]. Each set of data in the training dataset is a fuzzy set, and all the input values are then fuzzified into fuzzy membership functions. Next, the fuzzy output is calculated based on the rule base. Finally, according to the SOH of each fuzzy set, the estimated value of the SOH can be calculated using the weighted average of the SOH based on the fuzzy output [68].

A VRLA battery is a complex electrochemical system. With the charging and discharging process, there are many uncontrollable factors, such as the structural change of lead ions in electrolytes, the evolution of oxygen and hydrogen and the variation of ambient temperature, which affect the internal resistance, life and residual capacity of the battery. These uncontrollable factors bring many uncertainties and difficulties to VRLA battery SOH prediction, and the fuzzy logic system is just the means to solve the SOH estimation of this uncertain and complex electrochemical system.

Some researchers have applied the fuzzy logic method to estimate the SOH of VRLA batteries. FAN et al. [69] developed a fuzzy logic method to estimate the SOH of a VRLA battery in a substation online. According to the linear relationship between the amount of charge (Q) and the open-circuit voltage (OCV) of the battery, a fuzzy logic system based on Q-VOC slope and an SOH rule base was established. The input variable of the fuzzy system is the slope of the Q-VOC diagram, and the output is the SOH. The SOH of the battery was estimated in both online and offline states. The experimental results showed that the SOH error of online method was only 2% compared with the traditional check discharge method, demonstrating that this online SOH estimation method has the advantages of simplicity and short test time. Pritpal [70] described how impedance measurements, combined with fuzzy logic data analysis, have been used to estimate the SOH of lead-acid batteries. In the present method, the combination of fuzzy logic and EIS provided a powerful estimation method for SOH prediction of VRLA batteries. The fuzzy logic method is highly precise, offers good reliability and strong adaptability and can be used in a cheap microcontroller to provide low-cost cell surveillance systems.

4.4. Fusion Methods

In recent years, model fusion technology has received extensive attention from many researchers. The idea of the fusion method is to integrate multiple models, including experimental methods, model-based methods and data-driven methods, to give full play to their respective advantages and achieve accurate, reliable and robust battery health state estimation. Fusion-based methods usually include different model-based mutual integration, the merging of model methods and data-driven methods and the convergence of different data-driven methods.

Zhong et al. [71] proposed SOH estimation based on a fusion model for lead-acid batteries used in substations. Two models were established to estimate the SOH of a VRLA battery. The first model evaluates the relationship between the average resistance and SOH. The other model assesses the decline rate of battery voltage and SOH. According to the proportion of the influence of the nuclear discharge and floating charge state on SOH, a fusion model was established to estimate the SOH of a lead-acid battery in a substation. An accelerated life test was used to verify the proposed arithmetic, and the experimental results showed that the arithmetic was accurate and reliable and can realize real-time estimation the SOH of lead-acid batteries used in substations. Therefore, timely detection of poor SOH can greatly improve the safety and reliability of the battery pack.

The fusion estimation method overcomes the shortcomings of the single-model method or data-driven method, such as low prediction accuracy, poor reliability or misjudgment. It is an important method for battery SOH estimation in the future and has good application prospects.

5. Conclusions

Due to the complexity of electrochemical systems, the accurate estimation of SOH for VRLA batteries is still a challenge. Scholars at home and abroad have carried out a lot of research on the estimation of battery SOH and made many research advancements. Some research methods have been preliminarily applied. However, there is still a lack of a more complete theoretical system for battery SOH prediction, especially under actual operating conditions of backup power supply.

In this paper, the latest developments of SOH estimation methods for VRLA batteries in power system were reviewed. The basic principles, advantages and disadvantages of various methods were introduced. The main SOH estimation methods include experimental methods, model-based methods, data-driven methods and fusion methods. The traditional single model has poor accuracy in estimating battery SOH, and some SOH estimation methods can only be obtained offline and therefore cannot meet the future demand for high-precision and rapid battery SOH estimation in smart power plants and substations. For instance, although the ampere-hour counting method is simple, deep discharge damages the battery to some extent. The open-circuit voltage method requires a long standing, making it difficult to apply in the online estimation environment. The data-driven method can avoid the establishment of complex battery models. However, the establishment of databases, the selection of characteristic factors and the updating of estimation models are also considerable challenges. At present, the combination of model-based methods and data-driven methods has been widely used and achieved remarkable results. In particular, fusion methods are accurate and reliable for practical applications for electric power backup lead-acid batteries, effectively avoiding the adverse effects of systematic error and accidental error on SOH estimation.

The future development of VRLA battery SOH estimation may focus on the following aspects. First, in order to improve the accuracy of the model, efficient and accurate parameter identification methods should be further studied and developed. Second, various factors affecting battery SOH should be considered. By comparing the size of various factors, a more accurate battery model should be established to achieve accurate SOH estimation. Finally, with the development of artificial intelligence and big data technology, the integration of data and models, the complementary coordination of offline and online methods and the realization of online application requirements will also be also an important research direction to improve battery SOH estimation.

Author Contributions: Conceptualization, R.Y.; methodology, C.H.; validation, H.W.; investigation, Y.M. and L.X.; resources, G.L.; writing—original draft preparation, H.W.; writing—review and editing, C.H.; visualization, Y.M.; supervision, R.Y.; project administration, G.L.; funding acquisition, L.X. All authors have read and agreed to the published version of the manuscript.

Funding: This research was funded by the science and technology project of Zhejiang Zheneng Jiahua Electric Power Generation Co., Ltd. (ZNKJ-2021-041).

Institutional Review Board Statement: Not applicable.

Informed Consent Statement: Not applicable.

Data Availability Statement: Not applicable.

Conflicts of Interest: The authors declare no conflict of interest.

Abbreviations

AC	alternating current
ANN	artificial neural network
BP	backpropagation
DC	direct current
DVA	differential voltage analysis
ECM	equivalent circuit model
EIS	electrochemical impedance spectroscopy
FIS	fuzzy logic system
ICA	incremental capacity analysis
ICP	Inductively coupled plasma technology
KF	Kalman filter
OCV	open-circuit voltage
PAM	positive active material
RC	resistor–capacitor
SEM	scanning electron microscope
SOC	state of charge
SOH	status of health
SVM	support vector machine
VRLA	valve-regulated lead acid
XRD	X-ray diffraction technology
WNN	wavelet neutral network

References

1. Li, D.M.; Xu, Y.L.; Zhao, G.J.; Wang, Y.; Guo, Y. Study on the harmless treatment of valve regulated lead-acid batteries in substations. *Appl. Mech. Mater.* **2012**, *217*, 801–804. [CrossRef]
2. Liu, X.; Yang, Y.; He, Y.; Zhang, J.; Zheng, X.; Ma, M.; Zeng, G. A new dynamic SOH estimation of lead-acid battery for substation application. *Int. J. Energy Res.* **2017**, *41*, 579–592. [CrossRef]
3. Nascimento, R.; Ramos, F.; Pinheiro, A.; Junior, W.d.A.S.; Arcanjo, A.M.C.; Filho, R.F.D.; Mohamed, M.A.; Marinho, M.H.N. Case Study of Backup Application with Energy Storage in Microgrids. *Energies* **2022**, *15*, 9514. [CrossRef]
4. Chen, D.; Li, W.; Tian, J.-G. Failure Analysis and Countermeasures of Lead Acid Battery in Substation. *Distrib. Util.* **2016**, *33*, 10–14+9.
5. de Araujo Silva Júnior, W.; Vasconcelos, A.; Arcanjo, A.C.; Costa, T.; Nascimento, R.; Pereira, A.; Jatobá, E.; Filho, J.B.; Barreto, E.; Dias, R.; et al. Characterization of the Operation of a BESS with a Photovoltaic System as a Regular Source for the Auxiliary Systems of a High-Voltage Substation in Brazil. *Energies* **2023**, *16*, 1012. [CrossRef]
6. Costa, T.; Arcanjo, A.; Vasconcelos, A.; Silva, W.; Azevedo, C.; Pereira, A.; Jatobá, E.; Filho, J.B.; Barreto, E.; Villalva, M.G.; et al. Development of a Method for Sizing a Hybrid Battery Energy Storage System for Application in AC Microgrid. *Energies* **2023**, *16*, 1175. [CrossRef]
7. Lipu, M.H.; Hannan, M.; Hussain, A. A review of state of health and remaining useful life estimation methods for lithiumion battery in electric vehicles: Challenges and recommendations. *J. Clean. Prod.* **2018**, *205*, 115–133. [CrossRef]
8. Yang, Y.-F.; Liu, X.-T.; He, Y. State of health and life forecast of backup batteries for substation. *Chin. J. Power Sources* **2018**, *42*, 877–881.

9. Cuma, M.; Koroglu, T. A comprehensive review on estimation strategies used in hybrid and battery electric vehicles. *Renew. Sustain. Energy Rev.* **2015**, *42*, 517–531. [CrossRef]
10. Waltari, P.; Suntio, T. Survey and evaluation of battery monitoring methods and results from user's viewpoint. In Proceedings of the 21st International Telecommunications Energy Conference. INTELEC '99 (Cat. No.99CH37007), Copenhagen, Denmark, 9 June 1999; p. 421.
11. Ouyang, M.-S.; Yu, S.-J. The status quo and development of capacity forecasting technology for VRLA batteries. *Chin. LABAT Man* **2004**, *2*, 59–66.
12. Du, X.-H.; Li, B.-Y.; Jia, B.-Y.; Wang, L. Mechanism and experimental study on life attenuation of VRLA floating charge. *Chin. J. Power Sources* **2022**, *46*, 514–517.
13. Yan, J.-Y.; Li, W.-S.; Zhan, Q.-Y. Failure mechanism of valve-regulated lead–acid batteries under high-power cycling. *J. Power Sources* **2004**, *133*, 135–140. [CrossRef]
14. Chen, C.-M.; Chen, R.-X.; Yu, J.-X.; Liu, S.-T. Research progress of failure modes and recovery technology of lead acid batteries in substations. *Chin. J. Power Sources* **2022**, *46*, 842–844.
15. Pour-Ali, S.; Aghili, M.M.; Davoodi, A. Electrochemical corrosion behavior of Pb–Ca–Sn–Sm grid alloy in H_2SO_4 solution. *J. Alloys Compd.* **2015**, *652*, 172–178. [CrossRef]
16. Ruetschi, P. Aging mechanisms and service life of lead–acid batteries. *J. Power Sources* **2004**, *127*, 33–44. [CrossRef]
17. Guo, Y.; Tang, S.; Meng, G.; Yang, S. Failure modes of valve-regulated lead-acid batteries for electric bicycle applications in deep discharge. *J. Power Sources* **2009**, *191*, 127–133. [CrossRef]
18. Tong, P.Y.; Zhao, R.; Zhang, R.B.; Yi, F.; Shi, G.; Li, A.; Chen, H. Characterization of lead (II)-containing activated carbon and its excellent performance of extending lead-acid battery cycle life for high-rate partial-state-of-charge operation. *J. Power Sources* **2015**, *286*, 91–102. [CrossRef]
19. Lam, L.T.; Haigh, N.P.; Phyland, C.G.; Urban, A.J. Failure mode of valve-regulated lead-acid batteries under high-rate partial-state-of-charge operation. *J. Power Sources* **2004**, *133*, 126–134. [CrossRef]
20. Zou, X.; Kang, Z.; Shu, D.; Liao, Y.; Gong, Y.; He, C.; Hao, J.; Zhong, Y. Effects of carbon additives on the performance of negative electrode of lead carbon battery. *Electrochim. Acta* **2015**, *151*, 89–98. [CrossRef]
21. Hao, Z.-D.; Xu, X.-L.; Hao, W. Review on the roles of carbon materials in lead-carbon batteries. *Int. J. Ion. Sci. Technol. Ion. Motion* **2018**, *24*, 951–965. [CrossRef]
22. Pavlov, D.; Dimitrov, M.; Petkova, G.; Giess, H.; Gnehm, C. The Effect of Selenium on the Electrochemical Behavior and Corrosion of Pb-Sn Alloys Used in Lead-Acid Batteries. *J. Electrochem. Soc.* **1995**, *142*, 2919–2927. [CrossRef]
23. Jiang, S.-D.; Song, Z.-X. A review on the state of health estimation methods of lead-acid batteries. *J. Power Sources* **2022**, *517*, 230710. [CrossRef]
24. Pavlov, D. *Lead-Acid Batteries: Science and Technology*; Elsevier: Amsterdam, The Netherlands, 2011.
25. Pavlov, D. The Lead-Acid Battery Lead Dioxide Active Mass: A Gel-Crystal System with Proton and Electron Conductivity. *J. Electrochem. Soc.* **1992**, *139*, 3075–3080. [CrossRef]
26. Zhong, G.-B.; Liu, S.; Xu, K.-Q.; Chen, D.; Wei, Z.-F. Analysis of the typical failure modes and risk of the VRLA batteries for substations. *Chin. LABAT Man* **2017**, *54*, 160–194.
27. Culpin, B. Thermal runaway in valve-regulated lead acid cells and the effect of separator structure. *J. Power Sources* **2004**, *133*, 79–86. [CrossRef]
28. Song, Y.; Su, L.-S.; Wang, C.-J.; Zhang, Y.-k.; Zhang, J.-B. Research progress of aging for lithium ion batteries. *Chin. J. Power Sources* **2018**, *142*, 1578–1581.
29. Moss, P.L.; Au, G.; Plichta, E.J.; Zheng, J.P. Study of capacity fade of lithium-ion polymer rechargeable batteries with continuous cycling. *J. Electrochem. Soc.* **2010**, *157*, A1–A7. [CrossRef]
30. Ge, M.-F.; Liu, Y.-B.; Jiang, X.-X.; Liu, J. A review on state of health estimations and remaining useful life prognostics of lithium-ion batteries. *Measurement* **2021**, *174*, 1–27. [CrossRef]
31. Ge, L.-J.; Song, Z.-X.; Zhang, G.-G. Study on Float Life of Valve Regulated Lead Acid Batteries for Substation. *Power Capacit. React. Power Compens.* **2020**, *41*, 191–195.
32. Ng, K.S.; Moo, C.-S.; Chen, Y.-P. Enhanced coulomb counting method for estimating state-of-charge and state-of-health of lithium-ion batteries. *Appl. Energy* **2009**, *86*, 1506–1511. [CrossRef]
33. Ungurean, L.; Cârstoiu, G. Battery state of health estimation: A structured review of models, methods and commercial devices. *Int. J. Energy Res.* **2017**, *41*, 151–181. [CrossRef]
34. Jiang, J.-C.; Gao, Y.; Zhang, C.-P.; Wang, Y.-B.; Zhang, W.-G.; Liu, S.-J. Online Diagnostic Method for Health Status of Lithium-ion Battery in Electric Vehicle. *J. Mech. Eng.* **2019**, *55*, 60–84.
35. Aylor, J.H.; Thieme, A.; Johnson, B.W. A Battery State-of-Charge Indicator for Electric Wheelchairs. *Trans. Ind. Electron.* **1992**, *39*, 398–409. [CrossRef]
36. Shahriari, M.; Farrokhi, M. Online State-of-Health Estimation of VRLA Batteries Using State of Charge. *IEEE Trans. Ind. Electron.* **2013**, *60*, 191–202. [CrossRef]
37. Remmlinger, J.; Buchholz, M.; Meiler, M.; Bernreuter, P.; Dietmayer, K. State-of-health monitoring of lithium-ion batteries in electric vehicles by on-board internal resistance estimation. *J. Power Sources* **2011**, *196*, 5357–5363. [CrossRef]

38. Noura, N.; Boulon, L.; Jemeï, S. A Review of Battery State of Health Estimation Methods: Hybrid Electric Vehicle Challenges. *World Electr. Veh. J.* **2020**, *11*, 66. [CrossRef]
39. Jiang, J.-C.; Lin, Z.-S.; Ju, Q. Electrochemical impedance spectra for lithium-ion battery ageing considering the rate of discharge ability. *Energy Procedia* **2017**, *105*, 844–849. [CrossRef]
40. Huet, F. A review of impedance measurements for determination of the state-of-charge or state-of-health of secondary batteries. *J. Power Sources* **1998**, *70*, 59–69. [CrossRef]
41. Berndt, D.; Voss, E. 2—The voltage characteristics OF a lead–acid cell during charge and discharge. In Batter. 2; Collins, D.H., Ed.; Brighton: 1965; pp. 17–27.
42. Ry, P.A.; Kaczor, K.; Lipkowski, J.; Piszcz, M.; Biczel, P.; Siekiers, M. Coup de fouet effect in estimating battery state of health. *J. Power Technol.* **2021**, *101*, 112–126.
43. Delaill, A.; Perrin, M.; Huet, F.; Hernout, L. Study of the "coup de fouet" of lead-acid cells as a function of their state-of-charge and state-of-health. *J. Power Sources* **2006**, *158*, 1019–1028. [CrossRef]
44. Pascoe, P.E.; Anbuky, A.H. The behaviour of the coup de fouet of valve-regulated lead–acid batteries. *J. Power Sources* **2002**, *111*, 304–319. [CrossRef]
45. Yuan, S.-K.; Cheng, L. Estimation of SOH of battery based on Coup de fouet. *Chin. LABAT Man* **2018**, *55*, 65–68.
46. Wu, S.-J.; Yuan, X.-D.; Xu, Q.-S. Review on lithium-ion battery health state assessment. *Chin. J. Power Sources* **2017**, *41*, 1788–1792.
47. Liu, P.; Liang, X.-C.; Huang, G.-J. A Review of Lithium-ion Battery Models. *Chin. Battery Ind.* **2021**, *25*, 106–112.
48. Yang, J.; Wang, T.; Du, C.-Y.; Min, F.; Lyu, T. Overview of the modeling of lithium-ion batteries. *Energy Storage Sci. Technol.* **2019**, *8*, 58–64.
49. Gao, R.-J.; LV, Z.-Q.; Zhao, S.; Huang, X.-G. Health State Estimation of Li-Ion Batteries Based on Electrochemical Model. *Trans. Beijing Inst. Technol.* **2022**, *42*, 791–797.
50. Lyu, C.; Lai, Q.-Z.; Ge, T.-F. A lead-acid battery's remaining useful life prediction by using electrochemical model in the Particle Filtering framework. *Energy* **2017**, *120*, 975–984. [CrossRef]
51. Tran, M.; Fowler, M. A review of lithium-ion battery fault diagnostic algorithms: Current progress and future challenges. *Algorithms* **2020**, *13*, 62. [CrossRef]
52. Gregory, L.P. Extended Kalman filtering for battery management systems of LiPB-based HEV battery packs. Part 1. Background. *J. Power Sources* **2004**, *134*, 52–261.
53. Gregory, L.P. Extended Kalman filtering for battery management systems of LiPB-based HEV battery packs. Part 2. Modeling and identification. *J. Power Sources* **2004**, *134*, 262–276.
54. Gregory, L.P. Extended Kalman filtering for battery management systems of LiPB-based HEV battery packs Part 3. State and parameter estimation. *J. Power Sources* **2004**, *134*, 277–292.
55. Rezvanizaniani, S.M.; Liu, Z.-C.; Chen, Y.; Lee, J. Review and recent advances in battery health monitoring and prognostics technologies for electric vehicle (EV) safety and mobility. *J. Power Sources* **2014**, *256*, 110–124. [CrossRef]
56. Zhang, W.-Z.; Zhang, Y.-H.; Yang, R.-Z. Equivalent circuit model and parameter identification of VRLA batteries. *Chin. J. Power Sources* **2017**, *41*, 460–463.
57. Zhang, L.-G.; Li, Y.-J.; Liu, L.; Miao, G.-X.; Yu, D.-B. Health State Estimation of Battery Model Based on Parameter Analysis. *Process Autom. Instrum.* **2022**, *43*, 69–75.
58. He, H.; Xiong, R.; Guo, H. Online estimation of model parameters and state-of-charge of LiFePO$_4$ batteries in electric vehicles. *Appl. Energy* **2012**, *89*, 413–420. [CrossRef]
59. Li, Y.; Liuc, K.; Foleyd, A.M.; Zülke, A.; Berecibar, B.; Nanini-Maury, E.; Van Mierlo, J.; Hoster, H.E. Data-driven health estimation and lifetime prediction of lithium-ion batteries: A review. *Renew. Sustain. Energy Rev.* **2019**, *113*, 109254. [CrossRef]
60. Chen, M.; Bai, Y.-F.; He, Y. Comparison of data-·driven lithium battery state of health estimation methods. *Energy Storage Sci. Technol.* **2019**, *8*, 1204–1210.
61. Li, J.-L.; Xiao, H. Review on modeling of lithium-ion battery. *Energy Storage Sci. Technol.* **2022**, *11*, 697–703.
62. Talha, M.; Asghar, F.; Kim, S.H. A neural network-based robust online SOC and SOH estimation for sealed lead–acid batteries in renewable systems. *Arabian J. Sci. Eng.* **2019**, *44*, 1869–1881. [CrossRef]
63. Mei, C.-L.; Wang, W.-Q.; Liu, B.-F.; Li, T. Prediction of operating life of substation VRLA battery by wavelet neutral network. *Battery Bimon.* **2014**, *44*, 351–353. (In Chinese)
64. Cao, Y. Health Evaluation of Storage Battery in Substation Based on Support Vector Machine. *Electr. Drive Autom.* **2020**, *43*, 1–3.
65. Chang, L.; Xiaoluo, J. Kalman Filter Based on SVM Innovation Update for Predicting State-of Health of VRLA Batteries. In *Applied Informatics and Communication*; Zeng, D., Ed.; Springer: Berlin/Heidelberg, Germany, 2011; pp. 455–463.
66. Salkind, A.J.; Fennie, C.; Singh, P. Determination of state-of-charge and state-of-health of batteries by fuzzy logic methodology. *J. Power Sources* **1999**, *80*, 293–300. [CrossRef]
67. Singh, P.; Reisner, D. Fuzzy logic-based state-of-health determination of lead acid batteries. In Proceedings of the 24th Annual International Telecommunications Energy Conference, Montreal, QC, Canada, 29 September–3 October 2002; pp. 583–590.
68. Pascoe, P.; Anbuky, A. Adaptive fuzzy coup de fouet based VRLA battery capacity estimation. In Proceedings of the 2001 IEEE International Conference on Systems, Man and Cybernetics. e-Systems and e-Man for Cybernetics in Cyberspace (Cat.No.01CH37236), Tucson, AZ, USA, 7–10 October 2001; Volume 4, pp. 2157–2162.

69. Fan, X.-X.; Ding, H.; Chen, X.-G.; Wang, J.-B.; Yan, H.-M. On-Line Health Assessment of Substation Battery Based on Fuzzy Logic. *Chin. J. Electron Devices* **2021**, *44*, 136–140.
70. Singh, P.; Vinjamuri, R.; Wang, X. Fuzzy logic modeling of EIS measurements on lithium-ion batteries. *Electrochim. Acta* **2006**, *51*, 1673–1679. [CrossRef]
71. Zhong, G.-B.; Liu, X.-T.; He, Y. Estimation of SOH of lead acid batteries in substation. *Chin. J. Power Sources* **2016**, *40*, 2407–2410.

Disclaimer/Publisher's Note: The statements, opinions and data contained in all publications are solely those of the individual author(s) and contributor(s) and not of MDPI and/or the editor(s). MDPI and/or the editor(s) disclaim responsibility for any injury to people or property resulting from any ideas, methods, instructions or products referred to in the content.

Review

Recent Advances in the Structural Design of Silicon/Carbon Anodes for Lithium Ion Batteries: A Review

Yanan Mei [1], Yuling He [1], Haijiang Zhu [1], Zeyu Ma [2], Yi Pu [1], Zhilin Chen [1], Peiwen Li [1], Liang He [2], Wenwu Wang [2] and Hui Tang [1,*]

[1] School of Materials and Energy, University of Electronic Science and Technology, Chengdu 611731, China
[2] School of Mechanical Engineering, Sichuan University, Chengdu 610065, China
* Correspondence: tanghui@uestc.edu.cn; Tel.: +86-18782917701

Abstract: As the capacity of lithium-ion batteries (LIBs) with commercial graphite anodes is gradually approaching the theoretical capacity of carbon, the development of silicon-based anodes, with higher energy density, has attracted great attention. However, the large volume variation during its lithiation/de-lithiation tends to lead to capacity decay and poor cycling performance. While rationally designed silicon/carbon (Si/C) anodes can exhibit higher specific capacity by virtue of silicon and high electrical conductivity and volume expansion suppression by virtue of carbon, they still show poor cycling performance with low initial coulombic efficiency. This review focuses on three strategies for structural design and optimization of Si/C anodes, i.e., carbon-coated structure, embedded structure and hollow structure, based on the recent researches into Si/Canodes and provides deeper insights into the problems that remain to be addressed.

Keywords: lithium-ion batteries; anode; silicon/carbon; structural design; electrochemical performance

1. Introduction

Over the last decade, lithium-ion batteries (LIBs) have become an integral part of all types of electronic devices [1–4]. While the specific capacity of commercial graphite in LIBs is approaching its theoretical specific capacity (372 mAh g^{-1}) [5,6], there is a growing demand for higher energy density. Currently, the main anodes available on the market include graphite, metal oxide and alloy, etc. Recently, numerous high-performance anodes are reported, such as spinel ($Li_4Ti_5O_{12}$) composites [7], laminated $Ti_3C_2T_x$ (MXene) [8], and silicon-based anode [9].

Silicon stands out due to its ultrahigh theoretical specific capacity (4200 mAh g^{-1} [10]), which is 10 times that of a commercial graphite anode [11]. The silicon anode has a much richer source, lower de-lithiation potential, and better environmental friendliness compared with other anodes. However, silicon-based anodes face numerous limitations. For instance, during lithiation/de-lithiation [12], the huge volume expansion (up to 400% [13]) can generate great mechanical stress, fractures, and pulverization of the silicon particles after several cycles which seriously hinders the transport of lithium ions in the anode. This leads to electrode polarization, solid electrolyte interphase (SEI) film's reconstruction [14,15], low coulombic efficiency, and continuous capacity decay. Additionally, the SiO_x electrode [16] as LIB anode can generate an inactive lithium silicate phase during the first cycle of irreversible charging and discharging, which can severely reduce the initial coulombic efficiency of the LIB and affect the electrical conductivity of the whole electrode [17]. Furthermore, the existing preparation methods of silicon-based anodes are complicated, costly, and cannot be used in industry [18–20].

Strategies such as silicon/carbon (Si/C) compositing, nano-structuring silicon [21–23], and manufacturing new binders and electrolytes [24,25] can help to overcome the limitations of the silicon-based anodes. Among them, the Si/C composite is widely studied due to its high operability. Carbon materials show high cycling stability and high electrical conductivity. Therefore, compounding silicon with carbon will provide high specific capacity, and obtain significantly improved electrical conductivity of anode [26–28]. Additionally, as a buffer for the silicon anode, the outer carbon shell can prevent direct contact between electrolyte and silicon [29], forming a stable SEI film and reducing the volume changes during charging and discharging. Thus, the silicon-based structure is less likely to be destroyed and the structural integrity of the electrode is maintained [30]. The combination of silicon and carbon is very promising because they mutually compensate for the low specific capacity of carbon, and the mechanical instability and poor electrical contact of silicon [31]. Thus, the Si/C anodes show excellent lithiation and cycling stability [32].

The development of Si/C anodes began as early as 1995, when Dahn and co-workers used 11% atomic silicon embedded in pre-graphitic carbon [33]. Since then, significant researches were performed and the performance of the Si/C anodes was greatly improved. Following the development of silicon nanowires [34], Y. Cui and co-workers made an outstanding contribution to the structural design of Si/C anodes. In 2009, they developed nanowires with a Si/C core-shell structure [35]. In 2010, they developed carbon nanotube (CNT)-based silicon thin films [36], carbon nanofibers, and silicon nanowires as composite anodes [37] that lowered the weight of the electrode. In the following year, a silicon and carbon nano-coaxial sponge [38] was developed to increase the mass loading per unit area. To improve the industrial scalability, a yolk-shell structure that defined the void between "yolk" and "shell" [39] was developed in 2012, based on which a pomegranate structure [40] was also developed in 2014. Additionally, in 2012, a double-walled silicon nanotube anode [41] was developed based on silicon nanotubes, which significantly increased the charging rate. A carbon cladding structure [42] was developed in 2015. New structures were developed in 2016 like graphene and graphite in cladding [43] and embedded structures [44]. In addition, a conformal graphitic carbon/silicon/self-healing elastic polymer foam structure [45] was developed to meet the needs of stretchable materials. The main known methods to improve Si/C anodes are the structural design's optimization and the preparation of high-performance binders [46,47] and electrolytes [48,49]. Structural optimization is favored by researchers because of the potential for diversity [50–53]. However, there are still some bottlenecks in the widespread commercialization of Si/C anodes. Therefore, to obtain a clear understanding of the optimal Si/C anodes, it is necessary to summarize the current status of the performance improvement in Si/C anodes by structure optimization and presenting a perspective.

This review summarizes the research development of three main structures, carbon-coated structure, embedded structure, and hollow structure (Scheme 1). Moreover, this review analyzes the problems of each structure and provides an outlook on the development direction of the Si/C anodes.

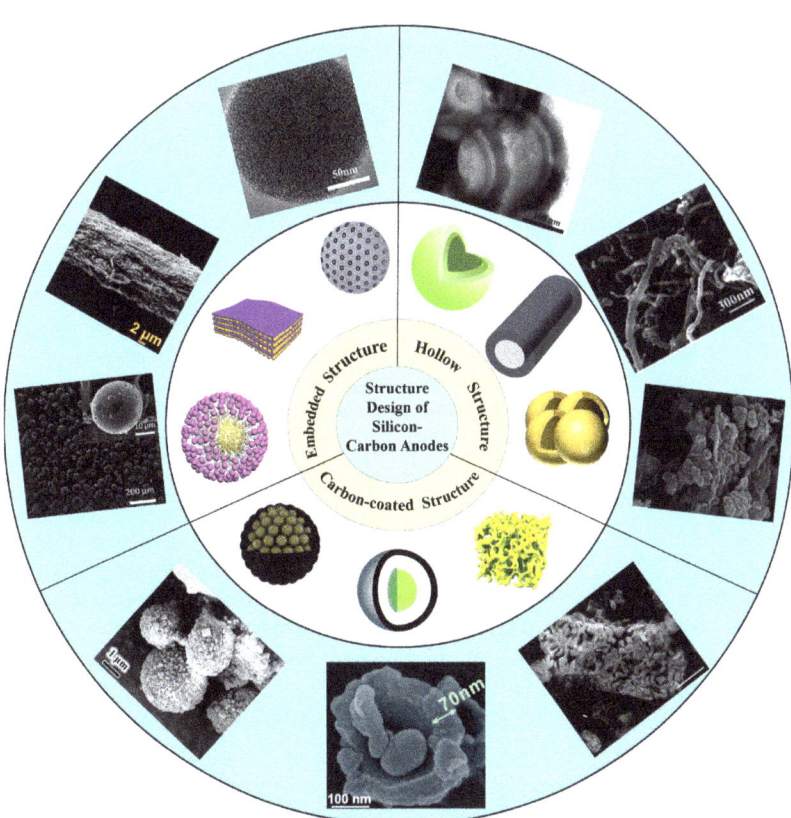

Scheme 1. Schematic diagram of different structures of Si/C composite anodes for lithium-ion batteries.

2. Carbon-Coated Structure

Since carbon-coated silicon-based anodes were proposed in the early 21st century [54], there are some important research results reported [55]. The main contribution of carbon coating is to enhance the electrical contact between the silicon particles, reduce electrode polarization, and improve the electrochemical performance of the assembled cell [56]. The carbon coating can also reduce the probability of direct contact between electrolyte and anode, inhibit the SEI's overgrowth, stabilize the interface, improve the coulombic efficiency, and prevent aggregation of silicon. The following subsections discuss the three main types of carbon-coated structures: core-shell structure, yolk-shell structure, and porous structure.

2.1. Core-Shell Structure

As early as 1999, Si/C anodes were reported as a core-shell structure composed of nano-silica powder and carbon black [57]. Conventional shell-layer structures usually have silicon nanoparticles (SiNPs) as the spherical core, with a layer of carbon encapsulated on the surface. However, in recent reports, various core-shell structures are reported such as those involving dimensional differences in the core, the multiple species of the shell, and heteroatom doping involved in the modification.

A new nanostructured Si/C anodes with enhanced lithium storage capacity and porous silicon microsphere@C core-shell structure (pSiMS@C) was proposed by Wang et al. [58]. The SiAl alloy was reacted with an organic acid under hydrothermal conditions by a facile self-corrosion method (Figure 1a). The SiAl/Al-MOF core-shell precursor was prepared, and the target product was synthesized by a series of annealing and etching processes. The novel pSiMS@C core-shell structure is consisted of an amorphous carbon shell encapsulated by interconnected nanowires. This anode exhibited high electrochemical performance and provided a reversible capacity of 1027.8 mAh g^{-1} after 500 cycles at 1 A g^{-1}, with a high capacity retention of 79%. The pSiMS@C anode was an in-situ MOF-derived carbon shell, which could effectively improve the electrical conductivity of the anode and mitigate its volume expansion.

Li et al. [59] abandoned the conventional method of carbon shells encapsulating unmodified SiNPs. They designed a silicon/nitrogen-doped carbon layer/carbon skeleton microsphere (SCM) using a facile and economical electrospray technique. Each SiNP within the SCM was encapsulated by carbon and connected by a PAN skeleton (Figure 1b). Notably, in the microscopic morphological characterization result, more paths for electron transport are found, which were provided by the nitrogen-doped carbon layer on the surface of the SiNPs and the PAN-derived nitrogen-doped carbon skeleton (Figure 1c), which allowed the material to have high electrical conductivity. Additionally, the carbon layer sustained the stress of SiNPs during (de)lithiation, and the robust three-dimensional carbon skeleton maintained overall integrity. This material is similar to the Si/C anodes with a pomegranate-like shell structure developed by Cui et al. [40] in 2014. Both anodes exhibited excellent electrochemical performance, i.e., a high reversible specific capacity of 746 mAh g^{-1} after 200 cycles, thus demonstrating the feasibility of this structure and potential for commercial applicability since the low cost of electrospray technology.

While the abovementioned core-shell structures exhibited high electrochemical performance, they still face critical challenges. For instance, the proper thickness of the carbon shell poses an important effect on buffering the expansion/extraction of the silicon core during the cycling process. Increasing the thickness helps reducing internal stress and avoid outer layer rupture. An excessively thick layer increases the weight of the anode, and the capacity of the nanocomposite anode is lowered due to the low lithium storage capacity of carbon [60]. Therefore, non-cracking and highly conductive shell layers are essential for core-shell structured Si/C anodes. A flexible N-doped freestanding core-shell Si/C nanofiber (SC-NF) anode reported by Li et al. [61] was prepared using a dual-coaxial electrostatic spinning and carbonization. The transmission electron microscope (TEM) image of this fiber structure (Figure 1d) shows that the Si particles are encapsulated by the carbon shell of the fiber, thus solving the problem of electrode's structural failure and alleviating the volume expansion of silicon. Notably, the binder-free core-shell structure has advantages over commercial binders, i.e., polyvinylidene fluoride and sodium carboxymethylcellulose, which are unable to suppress the drastic change in the volume of SiNPs during repeated cycling. Additionally, such a structural design simplifies the battery's assembly process and allows the homogenization process to be omitted when assembling coin-type batteries. In addition, such a one-dimensional structure brings high electrochemical performance to this anode. In the cycling curves of SC-NF-0, SC-NF-0.18, SC-NF-0.24, and SC-NF-0.30 electrodes at 0.5 A g^{-1}, the SC-NF-0.24 has the best performance with an initial specific discharge capacity of 1441 mAh g^{-1}.

Therefore, different types of core-shell structures, in terms of core form, fabrication process, and dimensionality, are reported in recent studies. While the core-shell structure is the first Si/C composite structure to be used, the electrochemical performance is often affected by the nonuniformity of the carbon coating. Additionally, modified core-shell structures have started to make a breakthrough in electrochemical performance, showing reversible capacity over 1000 mAh g^{-1} after 500 cycles at 1 A g^{-1}.

2.2. Yolk-Shell Structure

Yolk-shell structure differs from the core-shell structure with an additional artificially formed gap between the core and shell. With the yolk-shell structure, the volume expansion of silicon particles inside the shell can be accommodated by the engineered cavity. Furthermore, the conductive shell allows the transport of lithium ions and electrons, and provides a stable interface for good contact between the various particles, leading to a stable SEI layer. Different yolk-shell structures with different electrochemical performances can be obtained using various synthesis methods. As early as 2012, Y. Cui and co-workers [39] had already conducted research on the yolk-shell structured Si/C (YS-Si/C) anode. In this structure, SiNPs (about 100 nm) were attached to the "yolk" of amorphous carbon (carbon layer with a thickness of 5–10 nm) as a fixed side of the "shell", with voids on the corresponding side. The yolk structure has several advantages. Firstly, the gap between the carbon shell and SiNPs is significantly increased, allowing volume variations without affecting the electrode performance. Secondly, the integrity of the carbon shell prevents the electrolyte from coming into direct contact with the SiNPs. Thirdly, the electrical conductivity of the carbon shell structure improves the electrochemical performance. Fourthly, this fixed structure maintains the integrity of the microstructure inside the electrode. However, further studies have shown that the large voids in this structure cause a decrease in the bulk energy density and that the outer carbon shell is brittle.

Recently, Zhang et al. [62] proposed a new YS-Si/C anodes containing carbon-encapsulated rigid SiO_2 as the shell, multiple SiNPs as the yolk, and a flexible CNT network embedded with Fe_2O_3 nanoparticles as the filler (Figure 1e). The gaps brought by the fillers (Fe_2O_3 nanoparticles and CNTs) in YS-Si/C anodes have additional anode's reversible capacity for the electrode and increase the tap density. The carbon-coating by rigid SiO_2 provides greater mechanical strength than a thin carbon shell, improving the safety of the electrode. This anode exhibited excellent electrochemical performance when assembled with a commercially available cathode (Figure 1f).

Hu et al. [63] proposed a new type of double-shell layer yolk-shell structure (TSC-PDA-B) by encapsulating many small-sized SiNPs into a bilayer porous silicon shell (Figure 1g) through the strong etching effect of HF. The double-shell layer exhibited higher mechanical strength and electrical conductivity than the single-shell layer. Notably, this structure resulted in a large improvement in the lithium storage performance, i.e., a high initial capacity of 2108 mAh g^{-1} at a current density of 100 mA g^{-1} and superior cycling performance of 1113 mAh g^{-1} over 200 cycles, which is attributed to the large specific surface area provided by many SiNPs.

However, HF is not environmentally friendly and the etching of the SiO_2 layer needs to be controlled very precisely to obtain satisfactory results. Huang et al. [64] proposed an HF-free strategy using calcination to achieve the carbonization of the inner polymer, polystyrene (PS), and the formation of voids on the surface of the outer polymer, polyaniline (PANI), by exploiting the difference in carbon yield between the two layers of polymers that act as carbon sources grown on the surface of SiNPs (Figure 1h). A thin carbon layer, attached to the surface of the external polymer (PANI), is employed as an eggshell, as shown in the scanning electron microscope (SEM) image of Si@C@void@C (Figure 1i). In Si@C@void@C, the low internal carbon yielding polymer acts as a pore provider and the high carbon yielding external polymer acts as a shell, both of which compensate each other for the formation of a complete yolk-shell structure. Thus, the cladding of PS and SiNPs is a promising strategy.

According to recent studies on the YS-Si/C anode, the gap can help alleviating the volume expansion of silicon. The electrochemical performance of the yolk-shell structure is further enhanced through various designs. However, some following important challenges remain, such as safety problems when the carbon eggshell is too thin, the increased resistance to lithium-ion transport when the eggshell is a rigid carbon layer, the toxic HF for etching, and the encapsulation of the inner polymer with SiNPs using polymers with different types of carbon obtained by calcination.

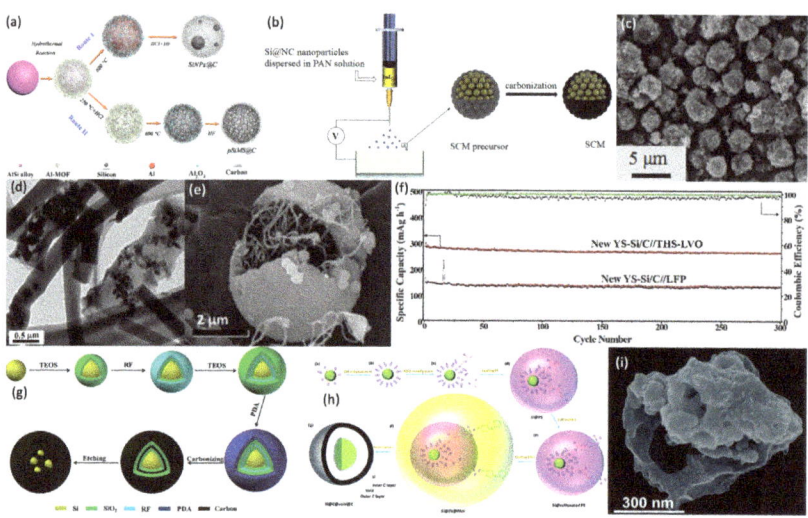

Figure 1. (**a**) Schematic illustration for the fabrication of SiNPs@C (Route I) and pSiMS@C (Route II) core-shell composites using SiAl@Al-MOF as the precursor. Reproduced with permission [58]. Copyright 2018, Elsevier. (**b**) Schematic diagram of SCM composite by electrospray technique, and (**c**) SEM image of SCM. Reproduced with permission [59]. Copyright 2022, Elsevier. (**d**) TEM image of SC-NF. Reproduced with permission [61]. Copyright 2021, ACS Publications. (**e**) SEM image of YS-Si/C anode, and (**f**) rate performance of the full cells of YS-Si/C anode//LFP. Reproduced with permission [62]. Copyright 2019, Wiley Online Library. (**g**) Schematic illustration of the preparation process of YS-Si/C composites. Reproduced with permission [63]. Copyright 2019, Elsevier. (**h**) Schematic illustration of the synthesis of the Si@C@viod@C, and (**i**) SEM image of Si@C@void@C. Reproduced with permission [64]. Copyright 2018, Royal Society of Chemistry.

2.3. Porous Structure

The mechanism of the porous structure is similar to that of the yolk-shell structure, and the silicon's volume expansion is relieved by the introduction of voids in the structure. Unusually, the porous structure allows for an increase in the specific surface area of the silicon anode, allowing for easier diffusion of lithium ions, and thus increasing the reactivity of the anode. However, larger voids and porous structures caused the compressive strength of the particles to be decreased significantly, leading to poor structural integrity and electrochemical performance. Therefore, it is necessary to rationally design a porous structure with optimal voids and densities, so that the Si/C anodes have a large energy density, high initial coulombic efficiency, and long cycling stability [65].

To solve the difficulties of scalable synthesis of silicon-based anodes, An et al. [66] reported an ant nest-like microscale porous silicon (AMPSi), consisting of three-dimensionally interconnected silicon nano-microspheres and bicontinuous nanopores. A reasonable nanopore structure was produced by a top-down scalable method (Figure 2a), which led to a higher electrochemical performance of the target product, as shown by the carbon-coated AMPSi anode that retains up to 90% capacity after 1000 cycles. It is worth noting that such a structure can achieve a self-volume inward expansion mechanism, combining the advantages of Si, both nanoscale and microscale. The expansion of the AMPSi@C electrode's thickness is decreased to 20% during cycling, which is a qualitative improvement compared with the volume expansion of 300% in pure silicon electrodes.

Chen et al. [67] successfully prepared a scalable mesoporous ultrathin two-dimensional silicon nanosheet by a magnesium thermal reduction method (Figure 2b), and overcame the experimental difficulties caused by the non-van der Waals structure of silicon (simultaneous implementation of two-dimensional and mesoporous structures). The unique mesoporous structure of the target product increased the specific surface area to 386.2 m² g⁻¹, thus reducing the magnitude of volume expansion and increasing the contact area between the anode and the electrolyte. Furthermore, the two-dimensional sheet structure accelerated the diffusion of lithium ions and reduced the structural stress. Thus, this method offers the possibility to attain economically scalable two-dimensional mesoporous materials.

The porous structure is also easily derived from the natural biomass porous structure. Xu et al. [68] used biomass derived from soybean as the carbon and nitrogen source to overcome the complicated preparation process and high cost of the Si/N doped carbon composites (Figure 2c), and the synthesized Si@NN-BC-25 composite anode showed excellent electrochemical performance. The SiNPs in Si@NNN-BC-25 were tightly wrapped in an amorphous carbon conductive network doped with nitrogen (Figure 2d), also known as a tofu gel matrix, resulting in "silicon tofu". Herein, nitrogen doping provides more active sites for lithium-ion adsorption, and the tofu structure has high porosity and low bulk density, which effectively mitigate the volume expansion of silicon and prevent direct contact with the electrolyte. The Si@NN-BC-25 composite maintained a reversible capacity of 731.6 mAh g⁻¹ at 1 A g⁻¹ after 300 cycles.

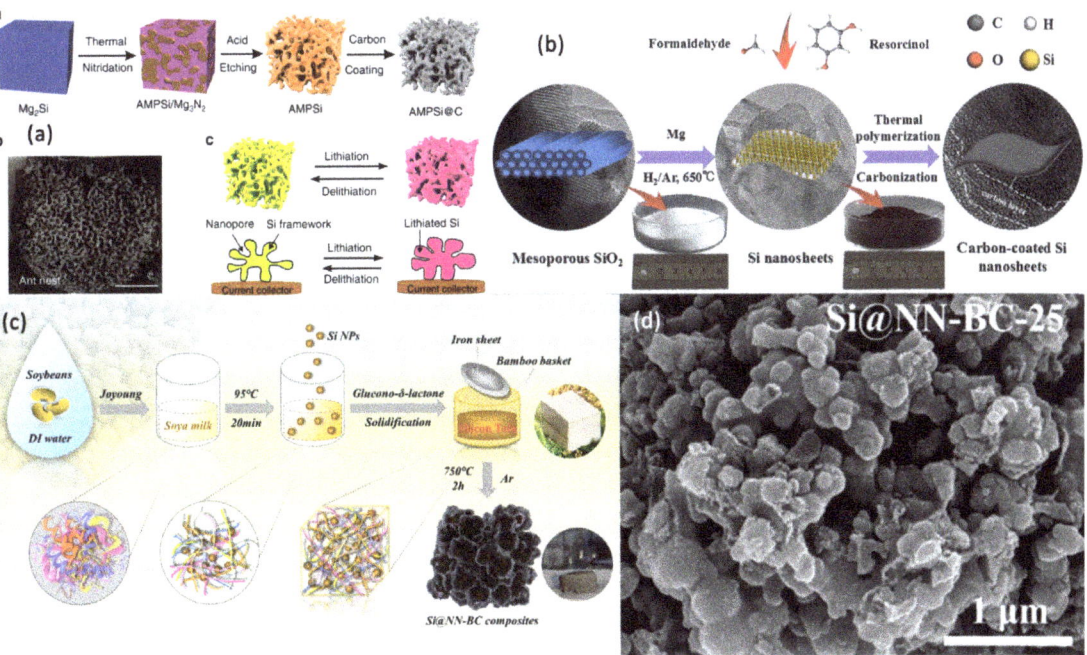

Figure 2. (**a**) Schematic diagram of AMPSi and AMPSi@C preparation and image of ant nests. Reproduced with permission [66]. Copyright 2019, Nature Communications. (**b**) Schematic illustration of the synthesis of nanosheets (inset: the large-scale products of SiO₂ precursors (1.5 g) and Si nanosheets (0.63 g) synthesized one time). Reproduced with permission [67]. Copyright 2018, Wiley Online Library. (**c**) Schematic synthesis of the Si@NN-BC composites derived from silicon tofu, and (**d**) SEM images of Si@NN-BC-25. Reproduced with permission [68]. Copyright 2021, ACS Publications.

Conventional porous structures can be classified as follows [69], carbon layer covered with porous silicon, SiNPs dispersed in the porous carbon base, and porous carbon covered with porous silicon. In recent studies, porous materials were developed through dopant modification [70], different scales and dimensions, types of wrapping agents [71], and difference in the production process [72]. Biomass-derived carbon also attracts attention because of its natural porous structure [73]. In addition to the "silica tofu" mentioned above, diatomaceous earth [74] was studied. However, some problems remain such as the complicated and expensive fabrication processes and environmental unfriendliness.

In summary, there are three types of carbon-coated structures, core-shell structure, yolk-shell structure, and porous structure. The carbon cladding provides reactive sites and alleviates volume expansion, which ensures the improved electrochemical activity of the silicon-based anodes. In a core-shell structure the outer carbon shell will relieve the volume expansion, and the thin carbon layer is prone to be broken during cycling. For the yolk–shell structure, the artificially added voids can compensate for the lack of the core-shell structure by allowing the expansion of the inner SiNPs to be largely relieved. The porous structure is unlike the others, which usually have a three-dimensional structure and a two-dimensional mesoporous structure. Such a porous structure has a large specific surface area, and the volume expansion of SiNPs is relieved [75,76]. Carbon-coated structure is one important structure applied to the design of Si/C anodes. Although the development of carbon coating in recent years presents a diversified trend, there are still some limitations such as uniform carbon coating, easy desorption of porous carbon coating, reasonable carbon content [77], and other limitations that still need to be solved [78]. Overall, carbon-coated structures have a high potential for commercialization. However, some challenges remain.

3. Embedded Structure

Embedded Si/C composites are generally SiNPs embedded in a continuous carbon matrix [79] or porous carbon skeleton [80], and the embedding of different nanoparticles into the carbon shell to achieve enhancement [81,82] The advantage of this structure is that it will successfully alleviate the agglomeration of SiNPs, which is observed in encapsulated structures. Additionally, the voids in the embedded structure provide space for the expansion of SiNPs, which can effectively alleviate the change in electrochemical performance due to volume expansion and provide a channel for the transport of lithium ions [83].

The existing commercial Si/C anodes, with the cycling performance of graphite-blended Si/C (Si/C@G), will be somewhat higher than those with core-shell Si/C [84]. Thus, embedded Si/C structures and silicon in continuous carbon matrices were studied widely. For instance, Kwon et al. [85] synthesized a high-performance Si/C anodes using a low-cost and scalable microemulsion method (Figure 3a), with SiNPs embedded in micro-scale amorphous carbon spheres (amorphous carbon spheres from corn starch), with an outer graphitic carbon coating prepared from corn starch as a biological precursor. The target product exhibited good extensibility and outstanding electrochemical performance when assembled with a commercially available cathode, showing a high capacity of 1800 mAh g^{-1}, high cycling stability of 80% capacity retention after 500 cycles (Figure 3b), and short charging time. These advantages broaden the upper limit of performance that can be achieved by SiNPs embedded in a continuous carbon matrix. Furthermore, Huang et al. [86] employed a biomass carbon source to make an embedded matrix, with SiNPs in a fluffy rice carbon/nano-Si@C (FRC/NP-Si@C) structure consisting of rice as the biomass carbon source. This anode showed high electrochemical performance even at high silicon content (60%).

Achieving uniform carbon distribution is challenging in Si/C anodes and can impact the coulombic efficiency during cycling. To achieve a uniform distribution of carbon at the atomic scale, Zhu et al. [87] selected a molecular precursor capable of containing both silicon and carbon precursors (one bridged organ alkoxysilane precursor $(R'O)_3Si-R-Si(OR')_3$ (R: incorporated organic group)) (Figure 3c). A silicon-based composite consisting of the homogeneous atomic-scale distribution of carbon (ASD-SiOC) was used, prepared from benzene-bridged mesoporous organosilicons (PBMOs), by a facile sol-gel method. Additionally, this structure led successfully to the optimization of the electrochemical performance of the Si/C anodes. This structure differs from conventional embedding, which does not embed SiNPs into a continuous carbon matrix or carbon skeleton. Conversely, the silicon-based material was used as a skeleton to achieve a uniform distribution of carbon at the atomic scale. This unique structure opens the door to a new field of embedded structures and provides guidance for subsequent, more comprehensive studies.

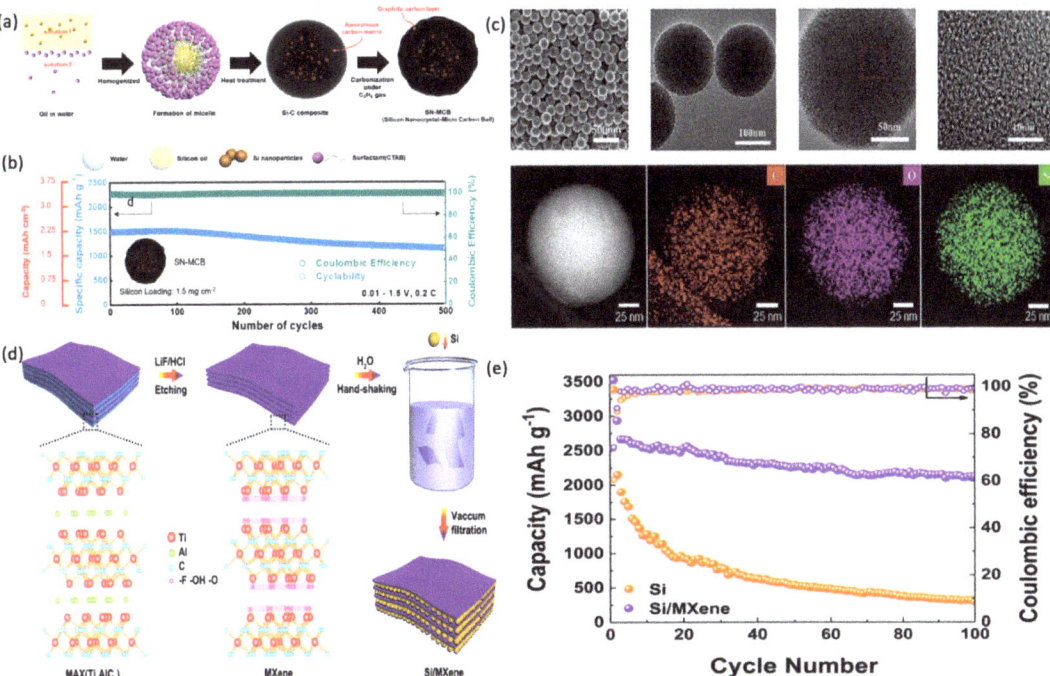

Figure 3. (a) Schematic Illustration of the synthesis of the proposed Si/C hybrid composite. Reproduced with permission [83]. Copyright 2019, ACS Publications. (b) Long-term cycling stability of the anodes in half-cells over a voltage range of 0.01–1.5 V at 0.2 C, and (c) the first row shows the field emission SEM images, TEM images and high-resolution TEM images of ASD-SiOC nanocomposites, and the second row shows the dark-field TEM images and corresponding elemental profiles of individual ASD-SiOC nanocomposite particles. Reproduced with permission [85]. Copyright 2019, Wiley Online Library. (d) The schematic diagram for preparing Si/MXene composite paper, and (e) cycling stability of Si/MXene and pure Si anodes at 200 mA·g^{-1}. Reproduced with permission [86]. Copyright 2019, ACS Publications.

The layered interlayers in the hierarchical structure provide natural sites for the embedding of SiNPs. Among such materials, the hierarchical structure favored by researchers in recent years is MXene, a new two-dimensional transition metal carbide with excellent electrical, electrochemical and mechanical performance. The large specific surface area of MXenes provides greater possibilities for embedding SiNPs. For instance, Tian et al. [88] embedded SiNPs into the laminate structure based on the structure of the flexible Si/MXene composite paper (Figure 3d). The advantages of two components in the materials are combined in this structure. The SiNPs embedded between the layers prevent the failure of the active sites due to the overlapping of the MXene sheets, while the flexible MXene sheets mitigate the volume expansion of the active material during cycling, thus, reducing the rupture of SEI film. This embedded structure of MXene allows the anode to have high lithium storage and electrochemical performance, showing 2118 mAh g^{-1} at 200 mAg^{-1} after 100 cycles (Figure 3e). Hui et al. [89] used the embedded structure of MXene composites with SiNPs. They obtained a better dispersion morphology of SiNPs and a chemical bond between MXene and Si, with favorable reaction kinetics, while preventing the disruption of the two-dimensional Ti$_3$C$_2$ MXene ultrathin layered structure. They developed an efficient low-temperature homogeneous in-situ growth method with relatively mild conditions for coupling Si and MXene. Notably, the Ti_3C_2 MXene shows a pseudocapacitive behavior, which contributes to the reversible capacity and makes the new layered porous MXene/Si nanocomposites with excellent electrochemical performance.

Conventional embedded Si/C anodes include embedding SiNPs into a continuous carbon matrix or carbon skeleton [90,91]. This embedding creates voids to achieve volume expansion mitigation, and the carbon matrix and the carbon skeleton are employed to obtain increased electrical conductivity, i.e., biomass-derived carbon as the carbon skeletons. However, embedded structures also present some challenges. Firstly, when SiNPs are embedded in the carbon skeleton after cycling, it is easy for the structure to break or detach, resulting in direct contact of the active material with the electrolyte which leads to continuous SEI growth and poor performance of the electrode. Secondly, the reversible capacity of the composite decreases when the Si: C ratio is increased [92]. In addition to conventional embedding structures, several embedding matrices can be investigated, such as the interlayer embedding mentioned above. However, more studies are needed because the fabrication processes involved in these strategies are complicated and commercially unavailable.

4. Hollow Structure

Unlike the yolk-shell structure with artificially introduced voids, hollow structure usually doesn't have a nucleus in the interior of the microstructure [93]. However, most preparation methods for hollow structure are similar to those of yolk-shell structure and involve the template method. The hollow structure was widely studied because of its large internal void volume, which provides considerable relief from the stress-strain caused by the embedding of lithium ions. The thin shell brings the active material closer to the electrolyte, shortening the transport path of lithium-ion and improving the electrochemical performance [94–96]. Depending on the template's morphology, templates can be classified into hard templates (i.e., porous silicon templates, silicon microspheres, and MgO) and soft templates (i.e., gels, anti-gels, and vesicles, formed by aggregation of surfactant molecules). Herein, the hollow-structured Si/C composite anodes are classified according to the type of template used in the hollow structure.

4.1. SiO_2 Template

The SiO_2 as a template could be used to obtain hollow structures through a magnesium thermal reduction reaction. However, the SiC produced in this process will decrease the electrochemical performance of the anode. Li et al. [97] developed a SiC-free hollow Si/C anodes using SiO_2 as a template through magnesium thermal reduction to obtain a porous Si@C ball-in-ball hollow structure. The porous Si@C hollow sphere structure generated by HCl etching of Al_2O_3, and the existence of reasonable voids between the carbon layer (Figure 4a) and the hollow structure are beneficial to alleviate the volume variation during cycling. The Si@C anode with this design showed excellent lithium storage performance, i.e., a capacity of 813.2 mAh g^{-1} at 0.1 C after 100 cycles.

Wang et al. [98] used natural reed extract as the Si source and CO_2 as the carbon source, leading to a low cost since biomass feedstock. The concentration of HCl and content of CO_2 had a significant impact on the composition, crystallinity, structure, electrical conductivity, specific surface area, pore volume, and particle distribution of the Si/C complexes. The optimal anode was obtained at 21 wt.% carbon and 3 M HCl, delivering a large reversible specific capacity of 1548 mAh g^{-1} at 100 mA g^{-1}. The development of this low-cost hollow structure paves a new path for the scalable preparation of Si/C composite anodes.

Jiang et al. [99] reported a Si/C composite prepared by SiO_2 template and obtained a unique structure of a silicon layer encapsulated in a bilayer of elastic carbon cushioned hollow shell by a simple layer-by-layer assembly and magnesium thermal reduction method (Figure 4b). In this structure, the elastic carbon shells inside and outside can make the volume expansion of silicon adaptive in both internal and external directions, while the hollow structure inside also stabilizes the mechanical performance. The electrode achieved a very high cycle life and reversible specific capacity.

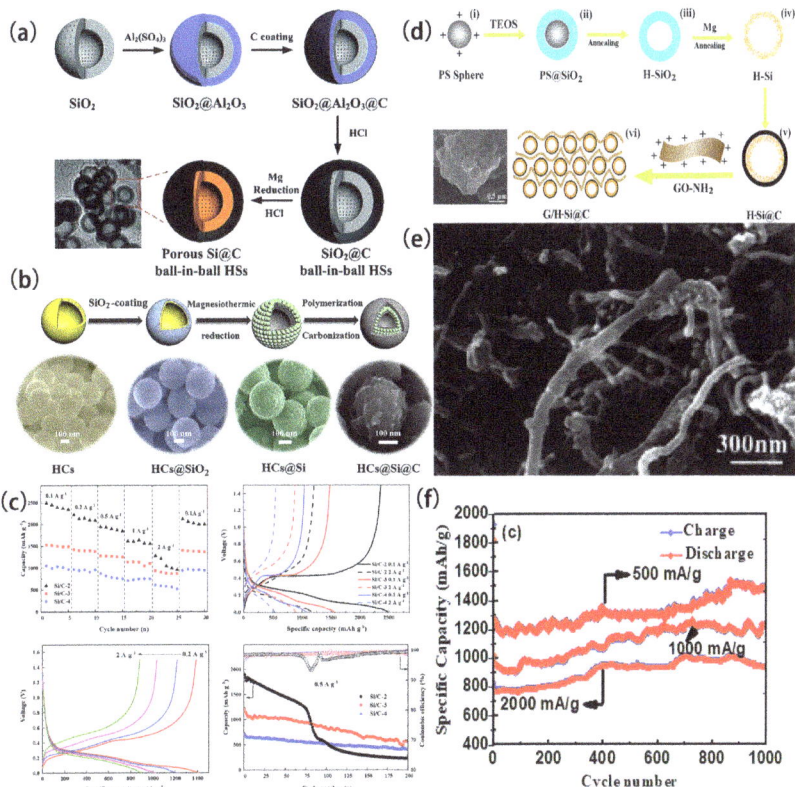

Figure 4. (a) Schematic fabrication process of porous Si@C ball-in-ball hollow spheres. Reproduced with permission [95]. Copyright 2018, Royal Society of Chemistry. (b) Schematic illustration of the synthesis processes of hollow carbon spheres (HCs)@Si@C "hollow triple-layer puff" structure. Reproduced with permission [97]. Copyright 2019, ACS Publications. (c) Rate capability, discharge-charge curve and cycling performance of different SiC hybrids. Reproduced with permission [98]. Copyright 2021, Elsevier. (d) Schematic illustration of producing G/H-Si@C bubble film composite structure. Reproduced with permission [99]. Copyright 2020, Elsevier. (e) SEM images of CNTs/Si/C, and (f) cycle performance of cells with CNTs/Si/C anodes at different current densities. Reproduced with permission [100]. Copyright 2018, Royal Society of Chemistry.

Preparing hollow Si/C structures using SiO_2 as a template has disadvantages. Firstly, the generation of SiC posed negative effects on the anode's electrochemical characteristics. Reducing the contact area between SiO_2 and the shell will prevent the generation of SiC. Secondly, the magnesium thermal reduction reaction requires a high temperature, causing difficulty in generating hollow structures. Thirdly, the gap between the hollow structure and the externally attached carbon layer needs to be moderate. The larger gap makes SiNPs agglomerate, which affects the electrochemical activity. In contrast, the smaller gap makes the volume expansion during cycling damage the carbon layer. The use of SiO_2 as a template for hollow structures, especially using cheaper biomass feedstock, is a feasible strategy. However, more studies are needed to achieve mass production.

4.2. Self-Sacrificing Templates

In addition to SiO_2 as a template to provide a silicon source to make hollow structures, there is also a self-sacrificing template. HF or HCl etching is typically used to sacrifice the material for obtaining the hollow structure. The materials often used as self-sacrificing templates to make cavities are MgO, amino PS, polyethyleneimine (PEI), etc. Among these, the most representative one is the MgO template. For instance, Zhang et al. [100] removed the in-situ-generated MgO templates by acid etching to achieve a hollow structure. They used excess Mg vapor and CO_2 (green carbon source) to achieve an interconnected carbon network between SiNPs, which significantly improved the electrical conductivity and electron transfer during cycling. This high-rate, long-cycle life interconnected hollow Si/C nanosphere anode (Figure 4c) contained no insulating SiC and exhibited excellent electrochemical performance.

Amino PSs are also used as self-sacrificing templates. For instance, Liu et al. [101] prepared a flexible graphene bubble film@thin mesoporous carbon shell, encapsulating submicron level Si@C hollow sphere structures and using PS as a sacrificing template through a thermal treatment and chemical vapor deposition (Figure 4d). Notably, the compact structure resulted in a significantly reduced transport distance of electrons and lithium ions. The effect of the double-shell layer alleviates volume expansion of the silicon, enhances lithium storage, and controls the SEI's formation. Thus, the composite anode has a high initial coulombic efficiency.

The hollow structure is also closely related to the suppression of pulverization caused by repeated rupture of the SEI in the silicon anode during cycling. Specifically, the inner and outer layers of the hollow structure will expand outward during the process of lithium, resulting in repeated breakage of the SEI film and thus thickening the SEI layer, establishing an insulating layer, and making the silicon anode activity ineffective, which is exactly the inhibition behavior we mentioned. As early as 2012, Y. Cui and co-workers [41] developed a silicon hollow nanofiber structure based on a self-sacrificing carbon template method to prepare a surface layer of silicon that was reduced to SiO_x during the annealing process, thus avoiding the problem of repeated breakage of the SEI film and leading to higher mechanical characteristics. In the last five years, hollow nanotube structures based on Si/C anodes have also been developed. For instance, Liu et al. [102] synthesized a sandwich-like carbon-coated coaxial core-shell CNTs/Si hollow nanofiber tube structure using CNTs as sacrificing template. In this structure, CNTs act as a perfect template for Si deposition, stabilizing the mechanical performance because of its sandwich-like hollow structure that can be observed in the microscopic form (Figure 4e). Both the CNTs and the carbon layer have a positive effect in achieving improved electrical conductivity of the CNTs/Si/C anode. Additionally, the target products still maintain good electrochemical characteristics. After long cycling, a stable reversible discharge capacity of CNTs/Si/C nanotubes of up to 1508.5 mA h g^{-1} was obtained. The gradual increase in capacity will be attributed to the formation of a stable polymer gel layer resulting from kinetically activated electrolyte degradation (Figure 4f)

Self-sacrificing templates are constructed by growing silicon on the surface of the template, followed by thermal decomposition or acid etching of its components to achieve hollow structures. The electrochemical behavior of the hollow structure prepared by this method is determined by the surface shell structure, which can alleviate the expansion of the active material. Additionally, the fabrication procedure is usually unstable and scalable.

The cavity in the hollow structure is the highlight of this structural design [103], which plays the role of relieving the volume variations during cycling internally. Different templates play different roles in the structural design of the hollow structure, some acting as the silicon source and some acting as the cavity obtained after self-sacrifice. However, during the cavitation process, acid etching is usually unavoidable, and thus, more environmentally friendly ways need to be developed [104]. In addition, the hollow structure of Si/C anodes also faces the inhibition behavior caused by the repeated breaking of SEI film on the surface shell. In a word, hollow structures have more possibilities for structural modification. It can also be found in the reported literature that the above problems can be solved by designing shells with different functional structures.

5. Conclusions

Silicon, a non-toxic material with abundant resources on earth and high theoretical capacity, is the ideal anode. However, commercialization of silicon anodes are limited due to their huge volume expansion and poor electrical conductivity. Hence, various strategies have been developed to overcome these limitations, such as designing electrolytes and Si/C anodes together through nano-structuring. In this review, based on the current state of research on structure optimization of Si/C anodes, we classified Si/C anodes into three categories, carbon-coated structure, embedded structure, and hollow structure. Moreover, the advantages and disadvantages of three types of Si/C anodes are summarized. However, these strategies still face challenges, which have been summarized as follows.

(1) Attaining uniform coatings in the carbon-coated structure is still challenging, and uneven coating impacts the lithium-ion diffusion rate, leading to poor cycle life.
(2) As the most developed structural optimization method, the carbon cladding includes a variety of strategies in terms of cladding content and cladding form. However, its impact on improving the overall electrochemical performance is still limited.
(3) The loss of SiNPs in the embedded structure reduces the anode content, which affects the performance.
(4) The fabrication of hollow structures usually involves templates and acid etching, which is not environmentally friendly.

The addition of carbon-based materials does not sufficiently address the challenges associated with the volume expansion of silicon-based anodes, the repeated fragmentation of SEI films, and the poor electrical conductivity. The following aspects should be attained in further studies.

(1) A uniform carbon coating and an optimal ratio of Si: C.
(2) More stable bonding when SiNPs are embedded.
(3) More environmentally friendly methods of template removal and more non-toxic precursors.
(4) Reasonable void and hole design, i.e., the more reasonable spacing between shell layers in hollow structures.

While the recently reported Si/C anodes have not reached the theoretical specific capacity of silicon (4200 mAh g^{-1}), the emergence of new processes and novel structures present promising electrochemical performance. As one of the silicon-based anodes most likely to be commercialized, Si/C anodes have the potential to replace commercial graphite anodes.

Author Contributions: Conceptualization, Y.M. and H.T.; investigation, Y.M., Y.H. and H.Z.; resources, Y.P. and Z.C.; writing—original draft preparation, Y.M.; writing—review and editing, Z.M., L.H. and W.W.; visualization, Y.P. and P.L.; supervision, H.T.; project administration, H.T.; funding acquisition, H.T. All authors have read and agreed to the published version of the manuscript.

Funding: This research was funded by the National Science and Technology Major Project (No. 2020YFB1506001), and the Department of Science and Technology of Sichuan Province (No. 2021YFG0231).

Institutional Review Board Statement: Not applicable.

Informed Consent Statement: Not applicable.

Data Availability Statement: All data that support the findings of this study are included within the article.

Conflicts of Interest: The authors declare no conflict of interest.

References

1. Grey, C.P.; Hall, D.S. Prospects for lithium-ion batteries and beyond-a 2030 vision. *Nat. Commun.* **2020**, *11*, 6279. [CrossRef]
2. Li, M.; Lu, J.; Chen, Z.; Amine, K. 30 years of lithium-ion batteries. *Adv. Mater.* **2018**, *30*, 1800561. [CrossRef] [PubMed]
3. Liu, J.; Bao, Z.; Cui, Y.; Dufek, E.J.; Goodenough, J.B.; Khalifah, P.; Zhang, J.G. Pathways for practical high-energy long-cycling lithium metal batteries. *Nat. Energy* **2019**, *4*, 180–186. [CrossRef]
4. Zubi, G.; Dufo-López, R.; Carvalho, M.; Pasaoglu, G. The lithium-ion battery: State of the art and future perspectives. *Renew. Sustain. Energy Rev.* **2018**, *89*, 292–308. [CrossRef]
5. Wu, F.; Maier, J.; Yu, Y. Guidelines and trends for next-generation rechargeable lithium and lithium-ion batteries. *Chem. Soc. Rev.* **2020**, *49*, 1569–1614. [CrossRef]
6. Zhang, W.J. A review of the electrochemical performance of alloy anodes for lithium-ion batteries. *J. Power Sources* **2011**, *196*, 13–24. [CrossRef]
7. Yi, T.F.; Yang, S.Y.; Xie, Y. Recent advances of $Li_4Ti_5O_{12}$ as a promising next generation anode material for high power lithium-ion batteries. *J. Mater. Chem. A* **2015**, *3*, 5750–5777. [CrossRef]
8. Xiong, D.; Li, X.; Bai, Z.; Lu, S. Recent advances in layered $Ti_3C_2T_x$ MXene for electrochemical energy storage. *Small* **2018**, *14*, 1703419. [CrossRef]
9. Feng, K.; Li, M.; Liu, W.; Kashkooli, A.G.; Xiao, X.; Cai, M.; Chen, Z. Silicon-based anodes for lithium-ion batteries: From fundamentals to practical applications. *Small* **2018**, *14*, 1702737. [CrossRef]
10. Wen, C.J.; Huggins, R.A. Chemical diffusion in intermediate phases in the lithium-silicon system. *J. Solid State Chem.* **1981**, *37*, 271–278. [CrossRef]
11. Wu, H.; Cui, Y. Designing nanostructured Si anodes for high energy lithium ion batteries. *Nano Today* **2012**, *7*, 414–429. [CrossRef]
12. Zuo, X.; Zhu, J.; Müller-Buschbaum, P.; Cheng, Y.-J. Silicon based lithium-ion battery anodes: A chronicle perspective review. *Nano Energy* **2017**, *31*, 113–143. [CrossRef]
13. Cui, Y. Silicon anodes. *Nat. Energy* **2021**, *6*, 995–996. [CrossRef]
14. Lucas, I.T.; Pollak, E.; Kostecki, R. In situ AFM studies of SEI formation at a Sn electrode. *Electrochem. Commun.* **2009**, *11*, 2157–2160. [CrossRef]
15. Jiang, J.; Zhang, H.; Zhu, J.; Li, L.; Liu, Y.; Meng, T.; Ma, L.; Xu, M.; Liu, J.; Li, C.M. Putting Nanoarmors on Yolk–Shell Si@C Nanoparticles: A Reliable Engineering Way To Build Better Si-Based Anodes for Li-Ion Batteries. *ACS Appl. Mater. Interfaces* **2018**, *10*, 24157–24163. [CrossRef]
16. Li, Z.; He, Q.; He, L.; Hu, P.; Li, W.; Yan, H.; Peng, X.; Huang, C.; Mai, L. Self-sacrificed synthesis of carbon-coated SiO_x nanowires for high capacity lithium ion battery anodes. *J. Mater. Chem. A* **2017**, *5*, 4183–4189. [CrossRef]
17. Chen, T.; Wu, J.; Zhang, Q.; Su, X. Recent advancement of SiO_x based anodes for lithium-ion batteries. *J. Power Sources* **2017**, *363*, 126–144. [CrossRef]
18. Ma, C.; Wang, Z.; Zhao, Y.; Li, Y.; Shi, J. A novel raspberry-like yolk-shell structured Si/C micro/nano-spheres as high-performance anode materials for lithium-ion batteries. *J. Alloys Compd.* **2020**, *844*, 156201. [CrossRef]
19. Xu, Q.; Sun, J.-K.; Li, J.-Y.; Yin, Y.-X.; Guo, Y.-G. Scalable synthesis of spherical Si/C granules with 3D conducting networks as ultrahigh loading anodes in lithium-ion batteries. *Energy Storage Mater.* **2018**, *12*, 54–60. [CrossRef]
20. Zhang, W.; Wang, D.; Shi, H.; Jiang, H.; Wang, C.; Niu, X.; Zhang, X.; Yu, L.; Zhang, Y. Industrial waste micron-sized silicon use for Si@C microspheres anodes in low-cost lithium-ion batteries. *Sustain. Mater. Technol.* **2022**, *33*, e00454. [CrossRef]
21. Xu, Z.-L.; Liu, X.; Luo, Y.; Zhou, L.; Kim, J.-K. Nanosilicon anodes for high performance rechargeable batteries. *Prog. Mater. Sci.* **2017**, *90*, 1–44. [CrossRef]
22. Kabashin, A.V.; Singh, A.; Swihart, M.T.; Zavestovskaya, I.N.; Prasad, P.N. Laser-Processed Nanosilicon: A Multifunctional Nanomaterial for Energy and Healthcare. *ACS Nano* **2019**, *13*, 9841–9867. [CrossRef]
23. Su, Y.; Feng, X.; Zheng, R.; Lv, Y.; Wang, Z.; Zhao, Y.; Shi, L.; Yuan, S. Binary Network of Conductive Elastic Polymer Constraining Nanosilicon for a High-Performance Lithium-Ion Battery. *ACS Nano* **2021**, *15*, 14570–14579. [CrossRef] [PubMed]
24. Wen, Z.; Wu, F.; Li, L.; Chen, N.; Luo, G.; Du, J.; Zhao, L.; Ma, Y.; Li, Y.; Chen, R. Electrolyte Design Enabling Stable Solid Electrolyte Interface for High-Performance Silicon/Carbon Anodes. *ACS Appl. Mater. Interfaces* **2022**, *14*, 38807–38814. [CrossRef] [PubMed]
25. Li, Y.; Zhou, X.; Hu, J.; Zheng, Y.; Huang, M.; Guo, K.; Li, C. Reversible Mg metal anode in conventional electrolyte enabled by durable heterogeneous SEI with low surface diffusion barrier. *Energy Storage Mater.* **2022**, *46*, 1–9. [CrossRef]
26. Zhang, X.; Weng, J.; Ye, C.; Liu, M.; Wang, C.; Wu, S.; Tong, Q.; Zhu, M.; Gao, F. Strategies for Controlling or Releasing the Influence Due to the Volume Expansion of Silicon inside Si−C Composite Anode for High-Performance Lithium-Ion Batteries. *Materials* **2022**, *15*, 4264. [CrossRef] [PubMed]
27. Hsieh, C.-C.; Liu, W.-R. Effects of nitrogen doping on Si/carbon composite anode derived from Si wastes as potential active materials for Li ion batteries. *J. Alloys Compd.* **2019**, *790*, 829–836. [CrossRef]
28. Liu, S.; Xu, J.; Zhou, H.; Wang, J.; Meng, X. B-Doped Si@C Nanorod Anodes for High-Performance Lithium-Ion Batteries. *J. Nanomater.* **2019**, *2019*, 6487156. [CrossRef]
29. Jiang, M.; Ma, Y.; Chen, J.; Jiang, W.; Yang, J. Regulating the carbon distribution of anode materials in lithium-ion batteries. *Nanoscale* **2021**, *13*, 3937–3947. [CrossRef]

30. Casimir, A.; Zhang, H.; Ogoke, O.; Amine, J.C.; Lu, J.; Wu, G. Silicon-based anodes for lithium-ion batteries: Effectiveness of materials synthesis and electrode preparation. *Nano Energy* **2016**, *27*, 359–376. [CrossRef]
31. Chae, S.; Xu, Y.; Yi, R.; Lim, H.; Velickovic, D.; Li, X.; Li, Q.; Wang, C.; Zhang, J. A Micrometer-Sized Silicon/Carbon Composite Anode Synthesized by Impregnation of Petroleum Pitch in Nanoporous Silicon. *Adv. Mater.* **2021**, *33*, 2103095. [CrossRef] [PubMed]
32. Tu, W.; Bai, Z.; Deng, Z.; Zhang, H.; Tang, H. In-Situ Synthesized Si@C Materials for the Lithium Ion Battery: A Mini Review. *Nanomaterials* **2019**, *9*, 432. [CrossRef] [PubMed]
33. Wilson, A.M.; Way, B.M.; Dahn, J.R.; van Buuren, T. Nanodispersed silicon in pregraphitic carbons. *J. Appl. Phys.* **1995**, *77*, 2363–2369. [CrossRef]
34. Chan, C.K.; Peng, H.; Liu, G.; McIlwrath, K.; Zhang, X.F.; Huggins, R.A.; Cui, Y. High-performance lithium battery anodes using silicon nanowires. *Nat. Nanotechnol.* **2008**, *3*, 31–35. [CrossRef]
35. Cui, L.F.; Yang, Y.; Hsu, C.M.; Cui, Y. Carbon−silicon core−shell nanowires as high capacity electrode for lithium ion batteries. *Nano Lett.* **2009**, *9*, 3370–3374. [CrossRef]
36. Cui, L.-F.; Hu, L.; Choi, J.W.; Cui, Y. Light-Weight Free-Standing Carbon Nanotube-Silicon Films for Anodes of Lithium Ion Batteries. *ACS Nano* **2010**, *4*, 3671–3678. [CrossRef]
37. Choi, J.W.; Hu, L.; Cui, L.; McDonough, J.R.; Cui, Y. Metal current collector-free freestanding silicon–carbon 1D nanocomposites for ultralight anodes in lithium ion batteries. *J. Power Sources* **2010**, *195*, 8311–8316. [CrossRef]
38. Hu, L.; Wu, H.; Gao, Y.; Cao, A.; Li, H.; McDough, J.; Xie, X.; Zhou, M.; Cui, Y. Silicon-Carbon Nanotube Coaxial Sponge as Li-Ion Anodes with High Areal Capacity. *Adv. Energy Mater.* **2011**, *1*, 523–527. [CrossRef]
39. Liu, N.; Wu, H.; McDowell, M.T.; Yao, Y.; Wang, C.; Cui, Y. A yolk-shell design for stabilized and scalable LIB alloy anodes. *Nano Lett.* **2012**, *12*, 3315–3321. [CrossRef]
40. Liu, N.; Lu, Z.; Zhao, J.; McDowell, M.T.; Lee, H.-W.; Zhao, W.; Cui, Y. A pomegranate-inspired nanoscale design for large-volume-change lithium battery anodes. *Nat. Nanotechnol.* **2014**, *9*, 187–192. [CrossRef]
41. Wu, H.; Chan, G.; Choi, J.W.; Yao, Y.; McDowell, M.T.; Lee, S.W.; Jackson, A.; Yang, Y.; Hu, L.; Cui, Y. Stable cycling of double-walled silicon nanotube battery anodes through solid-electrolyte interphase control. *Nat. Nanotechnol.* **2012**, *7*, 310–315. [CrossRef] [PubMed]
42. Lu, Z.; Liu, N.; Lee, H.-W.; Zhao, J.; Li, W.; Li, Y.; Cui, Y. Nonfilling Carbon Coating of Porous Silicon Micrometer-Sized Particles for High-Performance Lithium Battery Anodes. *ACS Nano* **2015**, *9*, 2540–2547. [CrossRef] [PubMed]
43. Li, Y.; Yan, K.; Lee, H.W.; Lu, Z.; Liu, N.; Cui, Y. Growth of conformal graphene cages on micrometre-sized silicon particles as stable battery anodes. *Nat. Energy* **2016**, *1*, 15029. [CrossRef]
44. Ko, M.; Chae, S.; Ma, J.; Kim, N.; Lee, H.-W.; Cui, Y.; Cho, J. Scalable synthesis of silicon-nanolayer-embedded graphite for high-energy lithium-ion batteries. *Nat. Energy* **2016**, *1*, 16113. [CrossRef]
45. Sun, Y.; Lopez, J.; Lee, H.-W.; Liu, N.; Zheng, G.; Wu, C.-L.; Sun, J.; Liu, W.; Chung, J.W.; Bao, Z.; et al. A Stretchable Graphitic Carbon/Si Anode Enabled by Conformal Coating of a Self-Healing Elastic Polymer. *Adv. Mater.* **2016**, *28*, 2455–2461. [CrossRef]
46. Chen, H.; Ling, M.; Hencz, L.; Ling, H.Y.; Li, G.; Lin, Z.; Liu, G.; Zhang, S. Exploring Chemical, Mechanical, and Electrical Functionalities of Binders for Advanced Energy-Storage Devices. *Chem. Rev.* **2018**, *118*, 8936–8982. [CrossRef]
47. Bresser, D.; Buchholz, D.; Moretti, A.; Varzi, A.; Passerini, S. Alternative binders for sustainable electrochemical energy storage –the transition to aqueous electrode processing and bio-derived polymers. *Energy Environ. Sci.* **2018**, *11*, 3096–3127. [CrossRef]
48. Ai, Q.; Fang, Q.; Liang, J.; Xu, X.; Zhai, T.; Gao, G.; Guo, H.; Han, G.; Ci, L.; Lou, J. Lithium-conducting covalent-organic-frameworks as artificial solid-electrolyte-interphase on silicon anode for high performance lithium ion batteries. *Nano Energy* **2020**, *72*, 104657. [CrossRef]
49. Pan, J.; Peng, H.; Yan, Y.; Bai, Y.; Yang, J.; Wang, N.; Dou, S.; Huang, F. Solid-state batteries designed with high ion conductive composite polymer electrolyte and silicon anode. *Energy Storage Mater.* **2021**, *43*, 165–171. [CrossRef]
50. Hou, G.; Cheng, B.; Yang, Y.; Du, Y.; Zhang, Y.; Li, B.; He, J.; Zhou, Y.; Yi, D.; Zhao, N.; et al. Multiscale Buffering Engineering in Silicon–Carbon Anode for Ultrastable Li-Ion Storage. *ACS Nano* **2019**, *13*, 10179–10190. [CrossRef]
51. Zhang, W.; Fang, S.; Wang, N.; Zhang, J.; Shi, B.; Yu, Z.; Yang, J. A compact silicon–carbon composite with an embedded structure for high cycling coulombic efficiency anode materials in lithium-ion batteries. *Inorg. Chem. Front.* **2020**, *7*, 2487–2496. [CrossRef]
52. Han, C.; Si, H.; Sang, S.; Liu, K.; Liu, H.; Wu, Q. Achieving fully reversible conversion in Si anode for lithium-ion batteries by design of pomegranate-like Si@C structure. *Electrochim. Acta* **2021**, *389*, 138736. [CrossRef]
53. Du, Y.; Yang, Z.; Yang, Y.; Yang, Y.; Jin, H.; Hou, G.; Yuan, F. Mussel-pearl-inspired design of Si/C composite for ultrastable lithium storage anodes. *J. Alloys Compd.* **2021**, *872*, 159717. [CrossRef]
54. Yoshio, M.; Wang, H.; Fukuda, K.; Umeno, T.; Dimov, N.; Ogumi, Z. Carbon-Coated Si as a Lithium-Ion Battery Anode Material. *J. Electrochem. Soc.* **2002**, *149*, A1598. [CrossRef]
55. Jang, J.; Kang, I.; Yi, K.-W.; Cho, Y.W. Highly conducting fibrous carbon-coated silicon alloy anode for lithium ion batteries. *Appl. Surf. Sci.* **2018**, *454*, 277–283. [CrossRef]
56. Ng, S.H.; Wang, J.; Wexler, D.; Chew, S.Y.; Liu, H.K. Amorphous Carbon-Coated Silicon Nanocomposites: A Low-Temperature Synthesis via Spray Pyrolysis and Their Application as High-Capacity Anodes for Lithium-Ion Batteries. *J. Phys. Chem. C* **2007**, *111*, 11131–11138. [CrossRef]

57. Li, H.; Huang, X.; Chen, L.; Wu, Z.; Liang, Y. A High Capacity Nano-Si Composite Anode Material for Lithium Rechargeable Batteries. *Electrochem. Solid State Lett.* **1999**, *2*, 547–549. [CrossRef]
58. Wang, K.; Pei, S.; He, Z.; Huang, L.-A.; Zhu, S.; Guo, J.; Shao, H.; Wang, J. Synthesis of a novel porous silicon microsphere@carbon core-shell composite via in situ MOF coating for lithium ion battery anodes. *Chem. Eng. J.* **2019**, *356*, 272–281. [CrossRef]
59. Li, W.; Peng, J.; Li, H.; Wu, Z.; Chang, B.; Guo, X.; Chen, G.; Wang, X. Architecture and performance of Si/C microspheres assembled by nano-Si via electro-spray technology as stability-enhanced anodes for lithium-ion batteries. *J. Alloys Compd.* **2022**, *903*, 163940. [CrossRef]
60. Dou, F.; Shi, L.; Chen, G.; Zhang, D. Silicon/Carbon Composite Anode Materials for Lithium-Ion Batteries. *Electrochem. Energy Rev.* **2019**, *2*, 149–198. [CrossRef]
61. Li, W.; Peng, J.; Li, H.; Wu, Z.; Huang, Y.; Chang, B.; Guo, X.; Chen, G.; Wang, X. Encapsulating Nanoscale Silicon inside Carbon Fiber as Flexible Self-Supporting Anode Material for Lithium-Ion Battery. *ACS Appl. Energy Mater.* **2021**, *4*, 8529–8537. [CrossRef]
62. Zhang, L.; Wang, C.; Dou, Y.; Cheng, N.; Cui, D.; Du, Y.; Liu, P.; Al-Mamun, M.; Zhang, S.; Zhao, H. A Yolk–Shell Structured Silicon Anode with Superior Conductivity and High Tap Density for Full Lithium-Ion Batteries. *Angew. Chem. Int. Ed.* **2019**, *58*, 8824–8828. [CrossRef] [PubMed]
63. Hu, L.; Luo, B.; Wu, C.; Hu, P.; Wang, L.; Zhang, H. Yolk-shell Si/C composites with multiple Si nanoparticles encapsulated into double carbon shells as lithium-ion battery anodes. *J. Energy Chem.* **2019**, *32*, 124–130. [CrossRef]
64. Huang, X.; Sui, X.; Yang, H.; Ren, R.; Wu, Y.; Guo, X.; Chen, J. HF-free synthesis of Si/C yolk/shell anodes for lithium-ion batteries. *J. Mater. Chem. A* **2018**, *6*, 2593–2599. [CrossRef]
65. Shen, X.; Tian, Z.; Fan, R.; Shao, L.; Zhang, D.; Cao, G.; Kou, L.; Bai, Y. Research progress on silicon/carbon composite anode materials for lithium-ion battery. *J. Energy Chem.* **2018**, *27*, 1067–1090. [CrossRef]
66. An, W.; Gao, B.; Mei, S.; Xiang, B.; Fu, J.; Wang, L.; Zhang, Q.; Chu, P.K.; Huo, K. Scalable synthesis of ant-nest-like bulk porous silicon for high-performance lithium-ion battery anodes. *Nat. Commun.* **2019**, *10*, 1447. [CrossRef]
67. Chen, S.; Chen, Z.; Xu, X.; Cao, C.; Xia, M.; Luo, Y. Scalable 2D Mesoporous Silicon Nanosheets for High-Performance Lithium-Ion Battery Anode. *Small* **2018**, *14*, 1703361. [CrossRef]
68. Xu, X.; Wu, F.; Yang, W.; Dai, X.; Wang, T.; Zhou, J.; Wang, J.; Guo, D. Silicon@Natural Nitrogen-Doped Biomass Carbon Composites Derived from "Silicon Tofu" as Green and Efficient Anode Materials for Lithium-Ion Batteries. *ACS Sustain. Chem. Eng.* **2021**, *9*, 13215–13224. [CrossRef]
69. Li, X.; Zhang, M.; Yuan, S.; Lu, C. Research Progress of Silicon/Carbon Anode Materials for Lithium-Ion Batteries: Structure Design and Synthesis Method. *Chemelectrochem* **2020**, *7*, 4289–4302. [CrossRef]
70. Li, L.; Deng, J.; Wang, L.; Wang, C.; Hu, Y.H. Boron-Doped and Carbon-Controlled Porous Si/C Anode for High-Performance Lithium-Ion Batteries. *ACS Appl. Energy Mater.* **2021**, *4*, 8488–8495. [CrossRef]
71. Chen, J.; Guo, X.; Gao, M.; Wang, J.; Sun, S.; Xue, K.; Zhang, S.; Liu, Y.; Zhang, J. Self-supporting dual-confined porous Si@c-ZIF@carbon nanofibers for high-performance lithium-ion batteries. *Chem. Commun.* **2021**, *57*, 10580–10583. [CrossRef] [PubMed]
72. Zeng, Y.; Huang, Y.; Liu, N.; Wang, X.; Zhang, Y.; Guo, Y.; Wu, H.-H.; Chen, H.; Tang, X.; Zhang, Q. N-doped porous carbon nanofibers sheathed pumpkin-like Si/C composites as free-standing anodes for lithium-ion batteries. *J. Energy Chem.* **2021**, *54*, 727–735. [CrossRef]
73. Muraleedharan Pillai, M.; Kalidas, N.; Zhao, X.; Lehto, V.P. Biomass-based silicon and carbon for lithium-ion battery anodes. *Front. Chem.* **2022**, *10*, 882081. [CrossRef] [PubMed]
74. Di, F.; Wang, N.; Li, L.; Geng, X.; Yang, H.; Zhou, W.; Sun, C.; An, B. Coral-like porous composite material of silicon and carbon synthesized by using diatomite as self-template and precursor with a good performance as anode of lithium-ions battery. *J. Alloys Compd.* **2021**, *854*, 157253. [CrossRef]
75. Qi, C.; Li, S.; Yang, Z.; Xiao, Z.; Zhao, L.; Yang, F.; Ning, G.; Ma, X.; Wang, C.; Xu, J.; et al. Suitable thickness of carbon coating layers for silicon anode. *Carbon* **2022**, *186*, 530–538. [CrossRef]
76. Zhou, S.; Fang, C.; Song, X.; Liu, G. The influence of compact and ordered carbon coating on solid-state behaviors of silicon during electrochemical processes. *Carbon Energy* **2020**, *2*, 143–150. [CrossRef]
77. Zhang, Y.; Pan, Y.; Chen, Y.; Lucht, B.L.; Bose, A. Towards reducing carbon content in silicon/carbon anodes for lithium ion batteries. *Carbon* **2017**, *112*, 72–78. [CrossRef]
78. Azam, M.A.; Safie, N.E.; Ahmad, A.S.; Yuza, N.A.; Zulkifli, N.S.A. Recent advances of silicon, carbon composites and tin oxide as new anode materials for lithium-ion battery: A comprehensive review. *J. Energy Storage* **2021**, *33*, 102096. [CrossRef]
79. Ma, Q.; Dai, Y.; Wang, H.; Ma, G.; Guo, H.; Zeng, X.; Tu, N.; Wu, X.; Xiao, M. Directly conversion the biomass-waste to Si/C composite anode materials for advanced lithium ion batteries. *Chin. Chem. Lett.* **2021**, *32*, 5–8. [CrossRef]
80. Peng, J.; Li, W.; Wu, Z.; Li, H.; Zeng, P.; Chen, G.; Chang, B.; Zhang, X.; Wang, X. Si/C composite embedded nano-Si in 3D porous carbon matrix and enwound by conductive CNTs as anode of lithium-ion batteries. *Sustain. Mater. Technol.* **2022**, *32*, e00410. [CrossRef]
81. Chen, S.; Zheng, S.; Shi, A.; Zheng, L.; Zhang, Y.; Wang, Z. Distinctive conductivity improvement by embedding Cu nanoparticles in the carbon shell of submicron Si@C anode materials for LIBs. *Sustain. Energ. Fuel.* **2022**, *6*, 2306–2313. [CrossRef]
82. Mu, T.; Lou, S.; Holmes, N.G.; Wang, C.; He, M.; Shen, B.; Lin, X.; Zuo, P.; Ma, Y.; Li, R.; et al. Reversible Silicon Anodes with Long Cycles by Multifunctional Volumetric Buffer Layers. *ACS Appl. Mater. Interfaces* **2021**, *13*, 4093–4101. [CrossRef] [PubMed]

83. Luo, H.; Wang, Q.; Wang, Y.; Xu, C.; Wang, B.; Wang, M.; Wu, H.; Zhang, Y. Nano-silicon embedded in MOFs-derived nitrogen-doped carbon/cobalt/carbon nanotubes hybrid composite for enhanced lithium ion storage. *Appl. Surf. Sci.* **2020**, *529*, 147134. [CrossRef]
84. Wu, S.J.; Wu, Z.H.; Fang, S.; Qi, X.P.; Yu, B.; Yang, J.Y. A comparison of core-shell Si/C and embedded structure Si/C composites as negative materials for lithium-ion batteries. *Rare Met.* **2021**, *40*, 2440–2446. [CrossRef]
85. Kwon, H.J.; Hwang, J.-Y.; Shin, H.-J.; Jeong, M.-G.; Chung, K.Y.; Sun, Y.-K.; Jung, H.-G. Nano/Microstructured Silicon–Carbon Hybrid Composite Particles Fabricated with Corn Starch Biowaste as Anode Materials for Li-Ion Batteries. *Nano Lett.* **2019**, *20*, 625–635. [CrossRef]
86. Huang, S.; Shan, W.; He, S.; Chen, H.; Qin, H.; Wang, S.; Hou, X. Scalable Synthesis of Si/C Microspheres with 3D Conducting Nanosized Porous Channels as High-Performance Anodes in LIBs. *Energ. Fuel.* **2020**, *34*, 13137–13143. [CrossRef]
87. Zhu, G.; Zhang, F.; Li, X.; Luo, W.; Li, L.; Zhang, H.; Wang, L.; Wang, Y.; Jiang, W.; Liu, H.K.; et al. Engineering the Distribution of Carbon in Silicon Oxide Nanospheres at the Atomic Level for Highly Stable Anodes. *Angew. Chem. Int. Ed.* **2019**, *58*, 6669–6673. [CrossRef]
88. Tian, Y.; An, Y.; Feng, J. Flexible and Freestanding Silicon/MXene Composite Papers for High-Performance Lithium-Ion Batteries. *ACS Appl. Mater. Interfaces* **2019**, *11*, 10004–10011. [CrossRef]
89. Hui, X.; Zhao, R.; Zhang, P.; Li, C.; Wang, C.; Yin, L. Low-temperature reduction strategy synthesized Si/Ti$_3$C$_2$ MXene composite anodes for high-performance Li-ion batteries. *Adv. Energy Mater.* **2019**, *9*, 1901065. [CrossRef]
90. Yang, S.H.; Kim, J.K.; Jung, D.-S.; Kang, Y.C. Facile fabrication of Si-embedded amorphous carbon@graphitic carbon composite microspheres via spray drying as high-performance lithium-ion battery anodes. *Appl. Surf. Sci.* **2022**, *606*, 154799. [CrossRef]
91. Chen, W.; Liu, H.; Kuang, S.; Huang, H.; Tang, T.; Zheng, M.; Yu, X. In-situ low-temperature strategy from waste sugarcane leaves towards micro/meso-porous carbon network embedded nano Si-SiO$_x$@C boosting high performances for lithium-ion batteries. *Carbon* **2021**, *179*, 377–386. [CrossRef]
92. Zhenjie, L.I.; Du Zhong, J.Z.; Jinwei, C.; Gang, W.; Ruilin, W. Silicon nanoparticles/carbon composites for lithium-ion battery. *Prog. Chem.* **2019**, *31*, 201.
93. Wang, W.; Gu, L.; Qian, H.; Zhao, M.; Ding, X.; Peng, X.; Sha, J.; Wang, Y. Carbon-coated silicon nanotube arrays on carbon cloth as a hybrid anode for lithium-ion batteries. *J. Power Sources* **2016**, *307*, 410–415. [CrossRef]
94. Han, P.; Sun, W.; Li, D.; Luo, D.; Wang, Y.; Yang, B.; Li, C.; Zhao, Y.; Chen, L.; Xu, J.; et al. Morphology-controlled synthesis of hollow Si/C composites based on KI-assisted magnesiothermic reduction for high performance Li-ion batteries. *Appl. Surf. Sci.* **2019**, *481*, 933–939. [CrossRef]
95. Tang, H.; Xu, Y.; Liu, L.; Zhao, D.; Zhang, Z.; Wu, Y.; Zhang, Y.; Liu, X.; Wang, Z. A Hollow Silicon Nanosphere/Carbon Nanotube Composite as an Anode Material for Lithium-Ion Batteries. *Coatings* **2022**, *12*, 1515. [CrossRef]
96. Chen, H.; He, S.; Hou, X.; Wang, S.; Chen, F.; Qin, H.; Xia, Y.; Zhou, G. Nano-Si/C microsphere with hollow double spherical interlayer and submicron porous structure to enhance performance for lithium-ion battery anode. *Electrochim. Acta* **2019**, *312*, 242–250. [CrossRef]
97. Li, B.; Li, S.; Jin, Y.; Zai, J.; Chen, M.; Nazakat, A.; Zhan, P.; Huang, Y.; Qian, X. Porous Si@C ball-in-ball hollow spheres for lithium-ion capacitors with improved energy and power densities. *J. Mater. Chem. A* **2018**, *6*, 21098–21103. [CrossRef]
98. Wang, J.; Wang, Y.; Jiang, Q.; Zhang, J.; Yin, H.; Wang, Z.; Gao, J.; Wu, Z.; Liang, J.; Zuo, S. Interconnected Hollow Si/C Hybrids Engineered by the Carbon Dioxide-Introduced Magnesiothermic Reduction of Biosilica from Reed Plants for Lithium Storage. *Energy Fuel.* **2021**, *35*, 10241–10249. [CrossRef]
99. Jiang, H.; Wang, S.; Shao, Y.; Wu, Y.; Shen, J.; Hao, X. Hollow Triple-Layer Puff-like HCs@Si@C Composites with High Structural Stability for High-Performance Lithium-Ion Battery. *ACS Appl. Energy Mater.* **2019**, *2*, 896–904. [CrossRef]
100. Zhang, J.; Zuo, S.; Wang, Y.; Yin, H.; Wang, Z.; Wang, J. Scalable synthesis of interconnected hollow Si/C nanospheres enabled by carbon dioxide in magnesiothermic reduction for high-performance lithium energy storage. *J. Power Sources* **2021**, *495*, 229803. [CrossRef]
101. Liu, X.; Shen, C.; Lu, J.; Liu, G.; Jiang, Y.; Gao, Y.; Li, W.; Zhao, B.; Zhang, J. Graphene bubble film encapsulated Si@C hollow spheres as a durable anode material for lithium storage. *Electrochim. Acta* **2020**, *361*, 137074. [CrossRef]
102. Liu, R.; Shen, C.; Dong, Y.; Qin, J.; Wang, Q.; Iocozzia, J.; Zhao, S.; Yuan, K.; Han, C.; Li, B.; et al. Sandwich-like CNTs/Si/C nanotubes as high performance anode materials for lithium-ion batteries. *J. Mater. Chem. A* **2018**, *6*, 14797–14804. [CrossRef]
103. Zhu, R.; Li, L.; Wang, Z.; Zhang, S.; Dang, J.; Liu, X.; Wang, H. Adjustable Dimensionality of Microaggregates of Silicon in Hollow Carbon Nanospheres: An Efficient Pathway for High-Performance Lithium-Ion Batteries. *ACS Nano* **2021**, *16*, 1119–1133. [CrossRef] [PubMed]
104. Man, Q.; An, Y.; Liu, C.; Shen, H.; Xiong, S.; Feng, J. Interfacial design of silicon/carbon anodes for rechargeable batteries: A review. *J. Energy Chem.* **2023**, *76*, 576–600. [CrossRef]

Disclaimer/Publisher's Note: The statements, opinions and data contained in all publications are solely those of the individual author(s) and contributor(s) and not of MDPI and/or the editor(s). MDPI and/or the editor(s) disclaim responsibility for any injury to people or property resulting from any ideas, methods, instructions or products referred to in the content.

MDPI
St. Alban-Anlage 66
4052 Basel
Switzerland
www.mdpi.com

Coatings Editorial Office
E-mail: coatings@mdpi.com
www.mdpi.com/journal/coatings

Disclaimer/Publisher's Note: The statements, opinions and data contained in all publications are solely those of the individual author(s) and contributor(s) and not of MDPI and/or the editor(s). MDPI and/or the editor(s) disclaim responsibility for any injury to people or property resulting from any ideas, methods, instructions or products referred to in the content.

www.ingramcontent.com/pod-product-compliance
Lightning Source LLC
LaVergne TN
LVHW070709100526
838202LV00013B/1061